Furfural

An Entry Point of Lignocellulose in Biorefineries to Produce Renewable Chemicals, Polymers, and Biofuels

Sustainable Chemistry Series

ISSN: 2514-3042

Series Editor: Nicholas Gathergood *(Tallinn University of Technology, Estonia)*

The concept of Green Chemistry was first introduced in 1998 with the publication of Anastas and Warner's "12 Principles of Green Chemistry". Today, these principles are becoming adopted as general practice in the chemical industries in order to reduce or eliminate the use and generation of hazardous materials, reduce waste, and make use of sustainable resources. New, safer materials and products are being released all the time. Alternative technologies are being developed to improve the efficiency of the chemical industry, while reducing its environmental impact. Sustainable resources are being investigated to replace our reliance on fossil fuels – not only as source of energy but also a source of chemicals — be they feedstock, bulk, or fine. Consideration is now given to the whole life cycle of a product or chemical — from design to disposal. And, as more of the Earth's resources become scarce so new alternatives must be found.

As the world works towards meeting the needs of the present generation without compromising the needs of the future, this series presents comprehensive books from leaders in the field of green and sustainable chemistry. The volumes will offer an excellent source of information for professional researchers in academia and industry, and postgraduate students across the multiple disciplines involved.

Published

Vol. 2   *Furfural: An Entry Point of Lignocellulose in Biorefineries to Produce Renewable Chemicals, Polymers, and Biofuels*
edited by Manuel López Granados and David Martín Alonso

Vol. 1   *Sorption Enhanced Reaction Processes*
by Alírio Egídio Rodrigues, Luís Miguel Madeira,
Yi-Jiang Wu and Rui Faria

Forthcoming

Functional Materials from Lignin: Methods and Advances
edited by Xian Jun Loh, Dan Kai and Zibiao Li

Sustainable
Chemistry
Series

Volume 2

Furfural
An Entry Point of Lignocellulose in Biorefineries to Produce Renewable Chemicals, Polymers, and Biofuels

Series Editor

Nicholas Gathergood
Tallinn University of Technology, Estonia

Editors

Manuel López Granados
Institute of Catalysis and Petrochemistry (CSIC), Spain

David Martín Alonso
University of Wisconsin-Madison, USA

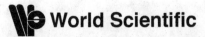

World Scientific

NEW JERSEY · LONDON · SINGAPORE · BEIJING · SHANGHAI · HONG KONG · TAIPEI · CHENNAI · TOKYO

Published by

World Scientific Publishing Europe Ltd.

57 Shelton Street, Covent Garden, London WC2H 9HE

Head office: 5 Toh Tuck Link, Singapore 596224

USA office: 27 Warren Street, Suite 401-402, Hackensack, NJ 07601

Library of Congress Cataloging-in-Publication Data
Names: López Granados, Manuel, editor. | Alonso, David Martín, editor.
Title: Furfural : an entry point of lignocellulose in biorefineries to produce renewable chemicals,
    polymers, and biofuels / edited by Manuel López Granados (Institute of Catalysis and
    Petrochemistry (CSIC), Spain), David Martín Alonso (University of Wisconsin-Madison, USA).
Description: New Jersey : World Scientific, 2018. | Series: Sustainable chemistry series ; volume 2 |
    Includes bibliographical references.
Identifiers: LCCN 2017054509 | ISBN 9781786344861 (hc : alk. paper)
Subjects: LCSH: Furfural. | Vegetable oils. | Biomass energy. | Lignocellulose. | Agricultural wastes.
Classification: LCC TP680 .F87 2018 | DDC 662/.88--dc23
LC record available at https://lccn.loc.gov/2017054509

British Library Cataloguing-in-Publication Data
A catalogue record for this book is available from the British Library.

For any available supplementary material, please visit
http://www.worldscientific.com/worldscibooks/10.1142/Q0142#t=suppl

Desk Editors: Herbert Moses/Jennifer Brough/Koe Shi Ying

Typeset by Stallion Press
Email: enquiries@stallionpress.com

Printed in Singapore

Preface

Our society is demanding for a sustainable production of chemicals, polymers and biofuels and, in this context, lignocellulosic biomass is called to partially replace oil as a primary feedstock. Consequently the search of economic and environmentally viable transformations of lignocellulose into valuable products is a vivid topic of research. Furfural features the advantage of being already a commercial reality and several chemicals and resins are already produced and commercialized from furfural. More interestingly, a myriad of many others, including biofuels, has been reported to be technically viable and even though more research is still necessary they may become commercially available in a near future.

This book has been always conceived and envisaged to become a solid foundation for those interested in fostering the deployment of a furfural bio-industry. To do so, the book critically reviews the most significant technical aspects and innovations of the main transformations of furfural, by pinpointing the crucial aspects that deserves more attention and more room for improvements. This is the object of the Chapter 3, obviously the most extended chapter of the book. The number of derivable products is overwhelming, so we decided to focus on those products that are directly derived from furfural. A subchapter has been included for each of these direct transformations, describing the multiple applications that these products have or may have and organising the relevant chemical and engineering aspects of their production. The later information was critically condensed to get a global vista of the process and to stress the bottle-necks for further technical, economic and environmental improvements. We have only included those products that have relevance in the current chemical industry or that we consider it may have relevance because either they can be transformed into drop-in products or further processed into new products that can replace those ones already with a market.

The book starts with a chapter overviewing the fundamentals of the chemical reactivity of the furan ring and the carbonyl group, the two functionalities of the furfural molecule. The book is called to serve to a wide audience, including industrial professional and academic researchers but also PhD and advanced undergraduate students, beginners in the world of the furfural transformations. So we decided to include a brief chapter that refreshes the organic chemistry of furfural. Annexes 1 and 2 (Introduction to Chemical Reactors and Introduction to Electrochemistry) were included under the same considerations but applied now to the techniques and equipment require for converting furfural and how to quantify its transformation.

Chapter 2 recapitulates the past and the future technologies of furfural production. The technical details of how furfural is produced and the new approximations to its manufacture must be known for those interested in its further transformation: the upgrading of furfural to valuable products may need a complete reconsideration of how furfural is produced and purified.

Finally we would like to express our gratitude to all the authors that have contributed so outstandingly to this collective book because of his/her rapid and enthusiastic response to join this enterprise and for finding time and devoting effort for writing their chapters. The gratitude is also extended to the exceptional editorial team that has smoothly but firmly guided and helped us to reach the end of this exciting and rewarding journey.

David Martín Alonso and Manuel López Granados

About the Authors

Alberto Marinas (Madrid, Spain, 1974) carried out his International Doctorate between the University of Córdoba (Spain) and École Centrale de Lyon (France, co-supervised by Prof Herrmann on photocatalysis). Then he made a postdoctoral stay with Prof Corma at ITQ (Valencia, Spain) working on in-situ monitoring of catalyzed reaction mechanisms and another one at ETH (Zurich, Switzerland) under the supervision of Prof Baiker, working on enantioselective catalysis. Since 2009, he is Associate Professor in Organic Chemistry at the University of Córdoba (Spain). He has supervised 10 PhD Theses, co-authored over 65 papers in SCI journals (Hirsch index $h = 22$), ca. 100 presentations to national and international congresses and participated in 22 research projects and contracts (six of them as the main researcher). Moreover, he was the Vice-Chair of COST Action CM0903 on "Utilisation of Biomass for Sustainable Fuels and Chemicals" (2009–2013).

His research interests deal with the application of heterogeneous catalysis to Sustainable/Green Chemistry (photocatalysis, selective reductions, biomass valorization, etc). Since October 2015, he is the Deputy Director of Research at University of Córdoba (Spain).

David Martín Alonso earned his Bachelor degree in Chemical Engineering at the University of Salamanca (Spain) and his PhD at Catalysis and Petroleochemistry Research Institute (Spain) under the supervision of Dr Rafael Mariscal working in the production of biodiesel using heterogeneous catalysis. In 2009, he moved to UW-Madison to work with Prof James A. Dumesic as a Research Associate at the University of Wisconsin-Madison studying new catalytic processes to convert lignocelluloses into valuable chemicals and fuels. He is (co)author of several patents related with biomass conversion and furfural production and more than 45 peer-reviewed manuscripts and book chapters. He continues his work with Prof Dumesic at UW-Madison as an Honorary Associate and he is part of a start-up company, Glucan Biorenewables LLC based on the production of chemicals from biomass, such as furfural and its derivatives.

Federica Menegazzo received her PhD degree in Chemical Sciences at the University of Ferrara in 2003. Currently she works as a postdoc researcher at CAT-MAT research group at the Department of Molecular Sciences and Nanosystems, Ca' Foscari University of Venice. Her principal scientific interests are in the field of heterogeneous catalysis with a focus on the development of innovative nanostructured catalysts and their use in industrial and sustainable chemistry. She has co-authored two book chapters and over 50 publications in peer-reviewed journals.

Francisco J. Urbano (Córdoba, Spain, 1964) is Professor at the Department of Organic Chemistry of the University of Córdoba (Spain). He obtained his PhD in 1991 and made a postdoctoral stay with Prof R. Burch at the University of Reading (UK) working on methane catalytic combustion over platinum group metal catalysts. He has supervised 11 PhD Theses and co-authored over 125 research papers with more than 3500 citations (Hirsch index $h = 33$). Since June 2014, he is Head of the Central Services for Research Support of the University of Córdoba (Spain).

His research interest is focused on the synthesis and characterization of heterogeneous catalysts and their application in green organic processes and biomass valorization. He is an expert in reduction processes of organic compounds with special focus on hydrogen transfer reduction and in photocatalytic processes such as hydrogen production by photoreforming of biomass-derived oxygenates, selective photo-oxidation of organic compounds into high added value products and destruction of organic contaminants in water by photocatalysis.

Francisco Vila received his Bachelor degree in Chemistry from the Autonomous University of Madrid in 2008. This year he joined Prof José Luis García Fierro research group at Catalysis and Petrochemistry Research Institute (ICP) from Spain. During his stay, he participated in six projects with European and Spanish funding. Finally, he obtained his PhD from Autonomous University of Madrid under the supervision of Dr Rafael

Mariscal (ICP researcher) in 2012. Since then he continues his research career within IBERCAT, a private Spanish start-up, as R&D Manager including the development of main research lines: Solid Catalysts applied to Biomass conversion to added value chemicals and biofuels.

Gianluca Marcotullio was born in L'Aquila, Italy, in 1979. After an MSc degree in Mechanical Engineering he joined Delft University of Technology in 2006 as Marie Curie fellow, where received a PhD degree in 2011 with a thesis on furfural production in modern biorefineries. He was actively involved between 2006 and 2010 in the EU FP6 research project Biosynergy on biorefinery development in Europe. He has authored several scientific papers and two patents in the field of furfural chemistry and process technology. He is presently responsible for energy auditing and R&D activities at SEA Servizi Energia Ambiente srl, Italy.

Iker Obregón obtained his degree in Chemical Engineering as best of his class at the University of the Basque Country (UPV/EHU) in 2012. He joined the Research Group of Prof Pedro L. Arias (UPV/EHU) for his master studies and completed his PhD thesis (2014–2017) on liquid-phase levulinic acid hydrogenation. During his PhD, he completed a 6-month research stage at the Institut für Technische und Makromolekulare Chemie

(ITMC) of the RWTH Aachen University (Germany) in Prof Regina Plalkovits' Research Group.

His research scope is the development of sustainable biorefinery processes for biofuel and fine chemicals production. His area of expertise is base metal heterogeneous catalyst synthesis, testing and characterization for liquid-phase hydrogenation reactions, area in which he co-authored five publications in international journals.

Inaki Gandarias obtained his degree in Chemical Engineering at the University of the Basque Country (UPV/EHU) in 2007. After 1 year working as Junior Engineer at the civil explosive manufacturing sector he joined the Research Group of Prof P.L. Arias (UPV/EHU). He developed his PhD thesis (2009–2012) on the aqueous phase glycerol hydrogenolysis and spent 6 months at the Leibniz Institute for Catalysis, Germany, at the research group of Prof Andreas Martin.

As a postdoctoral researcher, he spent the year 2013 at the research group of Prof Mark Mascal (University of California Davis, USA) and the year 2014 at the research group of Prof Graham Hutchings at the Cardiff Catalysis Institute (CCI), United Kingdom. In January 2015 he joined back the UPV/EHU as Assistant Professor at the Department of Chemical and Environmental Engineering.

His research work can be framed within the scope of biorefinery, with special focus on the development of heterogeneous catalysts that are active, selective and stable in liquid-phase reactions. He shows expertise in the valorization of biomass-derived platform chemicals through hydrogenation and oxidation processes. He is co-author of more than 25 publications in international journals and holds one patent.

Irantzu Sádaba received her PhD degree in 2012 at the Institute of Catalysis and Petrochemistry (Spanish National Research Council). Part of her research was conducted at the Colorado School of Mines (USA), the Center for Research in Ceramics and Composite Materials (Portugal) and the Technical University of Denmark (Denmark). After a 6 month-postdoc contract at the Centre for Catalysis and Sustainable Chemistry (Technical University of Denmark), she started as a research scientist at Haldor Topsøe A/S. Her current work focuses on the production of chemicals from biomass.

Jesús Hidalgo-Carrillo (Córdoba, Spain, 1983) is Researcher at the Department of Organic Chemistry of the University of Córdoba (Spain). He obtained his PhD in 2011 and made a postdoctoral stay with Prof R. Bulanek at the University of Pardubice (CZ) working on ODH reactions of propane over vanadium on zeolites. He has co-authored 10 research papers (Hirsch index, $h = 6$).

His research interest is focused on the synthesis and characterization of heterogeneous catalysts and their application in green organic processes and biomass valorization. He is an expert in reduction processes of organic compounds with especial focus on selective hydrogenations using noble metal and in photo-catalytic processes such as hydrogen production by photoreforming of biomass-derived oxygenates, selective photo-oxidation of organic compounds into high added value products and destruction of organic contaminants in water by photocatalysis.

José Miguel Campos-Martín earned his Bachelor degree in Chemistry (Industrial Chemistry) at the Universidad de Valencia (Spain) and his PhD at Institute of Catalysis and Petrochemistry (Spain) under the supervision of Dr A. Guerrero-Ruiz and Dr J.L.G. Fierro. After working as a postdoctoral student at REPSOL Technological Center, he was hired by CSIC to work in petrochemistry and sustainable chemistry applications. In 2004, he became Tenured Scientist at the Institute of Catalysis and Petrochemistry. In 2006, he was promoted to Scientific Researcher. He is co-author of more than 80 SCI publications (H-index of 30), 13 patents, four book chapters and supervisor of several PhD theses, Master and Degree works and is responsible for several projects. His areas of interest focus on the development of novel homogeneous and heterogeneous catalytic processes, with a strong emphasis on the efficient usage of raw materials and energy as well as on the minimal generation of unwanted by-products.

Keiichi Tomishige received his BS, MS and PhD from Graduate School of Science, Department of Chemistry, the University of Tokyo with Prof Y. Iwasawa. During his PhD course in 1994, he moved to Graduate School of Engineering, the University of Tokyo as a Research Associate and worked with Prof K. Fujimoto. In 1998, he became a Lecturer, and then he moved to Institute of Materials Science, University of Tsukuba as a Lecturer in 2001. Since 2004 he has been an Associate Professor, Graduate School of Pure and Applied Sciences, University of Tsukuba. Since 2010, he is a Professor at School of Engineering, Tohoku University. His research interests are the development of heterogeneous catalysts for

(1) production of biomass-derived chemicals, (2) direct synthesis of organic carbonates from CO_2 and alcohols, (3) steam reforming of biomass tar, and (4) syngas production by natural gas reforming. He is Associate Editor of Fuel Processing Technology (2014/2-), Editorial Board of Applied Catalysis A (2009/4-), Editorial Advisory Board of ACS Catalysis (2013/11-), International Advisory Board of ChemSusChem (2015/1-) and Advisory Board of Green Chemistry (2016/8-).

Manuel López Granados completed his undergraduate studies in Chemistry at the Universidad of Córdoba and obtained his PhD at the Universidad Complutense of Madrid in 1991. After two postdoctoral stays at the University of Utrecht in The Netherlands and the University of Notre Dame in USA, he got a permanent position at the Institute of Catalysis and Petrochemistry of CSIC, where he is currently Scientific Researcher. His area of interest currently focuses on the application of heterogeneous chemocatalysts to transformation of biomass into biofuels and chemicals. He is co-author of more than 100 peer-reviewed articles and four patents.

Manuel Ojeda obtained his PhD at the Institute of Catalysis and Petrochemistry (ICP, CSIC) in Spain. His doctoral thesis was devoted to the synthesis of higher oxygenates from synthesis gas using Rh-based catalysts. He then joined the Iglesia's group at the University of California (Berkeley, CA, USA). The postdoctoral studies focused on kinetic and mechanism studies of a variety of reactions, including

Fischer–Tropsch synthesis, CO oxidation and formic acid decomposition. He then returned to the ICP to investigate the use of lignocellulosic biomass to produce a range of biofuels and bioproducts by heterogeneous catalysis. Finally, in 2012 he joined to BP Chemicals Ltd, where he has been developing new catalytic technologies for the synthesis of fuels and chemicals.

Manuel Ojeda has co-authored more than 55 publications in highly-ranked journals, five book chapters and five patent applications. He has participated in more than 50 congress contributions, being also involved in the organization of three international conferences. He is also a regular reviewer of high-reputation scientific journals.

María Retuerto received her PhD degree at Universidad Complutense of Madrid. In 2010, she moved to Rutgers University in USA, and after 4 years, she became a Research Associate at the Niels Bohr Institute in Denmark. Since 2015 she is a Research Associate of the Sustainable Energy and Chemistry Group at the Institute of Catalysis and Petrochemistry, CSIC, Spain. Her research lines cover aspects of Material Science and Catalysis. At present, she is involved in the study of electrocatalysts for both electrolyzers and fuel cells reactions. Other areas of interest are magnetism, multiferroics, superconductivity and magnetoresistance. She has a strong background in diffraction techniques, especially Neutron Diffraction. She has published more than 70 scientific papers.

Masazumi Tamura is an Assistant Professor at School of Engineering in Tohoku University. He received his B.S. at School of Science from Kyoto University with Prof. K. Maruoka in 2003 and MS at Graduate School of Engineering from the University of Tokyo with Prof. M. Fujita in 2005. From 2005 to 2012 he was a researcher working at Kao Corporation, and during that period obtained PhD at Graduate School of Engineering from Nagoya University with Prof A. Satsuma in 2012. His current research interests focus on the development of heterogeneous catalysts for organic reactions including biomass conversion.

Michela Signoretto is Associate Professor of Industrial Chemistry at Ca' Foscari University of Venice. She leads CATMAT research group at the Department of Molecular Sciences and Nanosystems. She has co-authored seven book chapters and over 100 scientific publications on international journals in the field of heterogeneous catalysis and material sciences. Her research focuses on nanomaterials synthesis for energy and environmental applications. She is also involved in the formulation of new sustainable materials for pharmaceutical and cosmetic applications.

Pedro L. Arias is Professor and Head of the Chemical and Environmental Engineering Department of the University of the Basque Country (UPV/EHU). There, he is the leader of the Sustainable Process Engineering research group since its foundation in 1996. He has participated and led more than 50 research projects financially supported by the European Commission, the Spanish Research Institutions and several industrial companies. He has extensively published in peer-reviewed journals (more than 130 papers) in areas related to hydrogen production and utilization, industrial wastes treatment and recycling, catalysts and catalytic processes development for biorefineries and innovative catalytic reaction systems. He also holds two patents.

Prof Arias got his PhD in 1984 at the School of Engineering of Bilbao (Spain). He was a British Council postdoctoral researcher at the Imperial College of Science, Technology and Medicine (Department of Chemical Engineering and Chemical Technology) under the supervision of Prof Kandiyoti (1986–1987). He also worked as a visiting researcher the Department of Chemical Engineering of the MIT (1990, Prof Evans) and the Department of Chemical Engineering of the Carnegie Mellon University (1992, Prof Grossman).

Pedro Maireles-Torres received his PhD in 1991 at the University of Málaga (Spain). After a 3-month postdoctoral stay at the Petrochemical and Catalysis Institute (CSIC, Madrid, Spain), he carried out a 3-year postdoc at the University of Montpellier II (France). In 1995, he started his teaching career at the Department of Inorganic Chemistry, Crystallography and Mineralogy at University of

Málaga. Currently, he is Full Professor of Inorganic Chemistry and Head of the abovementioned department.

His current research interests include the development of mesostructured and porous materials for catalytic applications, mainly focused on the valorization of biomass by transformation in biofuels and chemicals, under heterogeneous conditions. He has co-directed nine doctoral theses, published more than 110 papers and five patents.

Rafael Mariscal studied Chemistry at the University of Córdoba, and obtained his PhD at the Complutense University of Madrid in 1991 on catalytic oxidative coupling of methane, work carried out in the Institute of Catalysis and Petrochemistry (ICP-CSIC). After a postdoctoral stay at the Technological University of Delft (The Netherlands), he returned to ICP-CSIC, where he is currently Scientific Researcher. He has carried out activities on catalytic process in energy and environment and for more than a decade his research focuses on applications of heterogeneous catalysts to obtain biofuels and renewable chemicals from biomass. He is co-author of more than 80 articles and four patents on this topic.

Sergio Rojas, graduated in Chemistry (Electrochemistry), he obtained his PhD in 1999 from the Universidad Autónoma de Madrid. After working as a postdoctoral student at KTH, he was hired by CSIC to work in electrocatalysis for sustainable energy applications. He holds a Tenured Scientist position at the Institute of Catalysis and Petrochemistry since 2006. His areas of interest focus on the development and understanding of catalysts and catalytic processes for clean energy, including biofuels production from syngas and electrocatalytic processes for energy conversion in fuel cells and electrolyzers. He is co-author of more than 100 scientific papers, several book chapters and he has delivered presentations in international conferences.

Sibao Liu received his BS and MS in Chemical Engineering from Dalian University of Technology in 2009 and 2012, respectively. He obtained his PhD from Tohoku University under the supervision of Prof Keiichi Tomishige in 2016. Then, he switched to work as a postdoctoral fellow with Prof Yoshinao Nakagawa in the same group for about 4 months. Now, he is a postdoctoral researcher with Prof Dionisios G. Vlachos at Catalysis Center for Energy Innovation, University of Delaware. His research focus is on developing heterogeneous catalysts and new catalytic process for the conversion of biomass to fuels and chemicals.

Sikander H. Hakim, PhD is Project Leader, Research and Development for Glucan Biorenewables LLC. Dr Hakim received his Bachelor's degree in Chemical Engineering at Indian Institute of Technology (IIT), Bombay (India) and his PhD at Iowa State University in Chemical Engineering under the supervision of Prof Brent Shanks, specializing in the synthesis and applications of catalytic materials for biomass conversion. Dr Hakim then joined Prof James Dumesic lab at the University of Wisconsin-Madison as a Postdoctorate Research Associate, specializing in catalytic processes for biomass conversion for the production of biorenewable fuels and chemicals.

Thomas J. Schwartz received BS degrees in Chemical Engineering and Biological Engineering from the University of Maine, and he completed his PhD in Chemical Engineering under the supervision of Prof James Dumesic at the University of Wisconsin-Madison. Dr Schwartz is currently Assistant Professor of Chemical Engineering at the University of Maine, where he is also affiliated with the Forest Bioproducts Research Institute and the Laboratory for Surface Science and Technology. His research group seeks to develop a molecular-level understanding of processes that occur on catalytic surfaces used for the upgrading of carbonaceous feedstocks to chemicals and fuels.

Yoshinao Nakagawa obtained his PhD in 2005 from Graduate School of Engineering, the University of Tokyo under the guidance of Prof N. Mizuno. After 4 years of postdoctoral research at the University of Tokyo, he joined the research group of Keiichi Tomishige at the University of Tsukuba. He moved to Tohoku University and became Assistant Professor in 2010. Since 2013, he has been Associate Professor. His current research interests are catalytic oxidations and reductions of biorelated chemicals.

Contents

Chapter 1

Chemistry of Furfural and Furanic Derivatives

Jesús Hidalgo-Carrillo, Alberto Marinas and Francisco J. Urbano*

Departamento de Química Orgánica,
Universidad de Córdoba, Campus de Rabanales,
Edificio Marie Curie, E-14071 Córdoba, España

1.1. Introduction

Furfural ($C_5H_4O_2$) and its derivatives have come to prominence within the last years as a source of renewable compounds with high applicability to different fields such as fuels and biochemical production. It is one of the top value-added chemicals that can be produced from biomass. As it will be shown in other chapters of this book, furfural is a natural precursor to a range of furan-based chemicals and solvents such as methyltetrahydrofuran, tetrahydrofuran, tetrahydrofurfuryl alcohol and furoic acid, among others. Thus, compounds derived from furfural are used as plastics, in the pharmaceutical industries, as agricultural fungicides or nematocides, lubricants, resins, bleaching agents, food and beverage additives, wood modifiers or book preservatives, among other uses.

This increase in the interest in these compounds is due to the flexibility in the production of furfural from biomass residues, being one of the top high value-added value products that can be obtained from residues of biomass [1,2].

---

*Corresponding author: fj.urbano@uco.es

Table 1.1.   Physical properties of main furan derivatives (adapted from Ref. [3]).

	CAS number	Mp (°C)	Bp (°C)	Density (20°C) (g/cm^3)	Solubility in water (25°C) (vol.%)
Furan	[110-00-9]	−85.6	31.4	0.938	1
Furfural	[98-01-1]	−36.5	161	1.160	8.3
Furfuryl alcohol	[98-00-0]	−14.6	170	1.129	Miscible
Tetrahydrofurfuryl alcohol	[97-99-4]	−80.0	178	1.051	Miscible
2-Methylfuran	[534-22-5]	−88.7	63.2	0.913	<0.3
2-Methyltretra- hydrofuran	[96-47-9]	−13.6	80.2	0.854	15.1
Furoic acid	[88-14-2]	128–132	230–232	0.550	Soluble

The principal physical properties of furfural and its main derivatives are shown in Table 1.1 [3]. The applications of products derived from furfural are multiple, Table 1.2 showing some of their main applications, as well as their synthetic route based on furfural. Further details will be given in Chapter 3.

Due to the two chemical functionalities present — aldehyde group and aromatic ring — furfural can undergo typical aldehyde reactions, such as nucleophilic additions, condensation reactions, oxidations or reductions, as well as others associated to the furan ring such as electrophilic aromatic substitution or hydrogenation.

The purpose of this chapter is to give an overview of furfural chemical reactivity based on the organic chemistry fundamentals, intending to help the reader to understand the multiple possible reactions that furfural has in function of the different functionalities of the molecule: the aromatic ring and the carbonyl group.

1.2. Furan and furan derivatives

1.2.1. *Furan structure, physical properties and synthesis*

Furan is a heterocyclic organic compound consisting of a five-membered aromatic ring with four carbon atoms and an oxygen

Table 1.2. Applications and synthetic procedure of the main furfural derivatives (adapted from Ref. [3]).

Derivative	Process reaction	Utilization
Furfural	Xylosans dehydration	Natural precursor to a range of furan-based chemicals and solvents
Furan	Furfural catalytic decarbonylation	Production of tetrahydrofuran and acetylfuran
Furfuryl alcohol	Furfural catalytic hydrogenation	Production of resins and tetrahydrofurfuryl alcohol; intermediate in fragrances production, lysine and vitamin C
Tetrahydrofurfuryl alcohol	Furfural catalytic hydrogenation	Solvent
2-Methylfuran	Furfural and 5-methylfurfural decarbonylation	Solvent and monomer
Furoic acid	Furfural oxidation	Synthesis of pharmaceuticals and perfumes

atom. It is a colorless, highly volatile and flammable liquid with a low-temperature boiling point (Table 1.1). It is slightly soluble in water but readily soluble in common organic solvents such as alcohol, ether or acetone.

1.2.1.1. *Furan aromaticity*

According to Huckel's rule, furan is aromatic because it is a flat ring having a six $(4n + 2)$ delocalized π electrons system. Thus, the four carbon atoms and oxygen have sp_2 hybridization, with an atomic p_z orbital perpendicular to the plane of the ring. Each carbon atom brings a p_z electron to the aromatic system, while the oxygen atom provides one of the lone pairs it possesses. The oxygen atom thus retains one of the lone electron pairs placed in a sp_2 hybrid orbital oriented away from the aromatic ring, but in the same plane as this one (Figure 1.1).

Furan has an aromatic resonance stabilization energy of $14.8 \, \text{kcal·mol}^{-1}$, energy which is lower than that of thiophene

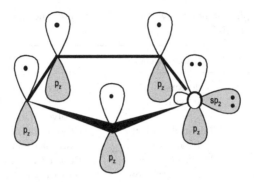

Figure 1.1. Structure of furan.

$(18.6\,\text{kcal·mol}^{-1})$ and pyrrole $(20.6\,\text{kcal·mol}^{-1})$. In other words, the aromatic character of furan is less marked than that of thiophene or pyrrole [4].

The oxygen atom of furan has an electrodonating mesomeric effect $(+M)$, associated with the ability to delocalize its lone electron pairs. Moreover, it also has an inductive effect $(-I)$ that is evident in the displacement towards the oxygen of the constituent electrons of the C–O bond due to the greater electronegativity of the oxygen.

Although both effects are globally opposed, oxygen exerts a net electron-donating effect, as a consequence of the mesomeric effect being greater than the inductive one $(+M > -I)$. Thus, the oxygen atom provides an additional electron density to the ring, which results in electron density values greater than 1 for each carbon atom.

As previously mentioned, furan as well as its homologs (thiophene and pyrrole) is a surplus π system due to the distribution of six π electrons delocalized in five atoms. This results in a greater electron density in the cycle than in its benzene analogues. This accounts for the higher reactivity of furan in *electrophilic aromatic substitution*, as compared to benzene.

On the other hand, contrary to the simple six elements aromatic rings, only one of its resonance structures is neutral, whereas the other forms exist as zwitterions (Figure 1.2). As result, although these charged forms contribute a minority to the resonance hybrid,

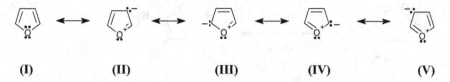

(I)   **(II)**   **(III)**   **(IV)**   **(V)**

Figure 1.2. Resonance structures for furan. Structures II–V are zwitterionic.

the chemistry of furan will be partly that of the aromatic compounds and partly that of the dienes.

1.2.1.2. *Furan synthesis*

Furan can be obtained by the Paal–Knorr synthesis, consisting in the dehydration of a 1,4-dicarbonyl compound by P_2O_5 or sulfuric acid (Figure 1.3(a)). However, at present, the main industrial process to obtain furan is decarbonylation of furfural using palladium or nickel catalysts (Figure 1.3(b)), by the copper-catalyzed oxidation of butadiene (Figure 1.3(c)) or by decarboxylation through oxidation with Ag_2O to furoic acid and subsequent elimination of CO_2 on copper catalysts (Figure 1.3(d)).

1.2.2. *Furan and furan derivatives reactivity*

As stated above, furan has an aromatic character that is associated to its *electrophilic aromatic substitution reactivity*, although as a result of charged resonance structures, it also behaves, in part, as a diene. Therefore, when establishing the reactivity of the furan ring, we must take into account both aspects.

1.2.2.1. *Electrophilic aromatic substitution reactions of furan*

Due to its aromatic character, furan undergoes *electrophilic aromatic substitution (EAS) reactions*, being considerably more reactive than benzene due to the electron-donating effect of the oxygen heteroatom [5]. The attack of the electrophile occurs mainly in carbon 2 (or 5), since the intermediate formed by the electrophilic attack at C2 is stabilized by charge delocalization to a greater degree than the intermediate from C3 attack (Figure 1.4). If the process is studied

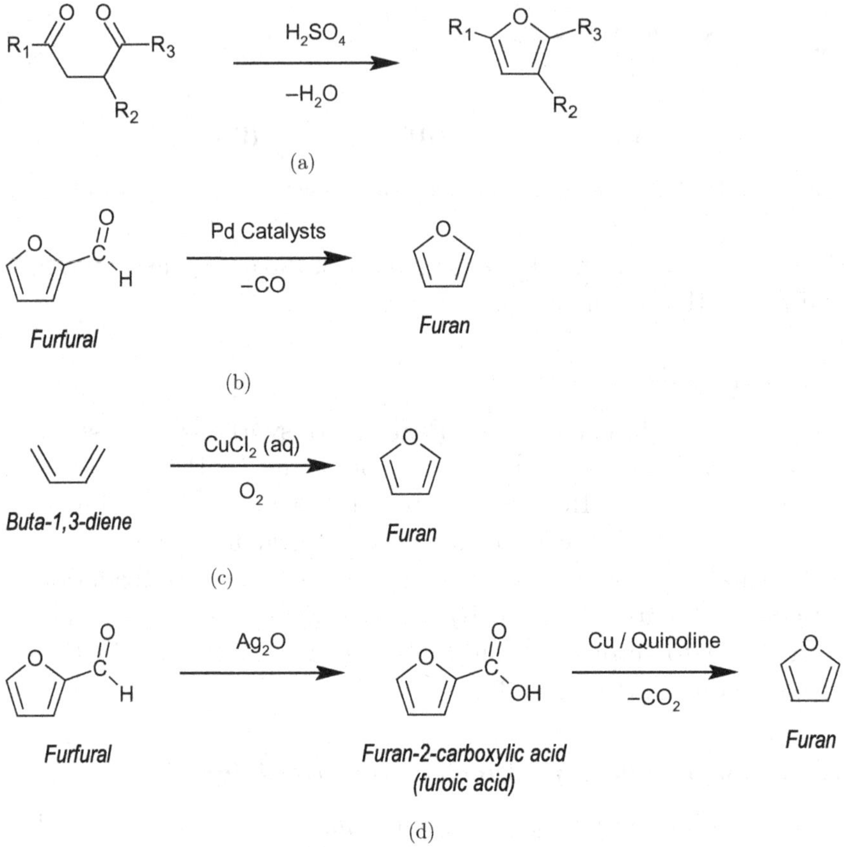

(a)

(b)

(c)

(d)

Figure 1.3. Some synthetic routes for obtaining furan. (a) Paal–Knorr synthesis of furans, (b) furan synthesis by decarbonylation of furfural, (c) furan synthesis by copper-catalyzed oxidation of butadiene, (d) furan synthesis by oxidation/decarboxylation of furfural.

by means of a diagram of energies, it is possible to observe how the attack to the C2 (or C5) is the most favored from the thermodynamic point of view, since the formed reaction intermediates are more stable, being necessary a smaller energy of activation.

It is noteworthy that, unlike benzene, furan is not resistant to strong mineral or Lewis acids and furan and alkyl furans are protonated in concentrated acid and, upon heating, the hydrolytic opening of the ring occurs (Figure 1.5(a)).

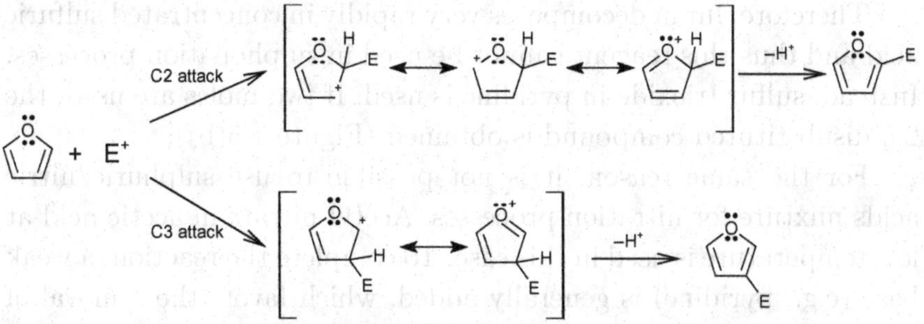

Figure 1.4. Electrophilic aromatic substitution of furan. Reaction mechanism.

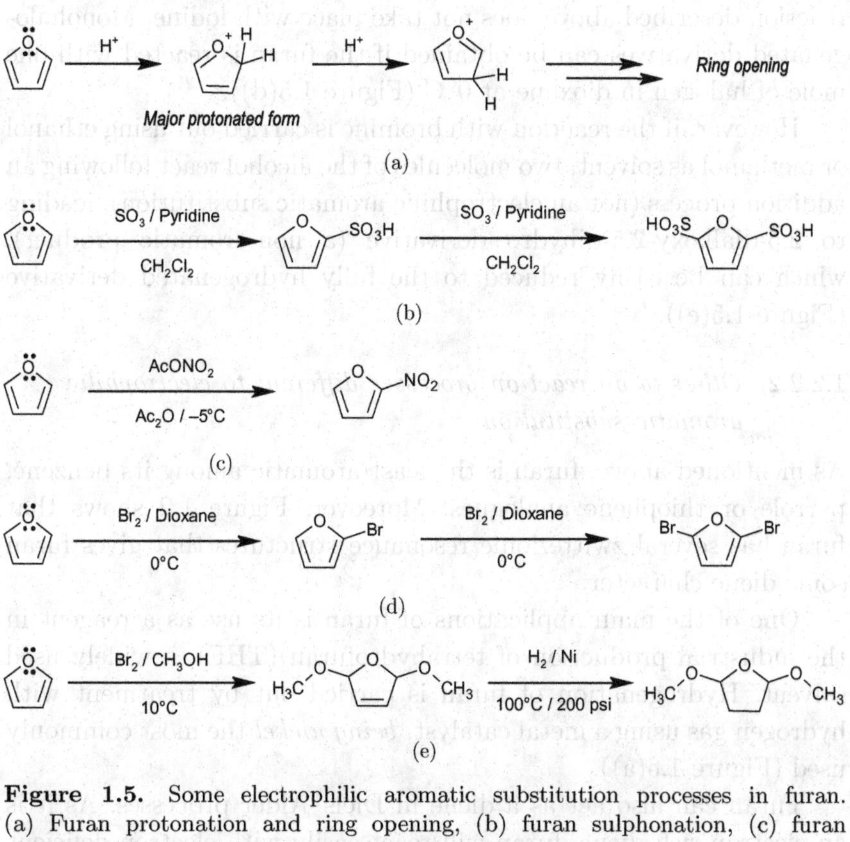

Figure 1.5. Some electrophilic aromatic substitution processes in furan. (a) Furan protonation and ring opening, (b) furan sulphonation, (c) furan nitration, (d) furan halogenation, (e) furan 2,5-dimethoxylation.

Therefore, furan decomposes very rapidly in concentrated sulfuric acid and thus this reagent cannot be used in sulphonation processes. Instead, sulfur trioxide in pyridine is used. If two moles are used, the 2,5-disubstituted compound is obtained (Figure 1.5(b)).

For the same reason, it is not possible to use sulphuric/nitric acids mixture for nitration processes. Acetyl nitrate in acetic acid at low temperature is used in this case. To complete the reaction, a weak base (e.g., pyridine) is generally added, which favors the removal of acetate in the form of acetic acid (Figure 1.5(c)).

Furans react vigorously with chlorine and bromine at room temperature, usually obtaining polyhalogenated compounds. The reaction described above does not take place with iodine. Monohalogenated derivatives can be obtained if the furan is reacted with one mole of halogen in dioxane at 0°C (Figure 1.5(d)).

However, if the reaction with bromine is carried out using ethanol or methanol as solvent, two molecules of the alcohol react following an addition process (not an electrophilic aromatic substitution), leading to 2,5-dialkoxy-2,5-dihydro derivative (a non-aromatic product), which can be easily reduced to the fully hydrogenated derivative (Figure 1.5(e)).

1.2.2.2. *Other furan reaction processes different to electrophilic aromatic substitution*

As mentioned above, furan is the least aromatic among its benzene, pyrrole or thiophene analogues. Moreover, Figure 1.2 shows that furan has several zwitterionic resonance structures that gives furan some diene character.

One of the main applications of furan is its use as a reagent in the industrial production of tetrahydrofuran (THF), a widely used solvent. Hydrogenation of furan is carried out by treatment with hydrogen gas using a metal catalyst, *being nickel* the most commonly used (Figure 1.6(a)).

Furan can also act as a diene in Diels–Alder processes. As it is an electron-rich diene, furan will react easily with electron-deficient dienophiles, following a concerted type mechanism that leads to the loss of aromaticity of the ring in the formed adduct (Figure 1.6(b)).

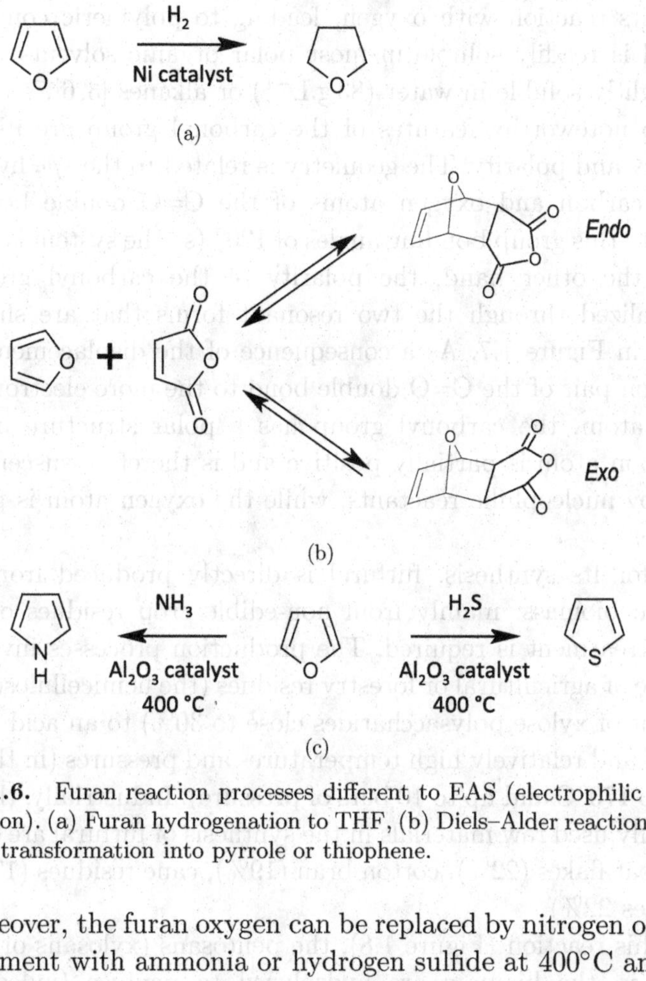

Figure 1.6. Furan reaction processes different to EAS (electrophilic aromatic substitution). (a) Furan hydrogenation to THF, (b) Diels–Alder reaction of furan, (c) furan transformation into pyrrole or thiophene.

Moreover, the furan oxygen can be replaced by nitrogen or sulfur, by treatment with ammonia or hydrogen sulfide at 400°C and using alumina as catalyst, thus leading to thiophene or pyrrole, respectively (Figure 1.6(c)).

1.3. Furfural

1.3.1. *Furfural structure, physical properties and synthesis (preparation)*

Furfural (furan-2-carbaldehyde) is a furan derived aromatic heterocyclic aldehyde. It is a colorless oily liquid with the smell of almonds in pure state. However, in contact with air, it quickly turns yellow due to its reaction with oxygen, leading to polymeric compounds. Furfural is readily soluble in most polar organic solvents but it is only slightly soluble in water $(83\,g.L^{-1})$ or alkanes [3, 6].

Two noteworthy features of the carbonyl group are its planar geometry and polarity. The geometry is related to the sp_2 hybridization of carbon and oxygen atoms of the C=O double bond that confers to this group bonding angles of 120° (so the system is planar).

On the other hand, the polarity of the carbonyl group can be visualized through the two resonant forms that are shown for furfural in Figure 1.7. As a consequence of the displacement of the π electron pair of the C=O double bond to the more electronegative oxygen atom, the carbonyl group has a polar structure in which the carbon atom is partially positive and is therefore susceptible to attack by nucleophilic reactants, while the oxygen atom is partially negative.

As for its synthesis, furfural is directly produced from lignocellulosic biomass, mainly from non-edible crop residues or wood. No pre-treatment is required. The production processes involve the exposure of agricultural or forestry residues (the hemicellulose having a content of xylose polysaccharides close to 30%) to an acid aqueous medium and relatively high temperatures and pressures (in the range of 150 to 170°C and up to 10 bars of pressure). Industrially, the most-commonly used raw materials in the synthesis of furfural are corn ear (23%), oat flakes (22%), cotton bran (19%), cane residues (17%) and rice flakes 22%).

In this reaction (Figure 1.8), the pentosans (xylosans or xylans) present in the biomass are hydrolyzed to pentose (xylose) in a first acid hydrolysis reaction, and subsequently, it undergoes an

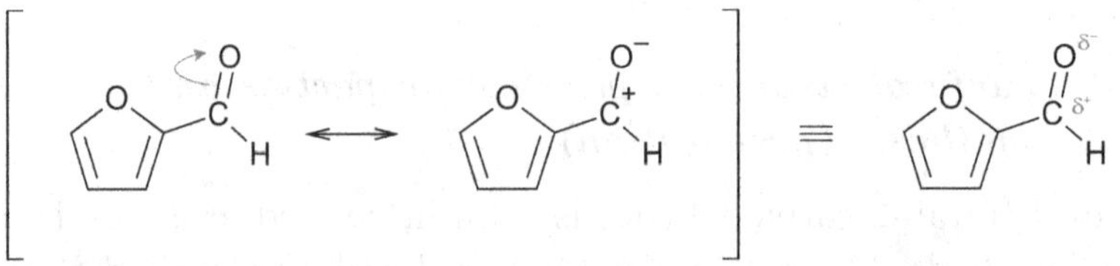

Figure 1.7.   Resonant structures of the furfural carbonyl group.

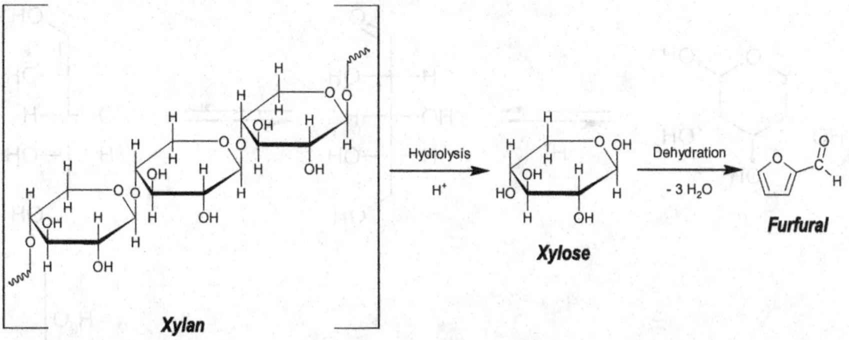

Figure 1.8. Hydrolysis of xylan in acid medium leading to the formation of xylose and subsequent dehydration of this to furfural.

acid-induced dehydration process from xylose to furfural [7, 8]. The aqueous solutions of furfural thus obtained must be distilled and purified in order to obtain furfural with a purity greater than 98%.

The second dehydration step is slower than the hydrolysis of xylan. These sequential reactions are acid catalyzed, the presence of an acid catalyst thus being necessary to convert the xylose to furfural. However, furfural can also be formed without addition of any catalyst; this is due to an autocatalytic reaction mechanism: the thermal decomposition of xylose leads to the formation of organic acids which can act as homogeneous catalysts in the production of furfural [9].

1.3.1.1. *Mechanism for furfural formation*

There are numerous studies in the literature on the production of furfural though the mechanism of their formation through dehydration of C5 sugars on different catalytic systems is still a matter of debate [10,11]. In this sense, two classic schemes have been proposed. The first mechanism (Figure 1.9) suggests the formation of a 1,2-enediol intermediate (open chain form of the sugar) which would initially dehydrate. Subsequent keto-enol tautomerisms would lead to the release of two new water molecules, eventually resulting in furfural formation.

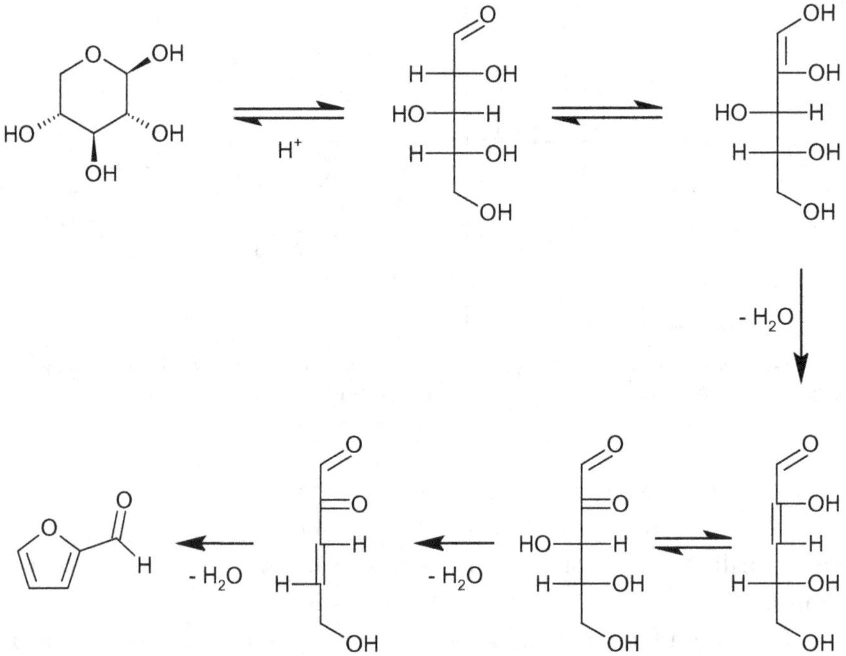

Figure 1.9. Mechanism of formation of furfural from the treatment of carbohydrates of five carbons in acid medium, through 1,2-enediol intermediates (adapted from Ref. [11]).

The second mechanism is based on the initial dehydration of a closed chain (cyclic) form of pentose sugars (Figure 1.10). Therefore, one hydrogen ion (H^+) would protonate one of the hydroxyl groups in the pentose, thus resulting in a positively charged oxygen atom. Due to the more electronegative character of oxygen as compared to carbon atom, the positive charge would delocalize to the carbon atom resulting in a release of one molecule of water. Subsequently, the two electrons from a neighboring C–O bond would migrate to form a C=C leading to the cleavage of one C–O bond (ring opening to form an aldehyde with migration of a hydrogen atom within the molecule). This results in hydrogen ion protonating another lone electron pair of hydroxyl group and liberation of a new water molecule. Finally, the trivalent carbon atom would combine and form a relatively stable

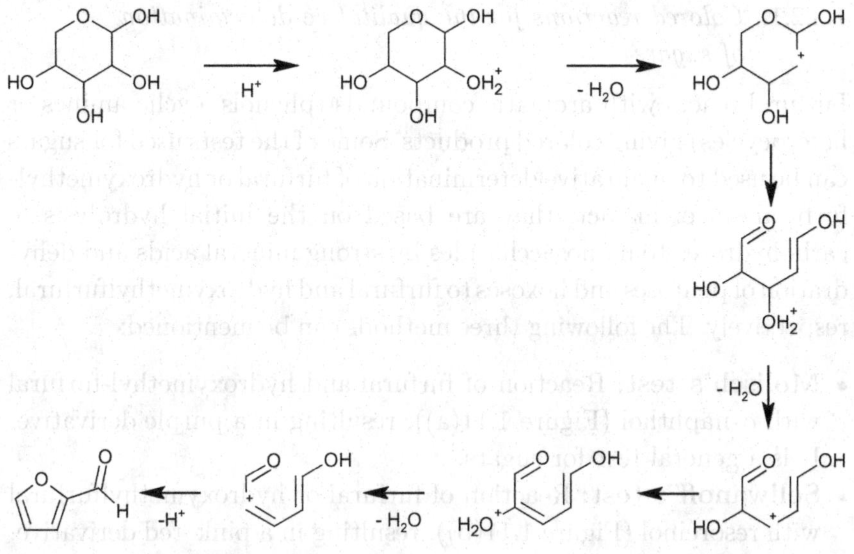

Figure 1.10. Furfural formation mechanism based on the initial addition of a proton to a hydroxyl group and subsequent release of a water molecule (adapted from Ref. [12]).

ring structure. The elimination of a H^+ would produce furfural in the end [12].

1.3.2. *Furfural and furfural derivatives reactivity*

As furfural holds two functional groups, an aromatic ring and an aldehyde, it is a versatile building block for several applications. The above-presented reactions of the furan aromatic ring could also take place in the ring of furfural. In this case, the carbonyl group acts as a substituent of the furan ring and therefore affects the *electrophilic aromatic substitution* (EAS) by means of its inductive (−I) and mesomeric (−M) effects. As an electron acceptor group (deactivating), the EAS of furfural takes place at lower reaction rates than for the furan itself.

Furthermore, the aldehyde group of furfural can undergo some reactions such as acylation, acetalization, aldol and Knoevenagel condensations, reduction to alcohols, reductive amination to amines, decarbonylation, and oxidation to carboxylic acids [12–18].

1.3.2.1. *Colored reactions for the qualitative determination of sugars*

Furfural reacts with aromatic compounds (phenols, cyclic amines or heterocycles) giving colored products. Some of the tests used for sugars can be used to qualitative determination of furfural or hydroxymethyl-furfural since, in fact, they are based on the initial hydrolysis of carbohydrates to monosaccharides by strong mineral acids and dehydration of pentoses and hexoses to furfural and hydroxymethylfurfural, respectively. The following three methods can be mentioned:

- **Molisch's test:** Reaction of furfural and hydroxymethyl-furfural with α-naphthol (Figure 1.11(a)), resulting in a purple derivative. It is a general test for sugars.
- **Seliwanoff's test:** Reaction of furfural or hydroxymethylfurfural with resorcinol (Figure 1.11(b)), resulting in a pink–red derivative. It is a ketose-specific test. Ketoses in acid medium dehydrate faster than the aldoses; therefore, both groups can be differentiated depending on the time the colored product takes to appear.
- **Bial's test:** Reaction of furfural with orcinol (Figure 1.11(c)) in an aqueous medium with HCl and in the presence of $FeCl_3$, resulting in a blue or green derivative. It is a specific test of pentoses. In case the medium contains hexoses, hydroxymethylfurfural would produce a muddy-brown or gray color.

1.3.2.2. *Furan ring based reactivity of furfural: Electrophilic aromatic substitution*

As commented above, in the case of substituted furans such as furfural, the influence of the substituent in the *electrophilic aromatic substitution* process is similar to that observed for substituted benzenes, the inductive and mesomeric effects of the substituents determining both the reaction rate and orientation.

Aldehyde group of furfural molecule is an electron withdrawing group due to its negative inductive and mesomeric effects $(-I; -M)$ and therefore should be considered as a deactivating substituent. *Electrophilic aromatic substitution* of furfural should lead preferentially to the 5- or 4-substituted derivatives as consequence of a combination of electric as well as steric factors.

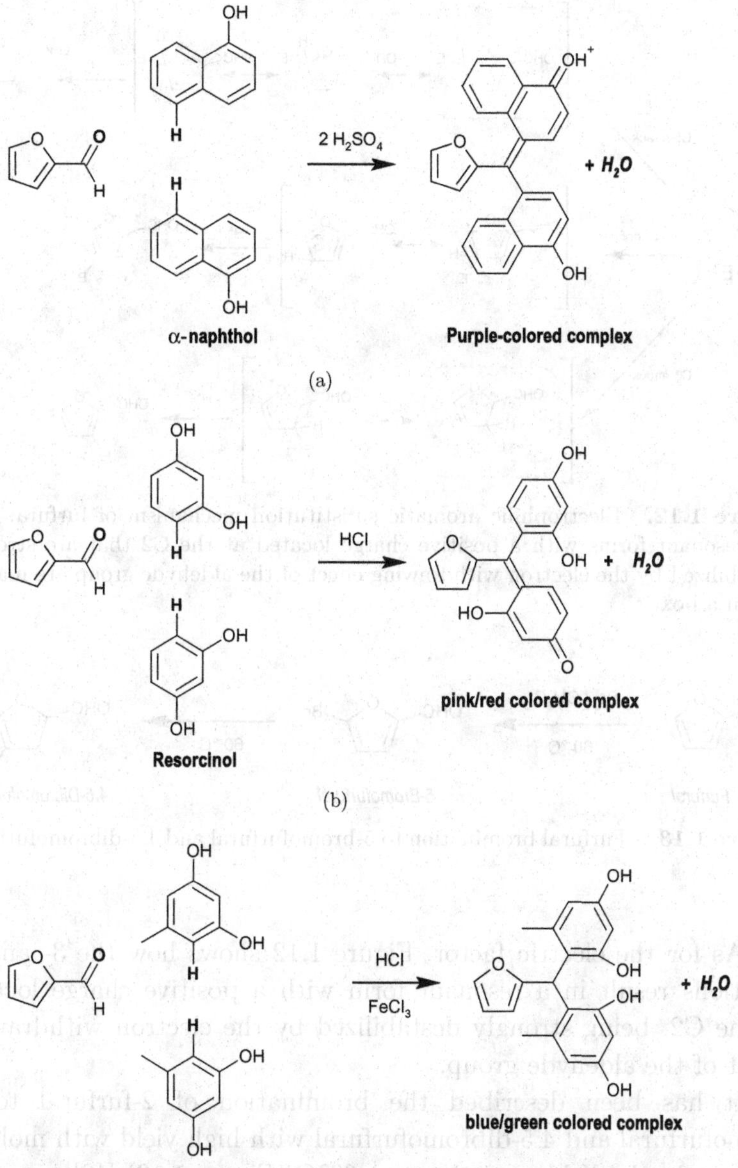

(a)

(b)

(c)

Figure 1.11. Colored reactions used as specific tests to determine furfural and furfural derivatives. (a) Molisch's test (general carbohydrates test), (b) Selivanoff's test (ketoses identification test), (c) Bial's test (pentoses identification test).

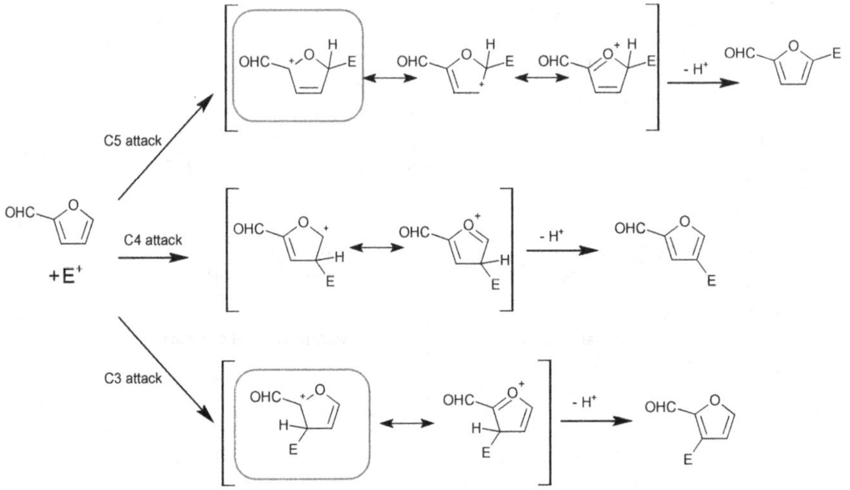

Figure 1.12. Electrophilic aromatic substitution mechanism of furfural. The two resonant forms with a positive charge located at the C2 that are strongly destabilized by the electron withdrawing effect of the aldehyde group are marked within a box.

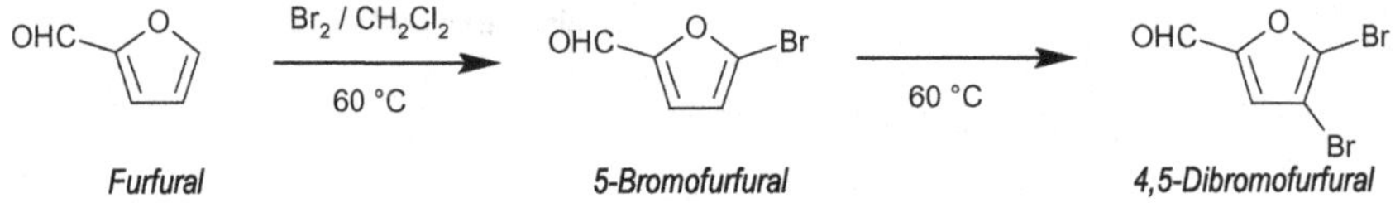

Figure 1.13. Furfural bromination to 5-bromofurfural and 4,5-dibromofurfural.

As for the electric factor, Figure 1.12 shows how the 3- and 5-positions result in a resonant form with a positive charge located at the C2, being strongly destabilized by the electron withdrawing effect of the aldehyde group.

It has been described the bromination of 2-furfural to 5-bromofurfural and 4,5-dibromofurfural with high yield with molecular bromine in dichloromethane at 60°C (Figure 1.13) [19].

Moreover, furan does not undergo Friedel–Crafts alkylation since the acid catalysts required cause polymerization because of the acid sensitivity of furan. However, furfural is less sensitive to acids due to the electron-withdrawing effect of the aldehyde group and, therefore,

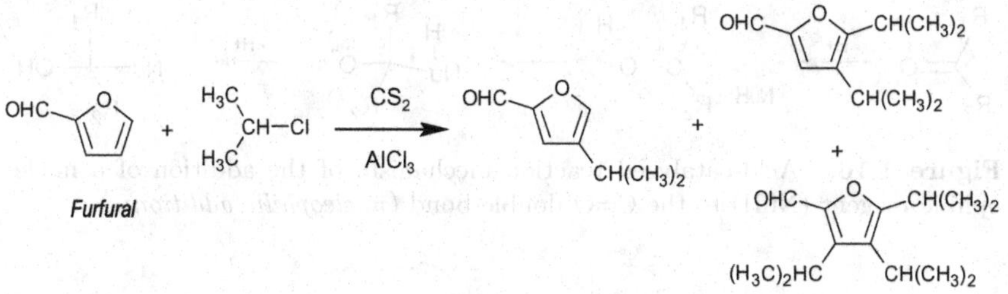

Figure 1.14. Friedel–Crafts alkylation of furfural at room temperature.

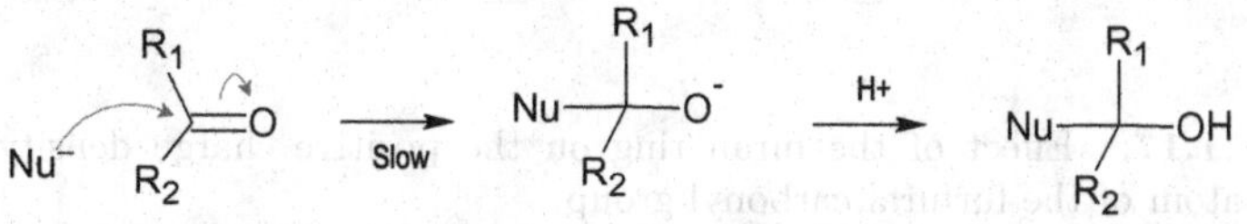

Figure 1.15. Reaction mechanism of the addition of a nucleophilic reagent (NüH) to the C=O double bond (*nucleophilic addition*).

can undergo Friedel–Crafts alkylation at room temperature yielding a mixture of alkylfurfurals (Figure 1.14) [20].

1.3.2.3. *Carbonyl group based reactivity*

One of the main type of reactions that suffers the carbonyl group is the addition to the carbon–oxygen double bond. Since C=O bond is strongly polar with the carbon atom being the positive end, nucleophilic species always attack to this atom while the electrophilic counterpart of the reactive (usually a proton) go to the oxygen.

These reactions named as *nucleophilic additions* lead to the formation of a new bond between the nucleophile and the carbon atom and to the disappearance of the C=O double bond, the rate-determining step being the one involving the nucleophilic attack (Figure 1.15).

In some cases, it is also possible an acid-catalyzed nucleophilic addition in which the oxygen atom protonates first thus creating a positively charged carbon atom which is more attractive to nucleophilic attack (Figure 1.16). Similar catalysis can be found with either mineral or solid acids (Brönsted or Lewis). Base catalysis is

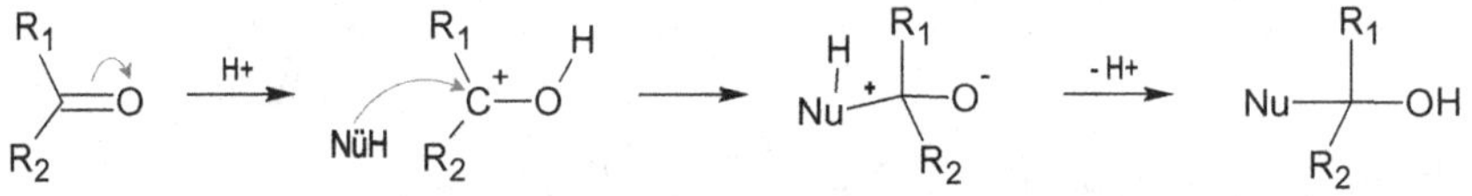

Figure 1.16. Acid-catalyzed reaction mechanism of the addition of a nucleophilic reagent (NüH) to the C=O double bond (*nucleophilic addition*).

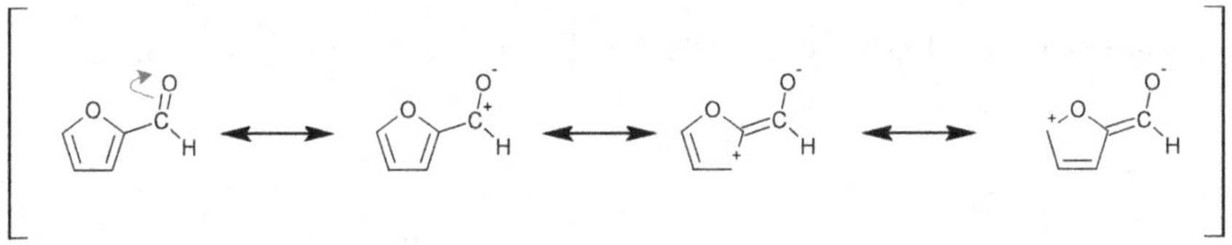

Figure 1.17. Effect of the furan ring on the positive charge density of the carbon atom of the furfural carbonyl group.

also possible but in this case the base allows the formation of a stronger negatively charged nucleophile (Nu$^-$) from a weaker neutral one (NüH).

Electron-donating groups attached to the carbonyl group cause a decrease in reaction rates while electron-attracting groups increase rates. In the case of furfural, because of the resonance effect (Figure 1.17) of the furan ring, the positive charge density of the carbon atom of the furfural carbonyl group is reduced and therefore nucleophilic addition to the carbonyl group of furfural requires more severe reaction conditions. Steric factors are also quite important and contribute to the decreased reactivity when bulky groups are present.

1.3.2.4. *Nucleophilic addition of alcohols: Furfural hemiacetal formation*

A first example of a nucleophilic addition process to furfural is shown in Figure 1.18, consisting of the acid-catalyzed addition of ethanol (alcohols in general) to give the corresponding furfural hemiacetal and if there is excess of alcohol a second step takes place leading to the acetal. This last step is a nucleophilic substitution of the hydroxyl group of the hemiacetal structure. Furfural hemiacetals and acetals were found as secondary products during the selective hydrogenation

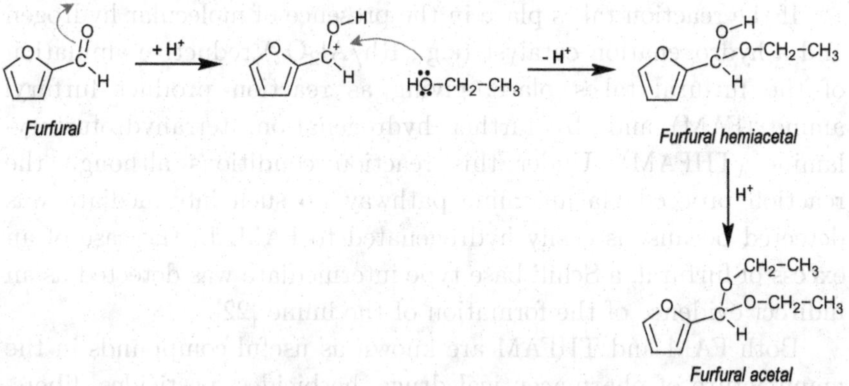

Figure 1.18. Acid-catalyzed nucleophilic addition of ethanol to the carbonyl group of furfural yielding the furfural hemiacetal and acetal.

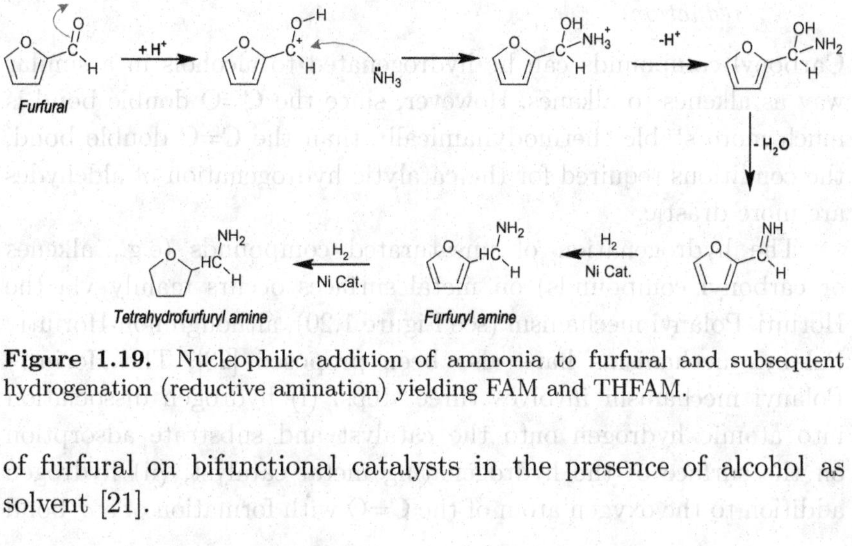

Figure 1.19. Nucleophilic addition of ammonia to furfural and subsequent hydrogenation (reductive amination) yielding FAM and THFAM.

of furfural on bifunctional catalysts in the presence of alcohol as solvent [21].

1.3.2.5. *Nucleophilic addition of ammonia: Furfural amination reactions*

The reaction of furfural with nitrogen-based nucleophilic reagents such as ammonia or primary amines leads in the first place to hemiaminals although these compounds are generally unstable and after a dehydration process yield the corresponding furfural-imines (Figure 1.19).

If the reaction takes place in the presence of molecular hydrogen and a hydrogenation catalyst (e.g., Rh/Al_2O_3) reductive amination of the furfural takes place, giving as reaction product furfuryl amine (FAM) and, by further hydrogenation, tetrahydrofurfurylamine (THFAM). Under this reaction conditions although the reaction proceed via an imine pathway no such intermediate was detected because is easily hydrogenated to FAM. In the case of an excess of furfural, a Schiff base type intermediate was detected as an indirect evidence of the formation of the imine [22].

Both FAM and THFAM are known as useful compounds in the manufacture of pharmaceutical drugs, herbicides, pesticides, fibers, perfumes, etc. [6]. Further details will be given in Chapter 3.8.

1.3.2.6. *Reduction to furfuryl alcohol (and further/deeper reduction)*

Carbonyl compounds can he hydrogenated to alcohols in a similar way as alkenes to alkanes. However, since the $C=O$ double bond is much more stable thermodynamically than the $C=C$ double bond, the conditions required for the catalytic hydrogenation of aldehydes are more drastic.

The hydrogenation of unsaturated compounds (e.g., alkenes or carbonyl compounds) on metal surfaces occurs mainly via the Horiuti–Polanyi mechanism (see Figure 1.20), although non-Horiuti–Polanyi mechanisms have also been proposed [23]. The Horiuti–Polanyi mechanism involves three steps: (i) hydrogen dissociation into atomic hydrogen onto the catalyst and substrate adsorption on the surface of the hydrogenating metal catalyst, (ii) hydrogen addition to the oxygen atom of the $C=O$ with formation of a σ-bond

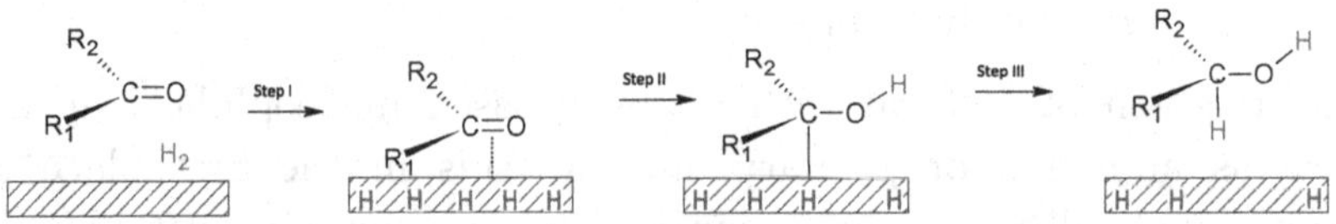

Figure 1.20.   Horiuti–Polanyi hydrogenation mechanism.

between the metal and the carbon atom, and finally (iii) reductive elimination of the free alcohol.

In the case of unsaturated aldehydes, the adsorption modes of the substrate (e.g., furfural) on the catalyst surface are determinant for obtaining a selectivity to a particular product, being those providing an interaction of the carbonyl group with the metal (second metal or additive) that lead successfully to furfuryl alcohol [24].

Under these severe reaction conditions, selective hydrogenation of furfural to furfuryl alcohol is complicated since different side reactions can take place like furan ring hydrogenation or the hydrogenolysis of the produced furfuryl alcohol leading to the methylfuran derivative. Figure 1.21 shows the reaction pathway for catalytic hydrogenation of furfural [25].

The liquid-phase selective catalytic hydrogenation of furfural to furfuryl alcohol has been widely investigated in the presence of Ni, Co, Cu, Rh, Ru, Pt and Pd based catalysts [24, 26–29]. In the past, the Cu–Cr based catalysts were commonly used in industry, but Cr had great impact on the environmental pollution, thus limiting its later applications [27]. In order to improve furfuryl alcohol selectivity, bimetallic catalysts have been used in the literature as well as the use of additives to avoid the above-mentioned secondary reactions [30].

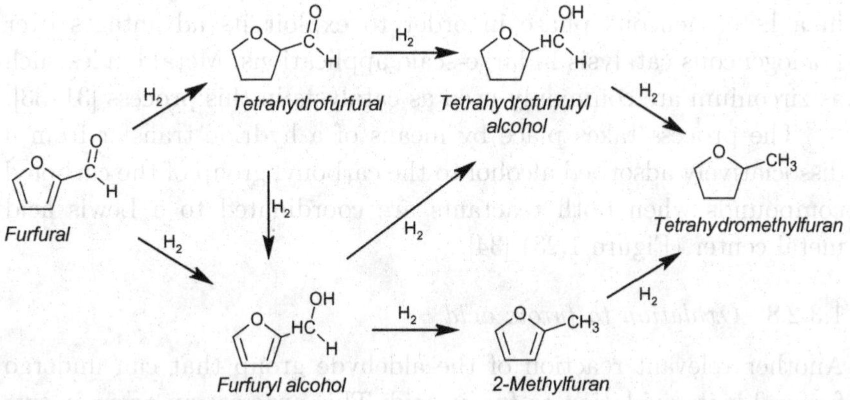

Figure 1.21. Reaction pathway for catalytic hydrogenation of furfural.

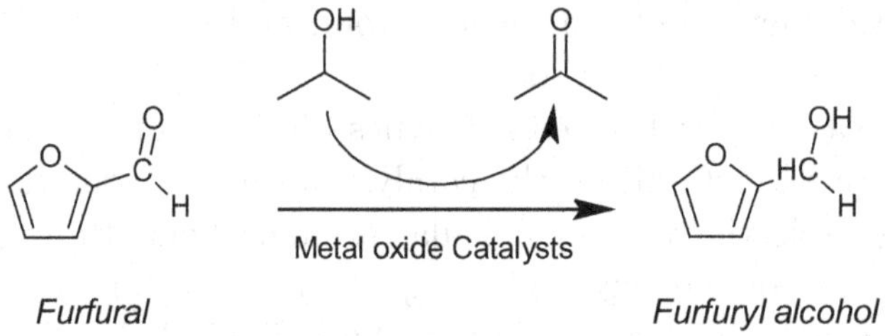

Furfural   *Furfuryl alcohol*

Figure 1.22. Meerwein–Ponndorf–Verley selective reduction of furfural to furfuryl alcohol.

1.3.2.7. *Hydrogen-transfer reduction of furfural to furfuryl alcohol*

The transformation of furfural into furfuryl alcohol can also be carried out by a hydrogen transfer catalytic process from a donor compound, mainly secondary alcohols such as 2-propanol, in the so called Meerwein–Ponndorf–Verley (MPV) process (Figure 1.22).

Traditionally, the MPV reaction has been conducted in a homogeneous medium containing an aluminium or zirconium alkoxide. The reaction mechanism in a homogeneous phase, which has been thoroughly studied, involves the formation of a six-membered cyclic intermediate where both alcohol and carbonyl are coordinated to the metal site in the alkoxide. Nevertheless, the last two decades have witnessed an increasing interest in conducting the process in a heterogeneous phase in order to exploit its advantages over homogeneous catalysis in large-scale applications. Metal oxides such as zirconium are commonly used as catalysts in this process [31–33].

The process takes place by means of a hydride transfer from a dissociatively adsorbed alcohol to the carbonyl group of the carbonyl compounds when both reactants are coordinated to a Lewis acid metal center (Figure 1.23) [34].

1.3.2.8. *Oxidation to furoic acid*

Another relevant reaction of the aldehyde group that can undergo furfural is its oxidation to furoic acid. This process can occur in two ways. On the one hand, since furfural does not possess α-hydrogens, its treatment with a base (e.g., NaOH aq.) leads to the *Cannizzaro*

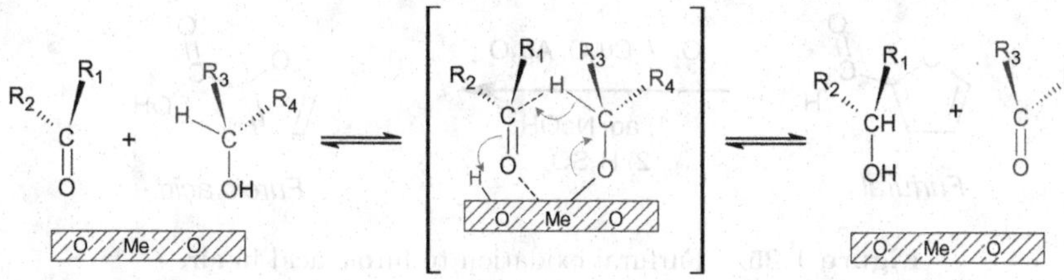

Figure 1.23. Meerwein–Ponndorf–Verley reaction mechanism through a six-membered cyclic intermediate (adapted from Ref. [34]).

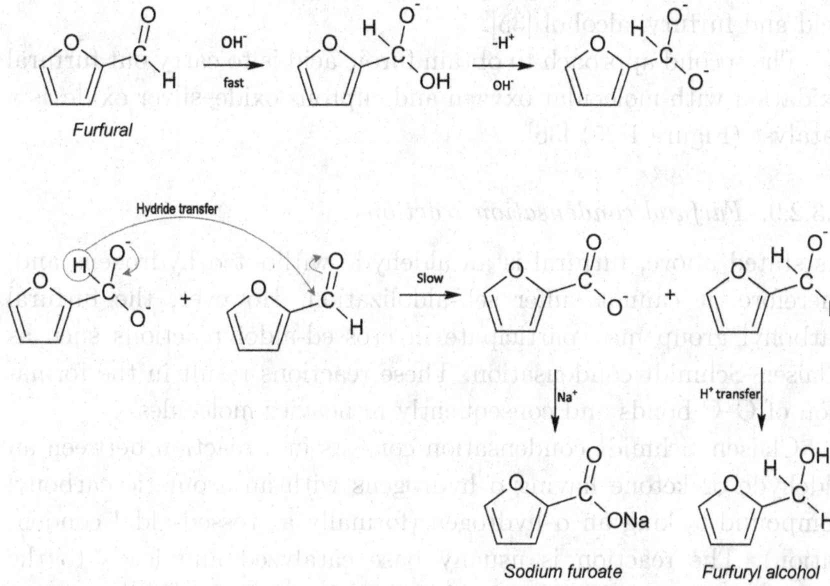

Figure 1.24. Reaction mechanism of the *Cannizzaro* reaction of furfural.

disproportionation reaction. In this process, two molecules of furfural are reacted, one of which transforms into furfuryl alcohol (reduction process) and the other yield furoic acid (oxidation process).

The reaction mechanism involves in the first instance the attack of a hydroxide ion to the C=O to give an intermediate that loses a proton in the basic solution to give the corresponding diion (Figure 1.24). The strong electron-donating character of both O– greatly facilitates the ability of the aldehydic hydrogen to leave

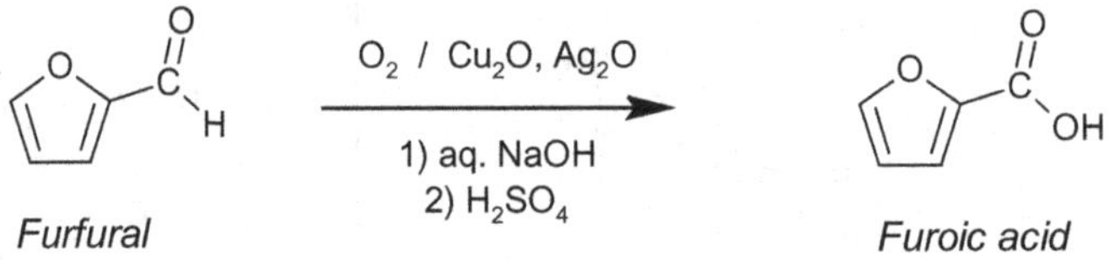

Figure 1.25. Furfural oxidation to furoic acid in air.

with its electron pair and thus a hydride is then transferred to another molecule of aldehyde yielding finally the sodium salt of furoic acid and furfuryl alcohol [35].

The second approach to obtain furoic acid is to carry out furfural oxidation with molecular oxygen and cuprous oxide–silver oxide as a catalyst (Figure 1.25) [36].

1.3.2.9. *Furfural condensation reactions*

As stated above, furfural is an aldehyde without α-hydrogens and, therefore, it cannot suffer self-aldolization. However, the furfural carbonyl group may participate in crossed-aldol reactions such as Claisen–Schmidt condensation. These reactions result in the formation of C–C bonds and consequently in heavier molecules.

Claisen–Schmidt condensation consists in a reaction between an aldehyde or ketone having α-hydrogens with an aromatic carbonyl compound lacking an α-hydrogen (formally a crossed-aldol condensation). The reaction is usually base catalyzed and leads to the formation of an α,β-unsaturated carbonyl compound [37].

Thus, the condensation of furfural (C5) with acetone (C3) (Figure 1.26) has been described to give rise to an adduct (C8) than can undergo a new condensation process with a second furfural molecule to yield 1,5-di(furan-2-yl)penta-1,4-dien-3-one (C13).

The acetone molecule losses a proton in basic medium and forms a stabilized enolate anion. The enolate ion is a good nucleophile that adds to the furfural molecule giving an intermediate that readily dehydrates to form the furfurylidene acetone.

Since there remain α-hydrogens in this molecule, the hydroxide ion facilitates the formation of a new enolate ion that adds to a

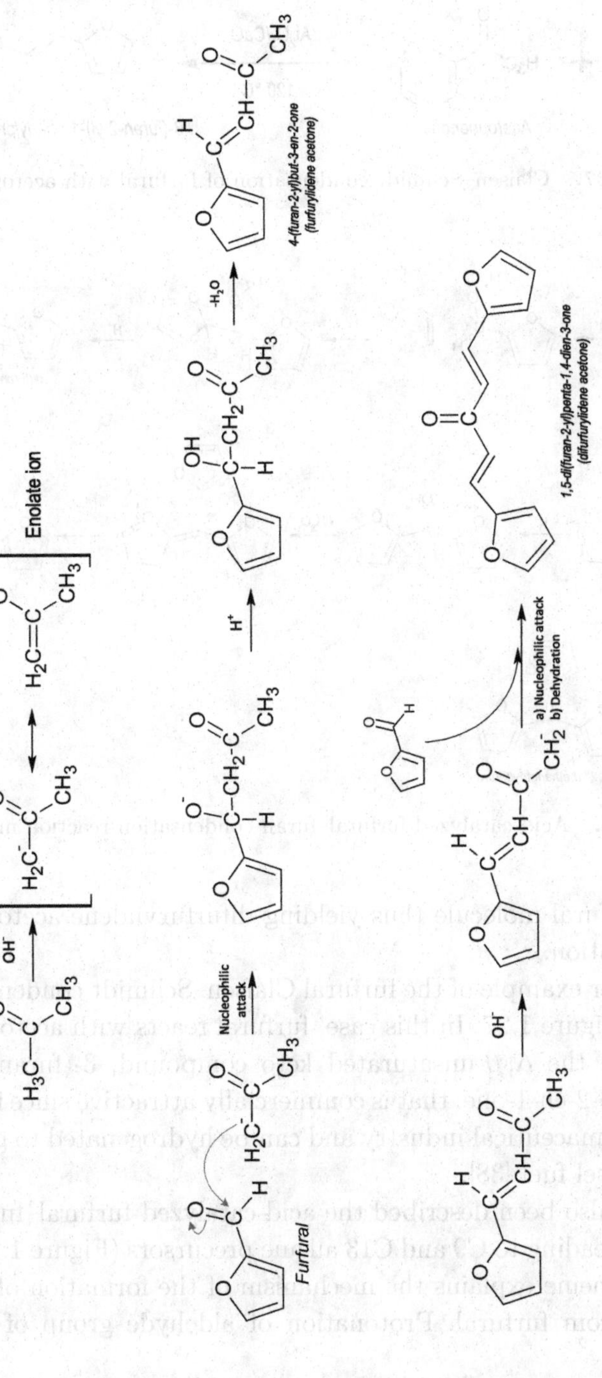

Figure 1.26. Reaction mechanism of the Claisen–Schmidt condensation of furfural and acetone.

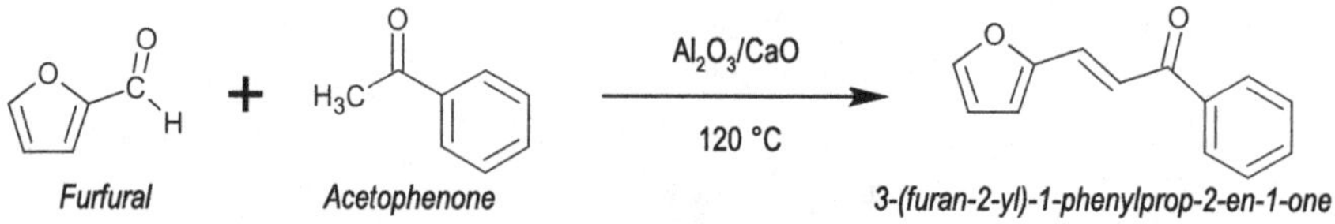

Figure 1.27. Claisen–Schmidt condensation of furfural with acetophenone.

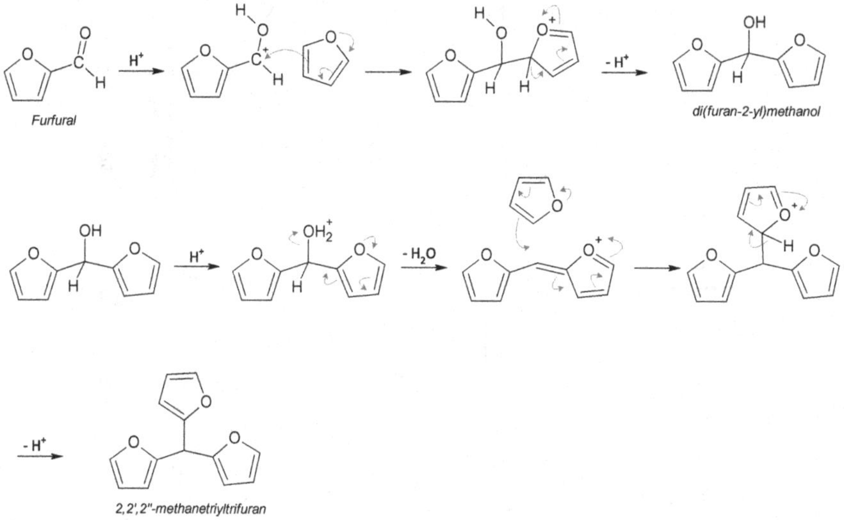

Figure 1.28. Acid-catalyzed furfural–furan condensation reaction mechanism.

second furfural molecule thus yielding difurfurylidene acetone upon its dehydration.

Another example of the furfural Claisen–Schmidt condensation is shown in Figure 1.27. In this case, furfural reacts with acetophenone to produce the α,β-unsaturated keto compound, 3-(furan-2-yl)-1-phenylprop-2-en-1-one, that is commercially attractive since it is used in the pharmaceutical industry and can be hydrogenated to give high quality diesel fuel [38].

It has also been described the acid-catalyzed furfural–furan condensation leading to C9 and C13 alkane precursors (Figure 1.28). The reaction scheme explains the mechanism of the formation of trifuryl methane from furfural. Protonation of aldehyde group of furfural followed by the addition of furan via oxonium ion intermediate led to product formation. The protonation of the resulting alcohol allows the attack of a second furan ring that finally yields the trifuryl methane. The reaction of furfural with furan requires catalytic amounts of acid to obtain good yields. Higher concentrations of acid and longer reaction times decreased the product formation due to the polymerization of furan under concentrated acid.

1.4. Summary

Furfural and structurally related compounds have received a growing interest in the context of the so-called green chemistry because they are platform molecules that can be easily obtained from inexpensive and non-edible biomass. Furfural itself has been pointed out as one of the most important biomass-derived chemicals in biorefineries. For the transformation of these molecules into high-added value products, it is necessary to understand both their physical properties and their chemical reactivity. In this sense, both furfural and other structurally related compounds have in their structure a furan ring, with an aromatic character, as well as other functional groups that contribute their own reactivity, as the carbonyl and alcohol groups of furfural and furfuryl alcohol, respectively. The aromatic character of furan ring justifies a series of reactions related to *electrophilic aromatic substitution* processes, in a similar way as benzene. Thus, furfural can be easily alkylated, nitrated or halogenated. Moreover, the furan molecule possesses a certain diene character that explains other reactions leading to the molecule aromaticity loss, such as reduction processes, Diels–Alder cycloaddition or ring opening reactions. In addition to the furan ring's own reactivity, the chemical properties of this type of compounds will also be related to the reactivity of the functional groups attached to it. Thus, furfural undergoes, for example, processes of *nucleophilic addition* to the carbonyl group, oxidation/reduction processes of the carbonyl or condensation reactions as the Claisen–Schmidt of furfural with acetone. In this chapter, there have been presented, from a simplified point of view, the basic concepts dealing with the reactivity furfural, as maximum exponents of this type of platform molecules.

Acknowledgments

The authors are thankful to *Fundación Ramón Areces* for financial support.

References

1. J.J. Bozell, G.R. Petersen, Technology development for the production of biobased products from biorefinery carbohydrates — the US Department of Energy's "Top 10" revisited, *Green Chemistry*, 12 (2010) 539–554.

2. S. Dutta, S. De, B. Saha, M.I. Alam, Advances in conversion of hemicellulosic biomass to furfural and upgrading to biofuels, *Catalysis Science & Technology*, 2 (2012) 2025–2036.

3. W.J. Mckillip, N.I. Sax, *Dangerous Properties of Industrial Materials*, 6th ed., Van Nostrand Rheinold, New York, 1984.

4. M.K. Cyrański, Energetic aspects of cyclic Pi-electron delocalization: Evaluation of the methods of estimating aromatic stabilization energies, *Chemical Reviews*, 105 (2005) 3773–3811.

5. P.Y. Bruice, *Organic Chemistry*, Pearson Prentice Hall, Upper Saddle River, NJ, 2007.

6. H.E. Hoydonckx, W.M. Van Rhijn, W. Van Rhijn, D.E. De Vos, P.A. Jacobs, *Furfural and Derivatives, Ullmann's Encyclopedia of Industrial Chemistry*, Wiley-VCH Verlag GmbH & Co., KGaA, 2000.

7. J.B. Binder, J.J. Blank, A.V. Cefali, R.T. Raines, Synthesis of furfural from xylose and xylan, *ChemSusChem*, 3 (2010) 1268–1272.

8. K. Yan, G. Wu, T. Lafleur, C. Jarvis, Production, properties and catalytic hydrogenation of furfural to fuel additives and value-added chemicals, *Renewable and Sustainable Energy Reviews*, 38 (2014) 663–676.

9. M.J. Antal, T. Leesomboon, W.S. Mok, G.N. Richards, Kinetic-studies of the reactions of ketoses and aldoses in water at high-temperature. 3. Mechanism of formation of 2-furaldehyde from D-xylose, *Carbohydrate Research*, 217 (1991) 71–85.

10. D.W. Harris, M.S. Feather, Evidence for a C-2→C-1 intramolecular hydrogen-transfer during the acid-catalyzed isomerization of D-glucose to D-fructose, *Carbohydrate Research*, 30 (1973) 359–365.

11. M.S. Feather, The conversion of D-xylose and D-glucuronic acid to 2-furaldehyde, *Tetrahedron Letters*, 11 (1970) 4143–4145.

12. K. Zeitsch, *The Chemistry and Technology of Furfural and Its Many Byproducts*, Elsevier, Amsterdam, 2000.

13. Y. Qiao, N. Theyssen, Z. Hou, Acid-catalyzed dehydration of fructose to 5-(hydroxymethyl)furfural, *Recyclable Catalysis*, 2 (2015) 36–60.

14. Y. Nakagawa, M. Tamura, K. Tomishige, Catalytic reduction of biomass-derived furanic compounds with hydrogen, *ACS Catalysis*, 3 (2013) 2655–2668.

15. S.G. Wettstein, D.M. Alonso, E.I. Gürbüz, J.A. Dumesic, A roadmap for conversion of lignocellulosic biomass to chemicals and fuels, *Current Opinion in Chemical Engineering*, 1 (2012) 218–224.

16. Q. Meng, C. Qiu, G. Ding, J. Cui, Y. Zhu, Y. Li, Role of alkali earth metals over Pd/Al2O3 for decarbonylation of 5-hydroxymethylfurfural, *Catalysis Science & Technology*, 6 (2016) 4377–4388.

17. A.S. Gowda, S. Parkin, F.T. Ladipo, Hydrogenation and hydrogenolysis of furfural and furfuryl alcohol catalyzed by ruthenium(II) bis(diimine) complexes, *Applied Organometallic Chemistry*, 26 (2012) 86–93.

18. K. Yan, J. Liao, X. Wu, X. Xie, A noble-metal free Cu-catalyst derived from hydrotalcite for highly efficient hydrogenation of biomass-derived furfural and levulinic acid, *RSC Advances*, 3 (2013) 3853–3856.

19. X. Wu, X. Peng, X. Dong, Z. Dai, Synthesis of 5-bromo-2-furfural under solvent-free conditions using 1-butyl-3-methylimidazolium tribromide as brominating agent, *Asian Journal of Chemistry*, 24 (2012) 927–928.

20. R.R. Gupta, M. Kumar, V. Gupta, *Heterocyclic Chemistry. Volume II: Five-Membered Heterocycles*, Springer, Berlin, Heidelberg, 1999.

21. P. Reyes, D. Salinas, C. Campos, M. Oportus, J. Murcia, H. Rojas, G. Borda, J.L. Garcia Fierro, Selective hydrogenation of furfural on Ir/TiO2 catalysts, *Quimica Nova*, 33 (2010) 777–780.

22. M. Chatterjee, T. Ishizaka, H. Kawanami, Reductive amination of furfural to furfurylamine using aqueous ammonia solution and molecular hydrogen: An environmentally friendly approach, *Green Chemistry*, 18 (2016) 487–496.

23. B. Yang, X.-Q. Gong, H.-F. Wang, X.-M. Cao, J.J. Rooney, P. Hu, Evidence to challenge the universality of the Horiuti–Polanyi mechanism for hydrogenation in heterogeneous catalysis: Origin and trend of the preference of a non-Horiuti–Polanyi mechanism, *Journal of the American Chemical Society*, 135 (2013) 15244–15250.

24. X. Chen, L. Zhang, B. Zhang, X. Guo, X. Mu, Highly selective hydrogenation of furfural to furfuryl alcohol over Pt nanoparticles supported on g-C3N4 nanosheets catalysts in water, *Scientific Reports*, 6 (2016) 28558.

25. A.B. Merlo, V. Vetere, J.F. Ruggera, M.L. Casella, Bimetallic PtSn catalyst for the selective hydrogenation of furfural to furfuryl alcohol in liquid-phase, *Catalysis Communications*, 10 (2009) 1665–1669.

26. S. Sang, Y. Wang, W. Zhu, G. Xiao, Selective hydrogenation of furfuryl alcohol to tetrahydrofurfuryl alcohol over Ni/γ-Al2O3 catalysts, *Research on Chemical Intermediates*, (2016) 1–17.

27. R. Rao, A. Dandekar, R.T.K. Baker, M.A. Vannice, Properties of copper chromite catalysts in hydrogenation reactions, *Journal of Catalysis*, 171 (1997) 406–419.

28. M.J. Taylor, L.J. Durndell, M.A. Isaacs, C.M.A. Parlett, K. Wilson, A.F. Lee, G. Kyriakou, Highly selective hydrogenation of furfural over supported

Pt nanoparticles under mild conditions, *Applied Catalysis B: Environmental*, 180 (2016) 580–585.

29. S.T. Thompson, H.H. Lamb, Palladium–Rhenium catalysts for selective hydrogenation of furfural: Evidence for an optimum surface composition, *ACS Catalysis*, 6 (2016) 7438–7447.

30. J. Wu, G. Gao, J.L. Li, P. Sun, X.D. Long, F.W. Li, Efficient and versatile CuNi alloy nanocatalysts for the highly selective hydrogenation of furfural, *Applied Catalysis B-Environmental*, 203 (2017) 227–236.

31. S. Axpuac, M.A. Aramendia, J. Hidalgo-Carrillo, A. Marinas, J.M. Marinas, V. Montes-Jimenez, F.J. Urbano, V. Borau, Study of structure-performance relationships in Meerwein–Ponndorf–Verley reduction of crotonaldehyde on several magnesium and zirconium-based systems, *Catalysis Today*, 187 (2012) 183–190.

32. G.K. Chuah, S. Jaenicke, Y.Z. Zhu, S.H. Liu, Meerwein–Ponndorf–Verley reduction over heterogeneous catalysts, *Current Organic Chemistry*, 10 (2006) 1639–1654.

33. F. Wang, Z. Zhang, Catalytic transfer hydrogenation of furfural into furfuryl alcohol over magnetic γ-Fe2O3@HAP catalyst, *ACS Sustainable Chemistry & Engineering*, 5 (2016) 942–947.

34. A. Corma, M.E. Domine, S. Valencia, Water-resistant solid Lewis acid catalysts: Meerwein–Ponndorf–Verley and Oppenauer reactions catalyzed by tin-beta zeolite, *Journal of Catalysis*, 215 (2003) 294–304.

35. J. March, *Advanced Organic Chemistry. Reactions, Mechanisms, and Structure*, John Wiley & Sons, New York, 1992.

36. Q. Tian, D. Shi, Y. Sha, CuO and Ag_2O/CuO catalyzed oxidation of aldehydes to the corresponding carboxylic acids by molecular oxygen, *Molecules*, 13 (2008) 948–957.

37. Z. Wang, *Claisen–Schmidt Condensation, Comprehensive Organic Name Reactions and Reagents*, John Wiley & Sons, 2010, pp. 660–664.

38. G.D. Yadav, A.R. Yadav, Novelty of Claisen–Schmidt condensation of biomass-derived furfural with acetophenone over solid super base catalyst, *RSC Advances*, 4 (2014) 63772–63778.

Chapter 2

Past, Current Situation and Future Technologies of Furfural Production

David Martín Alonso*,† and Gianluca Marcotullio‡,§

*Department of Chemical and Biological Engineering,
University of Wisconsin–Madison, Madison, WI 53706, USA

† Glucan Biorenewables LLC, 505 South Rosa Road,
Suite 112, Madison, WI 53719, USA

‡ Sea Servizi Energia Ambiente srl, via di Civita 2,
67100 L'Aquila, Italy

2.1. Industrial production of furfural — Past and current situation

Industrial furfural production has its roots in the effort done in the early 1920s by F.B. LaForge and G.H. Mains at the US National Bureau of Chemistry and, at about the same time, by C. Miner and H.J. Brownlee at The Quaker Oats Company.

The large experimental plant set up by LaForge and Mains was operated in Rosslyn, Virginia, between 1920 and 1921 with a capacity of roughly 100 lbs per day [1]. Optimized setting consisted in charging the digester with 200 lbs of corncobs and 550 lbs of cold water (no acid was added), letting steam inside the digester until reaching a pressure of about 9 bar(g) and thus starting withdrawing the furfural vapor at constant pressure for 2 h. 770 lbs of distillate were obtained containing 1.61 wt.% of furfural, around 6.2% of the initial weight

---

§Corresponding author: gianluca.marcotullio@sea.aq.it

of corncobs, i.e., roughly 30% of the theoretical yield. Remarkably, steam consumption accounted for nearly 100 ton/ton of furfural recovered.

On the other hand, in 1927 Brownlee reported the optimized Quaker Oats process conditions in terms of water and sulphuric acid use [2]. It implied charging the cylindrical rotating digester with 4500 lbs of oat hulls, 100 lbs of H_2SO_4 and 1200 lbs of water, heating it up with live steam up to 153°C (5 bar(g)) and thus starting the process of furfural vapor withdrawal for about 5 h. Reported furfural yield was 10.3 wt.% of initial dry biomass, slightly above 50% of theoretical. Furfural was recovered as a 5.8% aqueous solution and concentrated by distillation. Brownlee pointed out steam consumption as a critical part of the process, estimated under optimized conditions in around 25 ton/ton of furfural recovered, including heating-up, distillation and losses. This figure was confirmed during operations.

While the effort done at the US National Bureau of Chemistry provided a solid research basis for future developments, the Quaker Oats Company established a long lasting industrial production of furfural and derivatives, and its process evolved later into a series of large spherical digesters with a capacity of about 2.5 ton of oat hulls operated discontinuously, although the downstream phases of the process were continuous. At the Cedar Rapids mill furfural production reached around 11,000 tpa in 1943, and in 1951 the US production capacity topped 30,000 tpa [3]. At the same time, smaller production facilities were reported to operate in Sweden, France and Italy [4].

From the early 20s until the 50s the advent of furfural as a widely used industrial chemical stimulated the interest around new potential sources and production processes, and this effort is still continuing nowadays. In the early 50s the known literature accounted already for more than 50 among US, Russian, British, Austrian, German, French, Italian, Danish and Swedish patents attempting to improve the existing processes or describing new concepts [4].

In recent times, Karl J. Zeitsch, during the period spent as consultant to Illovo Sugar Ltd, made a remarkable effort tracking the progresses of the furfural industry up until the late 90s, when his well-known book was published [5]. Concerning the existing industrial processes for the production of furfural, Zeitsch reported of a continuous Quaker Oats process operated between the 60s and late 90s in Florida and then discontinued. This process consisted of long continuous cylindrical reactors with transport paddles for handling the solid biomass (namely bagasse), and equipped with multiple nozzles for the addition of sulphuric acid and superheated steam (10 bar, 650°C). The reported yield was in the order of 55%.

At the same time, other process concepts were developed and used, bringing about a substantial degree of innovation, namely the Agrifurane, Esher Wyss and Rosenlew, also well described by Zeitsch. These processes are now mostly abandoned [6], except from the production plant operated in South Africa by Illovo Sugar Ltd which is a modification of the Rosenlew process. The Rosenlew process improved the original concepts introducing a continuous reactor equipped at its top with two intermittently opening hydraulic shutters for feeding the coarse fraction of the bagasse, containing around 50% of moisture. No additional water or foreign acids are added, superheated steam enters the reactor from the bottom and flows upwards guaranteeing an increasing concentration of organic acids, which are generated inside the reactor during the hydrolysis of biomass. At the same time furfural is removed and recovered with the distillate. Illovo is the only furfural producer in South Africa and among the largest in the world and the process run at the Sezela plant has a remarkable degree of sophistication. Among other advantages, such as no use of strong mineral acids, the Illovo process favors the occurrence of diacetyl and 2,3-pentanedione, high added value food flavoring products, which are recovered by distillation and separated as low boilers. Although typical Rosenlew furfural yields have been reported to achieve 5.5 wt.% of the wet bagasse, roughly 60% of the theoretical yield [5], average yields experienced at the Sezela plant are around 3.6 wt.% [7]. The amount of steam required is substantial, around 30 ton/ton furfural recovered, and coal is used next to bagasse to balance the energy requirements of the furfural plant. With a full capacity of 20,000 tpa and a 2015 production of roughly 17,000 ton of furfural, beside significant amounts of furfuryl alcohol, diacetyl and other bioproducts [8], the downstream production plant at the Sezela sugar mill represents a world reference for furfural production.

A multitude of Chinese production plants, in most cases small in size and using corncobs as feedstock, make up for the large majority of global production and spare capacity (total capacity is estimated in up to 700,000 tpa, although mostly not in use). The process typically operated in China is a batch process resembling very closely the original Quaker Oats process both in terms of operating conditions and furfural yields, although employing vertical batch reactors [5]. Furfural yields are below 10 wt.% of the dry corncobs, and the residues are burned for steam production. For this reason, Chinese furfural production is strictly related to corncobs availability and price. A vast majority of the production is sold to furfuryl alcohol producers inside the country, the remainder being exported around the world.

The largest furfural production plant in the world is located in Dominican Republic, contiguous to a sugar mill owned by The Central Romana Corporation. This plant was started as a joined venture with the Quaker Oats Company in 1955 and expanded in 1973 to a full capacity of around 40,000 tpa [9]. The annual average production exceeds 30,000 tons, largely exported to the Belgian subsidiary Trans Furan Chemicals (TFC) for the production of furfuryl alcohol and other derivatives, and to the US based PennAKem, a specialty chemical company with a large portfolio of furfural-based molecules, incidentally located in the facilities formerly used by the Quaker Oats Chemicals in Memphis, TN.

The global production landscape is completed by small producers located in Thailand; in Argentina, where a subsidiary of the Italian vegetable extracts and tannins company Silvateam produces furfural from wood residues; in Iran, where Behran Oil Company produces less than 2000 tpa of furfural from bagasse, mostly for internal refining needs, using a process that resembles closely the typical Chinese batch process [10]; in Austria, where Lenzing operates a unique production facility recovering and purifying furfural that is formed during the pulping process of beech wood [11]; in Spain, where Nutrafur produces furfural from almond shells although, to the best of our knowledge, the production has been recently closed; and in Slovenia, where Tanin Sevnica produces furfural from wood since 1973 with a proprietary technology [12].

Many other production plants have closed in the past years, and furfural production disappeared from many countries, such as USA, France, India, former USSR, Kenya, Italy, Philippines and Turkey. All of them used agricultural or wood residues as feedstock, mainly corncobs, rice or oat hulls, bagasse, olive press cake or birch wood [13].

Plentiful of future concepts for the production of furfural have been described in literature, especially by Zeitsch [5] who directly contributed to some of those as inventor. Namely the SupraYield process concept held the potential to become a new standard for furfural production. Zeitsch suggested that steam stripping was incapable of bringing an aqueous furfural solution to boiling, making furfural removal ineffective to the detriment of yield. For this reason, the SupraYield process was conceived to operate at high temperature and pressure for allowing the reaction mixture to boil by gradual decompression. In comparison with traditional methods superior yields (70%) were obtained at pilot-scale. The patent was awarded to the International Furan Technology Ltd in 2004 and later the IP was transferred to the Proserpine Co-operative Sugar Milling Association (Australia) with the plan to build a full-scale plant and gain a major position in the furfural production landscape. Furthermore, in 2008 the technology was licensed to Arcoy Biorefinery Private Limited (India) who intended to produce furfural as feedstock for their new furfuryl alcohol manufacturing facility.

At Proserpine, the furfural plant was being commissioned in 2010 when there was unseasonal heavy rain. As a result, the sugar content of the cane supply fell below economic levels and the season ended prematurely. The lack of production placed great financial pressure on the company and the sugar factory including furfural plant was sold to Sucrogen in 2011. The new owners were not interested in the developing the furfural plant. The SupraYield reactors were operated for less than 10 runs with no chance for optimization, and yields were low [14].

The only plant to operate based on the SupraYield patent was Arcoy's biorefinery in Ankleshwar, India. Unfortunately, they could not achieve the yield obtained in the laboratory and closed as uneconomic [14]. As the plant was standalone an increase in the cost of bagasse also played a role. In summary, after almost 20 years from the first patent filing in 1999, the SupraYield concept is not yet fully proven at commercial scale, and presently no plans of new production plants are known.

In conclusion, active furfural production capacity is still strongly bound to the heritage from the very initial development efforts made in the first half of the 20th century, at least in terms of process principles. Low yields and high energy needs make production cost often too high and the whole process uneconomic. Current efforts are directed to new processes demonstrated at lab scale, and the hope is that any of those new technologies will be successfully scaled-up opening new possibilities for the conversion of furfural to renewable chemicals and fuels.

2.2. Future technologies for furfural production

Many new technologies to produce furfural are being developed at lab scale, demonstrating that the unique potential of furfural as building block molecule derived from biomass remains not only untouched but reinforced, even though no commercial progress has been done in the past 60 years. Most of the initial works to produce furfural at lab scale use xylose as feedstock. Other sugars such as arabinose or ribose have also been used, but they are typically less selective. Fortunately, more researchers are realizing that in order to bring their process to commercial scale and better understand the effect of biomass derived impurities, lignocellulosic biomass must be used as feedstock. Corncobs and bagasse are typically selected because of their high content of pentosane and availability. Corn stover is also an interesting feedstock, with a pentosane content over 20% and that is considered a very promising feedstock in a future biorefinery. Other suitable feedstock are oat hulls (however, the high content of arabinose limits its potential), and hardwoods (birch, beech). Softwood, even though a suitable feedstock for paper mills and future biorefineries, has a lower content of pentosane (typically 5–10%) limiting its interest as furfural feedstock. Related to paper mill applications, pre-hydrolysis liquid (C_5 sugars extracted from biomass by hot water treatments) [15, 16] are also a suitable feedstock for furfural production. Utilizing the hemicellulose to produce furfural has been a desire for paper mills for many years, as this possibility is seen as the key to improve the economics of a process over-optimized in terms of yields, energy efficiency, chemicals recovery and recycling. The main issue with these streams is the low concentration of pentosan and the high concentration of salts and impurities that limit the production of furfural. Another liquid stream that is gaining attention is the pre-hydrolysis liquid generated in the cellulosic ethanol facilities. The production of ethanol has been demonstrated successfully, but the price is higher than expected. Conversion of the hemicellulose to furfural instead to ethanol, can be an option to decrease the price of the cellulosic ethanol.

As the reaction mechanism to produce furfural is well known, new research related to furfural production at lab scale work is focused on maximizing the furfural yield as a way to improve the economics of the process. In this sense, the research can be categorized in four big groups.

1. **Stripping of furfural.** Quick removal of the furfural from the liquid solution where it is produced (normally water).
2. **Addition of salts.** Stabilization of reaction intermediates to prevent the furfural degradation.
3. **Biphasic systems.** Continuous removal of the furfural into an organic phase.
4. **Organic solvents.** Prevention of the furfural degradation by using polar aprotic solvents.

2.2.1. *New methods based on the stripping of furfural*

As it has been mentioned before, furfural is produced by acid dehydration of five carbon sugars. At the same reaction conditions used to produce furfural, it polymerizes with itself, xylose and other

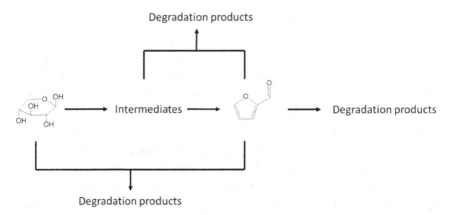

Figure 2.1. Reactions involved in the production and degradation of furfural.

reaction intermediates to produce degradation products, typically called "humins" reducing the selectivity to furfural (Figure 2.1).

The simplest method to avoid these degradation reactions is isolating the furfural from the reaction medium as quick as possible. This concept is the basis of the old Quaker Oats process, which uses steam to remove the furfural from the reaction and it is the method preferred to produce furfural commercially in batch mode in China and Dominican Republic (Central Romana). The low furfural yields obtained with this method (55–60%) have encouraged many researchers to find new methods to remove the furfural more effectively and increase the yields. Reactive distillation is one of the most studied methods. The process can work in continuous or batch and the feedstock utilized are normally aqueous solutions of xylose or pre-hydrolysis liquid. Mandalika *et al.* [17] reported that furfural yields over 85% could be achieved from hot water hydrolysates extracted from hybrid poplar, miscanthus, switchgrass and corn stover using sulfuric acid as catalyst. In the system reported, the dehydration takes places in an open batch reactor at 150–170°C, steam is continuously removed from the reactor through a controlled opened valve. As the furfural is formed, the low boiling point of the water–furfural azeotrope (98°C) allows the furfural to exit the reactor with the steam, preventing the degradation. The work successfully proves the concept of reactive distillation and other works have reported similar benefits contributing to a better understanding of the effect of impurities [18] on the reaction; however, in all cases the furfural produced was too diluted to implement the process commercially as large amount of energy would be necessary to separate the furfural from the water.

To increase the concentration of furfural and try to decrease the energy necessary to separate furfural and water, Arias *et al.* have extensively studied the use of nitrogen as carrier to strip the furfural. The furfural dehydration takes place in a batch reactor, using solutions of sugars in water as feedstock. Nitrogen is fed continuously to the reactor and act as a carrier for the furfural produced removing it quickly from the reactor. Upon condensation of the furfural and water, the nitrogen is easily separated from the liquid phase avoiding the dilution of the furfural that occurs when steam is used as the main gas-carrier. Using Amberlyst 70 as catalyst and diluted aqueous solutions of sugar as feedstock, they reported furfural yields over 75% [19, 20]. An interesting advantage of the process is that at high concentrations of xylose, enough furfural is produced that the ratio of water/furfural evaporated is over the solubility limit of furfural in water (around 8 wt.% at room temperature) allowing for the spontaneous separation into a rich layer of furfural (>90% furfural purity) and a poor layer (8 wt.% furfural in water). The system has been further improved by combining Lewis and Bronsted acid catalysts. It is well known that Lewis acid catalyst promote the isomerization of the xylose to xylulose, which is much more reactive and selective to furfural [21, 22]. Combining Nb_2O_5 and Amberlyst 70 with nitrogen stripping, Agirrezabal *et al.* increased the furfural yield over 80% [23, 24]. An economic analysis performed by the same authors [25] indicates that replacing steam by nitrogen to perform the stripping could lead to saving on the production of furfural. Further work is required to demonstrate the process at continuous mode and larger scale, where the diffusion and mixing of the nitrogen with the solution (mostly in the presence of solid biomass) could be a serious limitation compared with the steam stripping.

With the objective of reducing energy consumption yet improving furfural yield by providing abundant steam stripping, a process has been recently patented at Delft University of Technology [26–28] based on reactive distillation assisted by mechanical vapor compression. In this process, pentosans contained in an acidic aqueous solution are converted into furfural using a reactive distillation column where the top furfural containing vapors are compressed and used to heat the bottom reboiler of the same column. The energy input is thus limited to the power required by the steam compressor and, thanks to the very peculiar thermodynamics of the furfural–water system, specific energy consumption is reduced by a factor 10 compared to a traditional process. Claimed furfural yields are in the order of 80%, thanks to plentiful steam stripping in combination with the use of halides salts in a moderately acidic water solution. Claimed specific energy consumption is as low as 660 kWh/ton of furfural (excluding auxiliaries).

2.2.2. *Addition of salt to increase furfural selectivity*

Utilization of salts, mainly NaCl, has been reported as a way to improve furfural yields. Initially, the intention was to decrease the solubility of furfural in water, causing an effect of salting out that would help the evaporation of furfural via steam-stripping. Further work by Marcotullio *et al.* [29] points to a more important role of the Cl ions in the reaction mechanism. They reported that the Cl ions promote the formation of the 1,2-enediol from the acyclic form of xylose, and thus the subsequent acid catalyzed dehydration to furfural. Marcotullio *et al.* reported an increase in the furfural yields from 70% to >80% when HCl was used as catalyst and 1.7 M NaCl added to the solution [29]. Comparable results were observed when using K, Mg or Ca salts, indicating that the cation does not play a relevant role in the stabilization of the intermediates. In order to maximize the production of furfural, the presence of an acid catalyst is still necessary, even though weak acids, such as acetic and formic have been reported active in the presence of Cl salts [30]. Strong acids, such as HCl or H_2SO_4, are still preferred. Because the presence of Cl salts increases the conversion of sugars [31], if the dehydration reaction is not speed up by the acid catalyst, the sugars goes to degradation products. Other salts, such as $FeCl_3$ or

AlCl$_3$, add a different feature to the system. At controlled pH, they may act as a Lewis acid catalyzing the isomerization of xylose to xylulose and speeding up the reaction rates and the selectivity to furfural. Mao *et al.* [32] used a combination of FeCl$_3$ and acetic acid to produce furfural from corncobs obtaining furfural yields close to 70%. The advantage of that system is that a difference with the conventional process, where the cellulose is degraded and burned to produce energy, the acetic acid is not strong enough to depolymerize and degrade the cellulose so around 75% of it can be used for further chemical upgrading. Lopes *et al.* also reported a similar effect using formic acid as co-catalyst, in their case AlCl$_3$ resulted more effective than FeCl$_3$ [30] which indicates that further work is necessary to understand the exact role of the cation and the anion on the production of furfural as it is probably a combination of the Bronsted and Lewis acid sites, faster sugar conversion, reaction intermediate stabilization and salting out effects.

2.2.3. *Production of furfural in biphasic systems*

Utilization of biphasic systems (water/organic) is one of the most studied alternative to improve the production of furfural because of its simplicity at lab scale and the excellent results obtained. Similarly to the furfural stripping, the idea is to extract the furfural from the reaction phase as soon as possible to avoid its degradation. The sugars (highly soluble in water) and the acid catalyst (normally HCl or H$_2$SO$_4$) remains in the water phase. Ideally, neither the catalyst nor the sugars are soluble into the organic layer, where the furfural remains, minimizing the possibility of furfural condensation reactions and furfural degradation (due to the absence of catalyst).

The system has been further improved by adding salts. As mentioned in the previous section, the addition of salts has a double effect here, first it reduces the solubility of furfural into the aqueous phase increasing the partition coefficient of furfural into the organic phase, and second, they increase the selectivity to sugars (Figure 2.2).

Dumesic *et al.* have studied extensively the utilization of biphasic systems in dehydration reactions [33]. They stablished that the

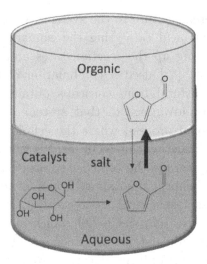

Figure 2.2.   Production of furfural in biphasic systems. The furfural is produced in the aqueous phase, where the sugars and the acid catalyst are soluble, and then it is extracted to the organic layer.

partition coefficient of the furfural in the organic solvent played the most relevant role to increase the furfural yield, as it is the key to prevent the degradation. Huber *et al.* continued the work including kinetic studies using methyl isobutyl ketone (MIBK) as organic solvent [34]. They demonstrated that, even thought there was a clear improvement in the selectivity and yield to furfural using the biphasic system, the fundamental kinetics of the reaction were similar than in the aqueous phase probing that the main effect of the solvent was the extraction of the furfural to prevent its degradation. In this system, it was essential that the xylose was not soluble in the organic solvent. Many other solvents have been tried for the reaction. Initially, the aim was to utilize low boiling point solvents, so they could be easily recovered by evaporation. Amiri *et al.* [35] reported that the use of butanol, propanol and THF significantly increased the yield of furfural when compared with a monophasic system. They didn't observe that effect when using acetone, where the furfural had a low partition coefficient (almost the same amount of furfural was found in the organic and the aqueous phase). Those results were confirmed by Li *et al.* [36], who used toluene, ethyl butyrate, butyl acetate, ethanol and butyl ether in their studies, those solvents with a better partition coefficient resulted in the highest furfural yield. Campos Molina *et al.* [37] explored the utilization of cyclopentyl-methyl-ether and confirmed the improved furfural yield and faster reaction kinetics observed in the presence of NaCl. This effect increased proportionally to the amount of salts, until 30 wt.% NaCl, where the aqueous solution is saturated. Huber group [38] performed a simple technoeconomic analysis to produce furfural using a THF/water/NaCl biphasic reactor to demonstrate the advantages of the process. Using as feedstock a 10 wt.% hemicellulose aqueous solution produced in a paper mill they obtained a furfural yield over 90% and a furfural production cost of $366 per MT. The low value was possible because of the co-production of acetic acid and formic acid, the high furfural yields obtained and the fact that the energy required to evaporate THF and purify the furfural produced was only 25% of the energy required in the current commercial process. In the study, the authors assume that the humins and degradation products will remain in the water phase after the THF evaporation. However, still many organic impurities (mostly phenolics compounds derived from lignin and sugar dehydration products such as HMF) remains soluble in the furfural and acetic acid contaminating the product, that is obtained in the reboiler of the distillation column. Gurbuz *et al.* [39] proposed the use of high boiling solvents, sec-butylphenol (SBP) for continuous extraction of furfural during the reaction. Using this system, the furfural can be easily separated and recovered in the top of a distillation column by taking advantage of the furfural–water azeotrope. They reported furfural yields up to 75% from corn stover, and 78% from xylose. By successive additions of the feedstock over the SBP, the furfural concentration could be increased to 5% reducing separation cost.

Biphasic systems have been clearly very successful at lab scale; however, none of them have achieved commercial or even demonstration scale. The main reasons are the difficulty to maintain an effective mixing at large scale and the formation of emulsions that complicate the phase separation. As mentioned before and demonstrated by several authors, effective extraction of furfural into the organic phase is critical to preserve the yields. While this is easily achieved in small reaction systems, the phase mixing is much more complex at large scale normally leading to mass transfer problems that minimize the diffusion of the furfural from the aqueous phase, where it is produced, to the organic phase. In addition, the accumulation of impurities in the organic phase may increase the solubility of water and sugars into the organic phase reducing the effectiveness of the system and increasing the formation of emulsions.

2.2.4. *Production of furfural in monophasic systems*

To solve the problems associated with biphasic reactors mixing Alonso *et al.* started using monophasic systems based on polar aprotic solvents [40]. Using a combination of gamma-valerolactone (GVL) and water as solvent and sulfuric acid as catalysts they obtained furfural yields up to 81%. Further work by Mellmer *et al.* [41] demonstrated that the xylose dehydration reaction turnover frequency (TOF) was 30× higher when GVL was used as solvent compared with water. They reported that the presence of GVL changed the activation energy for the reaction and suggested that the reason was the stabilization of the acidic proton relative to the protonated transition states, leading to accelerated reaction rates. Instead of preventing the degradation of furfural (polymerization or condensation reaction rates are similar in water and in GVL), the use of polar aprotic solvents allows to reduce the acid concentration, reaction time or reaction temperature leading to lower degradation of the furfural while maintaining the xylose conversion rates and selectivity to furfural. Gurbuz *et al.* [42] further improved the system by using Mordenite as catalyst. They demonstrated that the small pores of the Mordenite helped to decrease the condensation reaction between xylose, intermediates and furfural. Interestingly, they also reported that using GVL as solvent, up to 30% of C_6 sugars present in lignocellulosic biomass can be converted into furfural, which may be an interesting alternative to increase overall furfural yields. In addition, because the boiling point of furfural is lower than that of GVL, furfural could be continuously removed by vaporization adding the benefits of furfural stripping [40]. Further optimization of the reaction conditions have led to furfural yields close to 95% when using HCl as catalyst [43] (thanks to the advantage of the Cl ions effects reported in Section 2.2.2). The use of other chloride salts, such as FeCl$_3$, also resulted successful as reported by Zhang *et al.* [44] who obtained furfural yields close to 80% from corncobs. This demonstrated that the advantages to increase the furfural yields reported in the previous sections can be combined with the use of monophasic systems.

The idea of using monophasic systems to produce furfural was rapidly adopted by Dupont that has been particularly active in this subject, with many patents involving the use of several solvents such as sulfolane, PEG, propylene carbonate, GVL, gamma butyrolactone or diethylene glycol [45–51]. In particular, one of their processes combines the use of reactive distillation with solid acid catalyst, more selective than sulfuric acid [52]. The process works in continuous mode using pre-hydrolysate liquor (PHL) from wood chip digestion, an interesting feedstock that can be produced in paper mills and second-generation ethanol facilities. The liquid feedstock enters the distillation column, which is filled with the solid catalyst (Mordenite, H-Beta, ZSM-5), as the xylose is dehydrated to furfural, it evaporates providing an immediate furfural separation from the reaction zone minimizing the formation of undesired by-products. Sulfolane was using as a co-solvent (80 wt.% sulfolane, 15 wt.% water and 5 wt.% xylose) to increase the furfural yield following the work reported by Dumesic *et al.* [40]. All these features combined demonstrated that it is possible to produce furfural continuously at yields >75%. Using of xylose solutions presented no problems; however, when using pre-hydrolysate liquor as feedstock, the solid catalyst deactivated because of the presence of salts and other biomass derived impurities in less than 30–50 h. The problem could be solved by removing the salts using ion exchange resins and by regenerating the catalyst. Further work to demonstrate the effective recovering of the solvent and increase the stability of the catalyst will be needed before implementing the process commercially. In that sense, GVL presented some advantages compared with other solvents because of its boiling point (208°C), which is high enough to allow the recovery of furfural/water in the top of a distillation column, but low enough that the solvent still can be recovered by evaporation in a subsequent distillation column [43]. Cai *et al.* decided to use another polar aprotic solvent, THF, as solvent [53, 54] leading to similar kinetic effects to those observed for GVL. They reported furfural yields of up to 87% using sulfuric acid as a catalyst and THF/water as solvent. The furfural yield increased to 97% when $FeCl_3$ was added to the system. The main advantage of using a low boiling point solvent $(66°C)$ is that the solvent can be easily recovered by evaporation; however, the pressure inside of the reactor is considerably higher than in the previous systems and the furfural recovery is more complex because it ends at the bottom of the distillation column with other heavy and degradation products. Upgrading the furfural before its separation from the THF could led to improved systems.

An important field of monophasic systems is related to the use of ionic liquids as solvent. This approach has gained considerable attention in the past years and excellent results have been achieved in some cases. Zhang *et al.* [55] reported the use of [BMIM]Cl as solvent and solid acid catalyst to produce furfural obtaining yields over 93%. The ionic liquids, can not only be used as solvent, but as an acid catalyst itself, as demonstrated by Tao *et al.* [56] This dual behavior was utilized by Bogel–Lukasik [57] to improve the production of furfural using ionic liquids.

Finally, many authors have reported the use of solid acid catalysts, specially zeolites, to increase the furfural selectivity [41, 58–67]; however, until date all solid catalyst reported showed a fast deactivation requiring frequent calcination step to recover its activity and preventing them to be implemented at demonstration or commercial scale.

2.3. Challenges and prospective

When people with relevant roles in the industry of furfural were interviewed on the threats and opportunities, mostly agreed on a few aspects that summarizes the challenges and prospective [7, 10, 14]:

— Production costs are too high. Considering that feedstock has a major share of the total cost, yields should be improved. Notably,

yields around 50% of theoretical, often mentioned in literature as unsatisfactory, represent an optimistic target for many producers.
— Energy use is the other big cost item in furfural production. Energy efficient operation is critical to a long-term economic sustainability, and R&D efforts are being put on this topic.
— The potential of a new furfural-based chemical industry is seen as the biggest future opportunity. The expectation from the scientific world is the development of new furfural applications and furfural integration in new generation biorefineries, i.e., along with second generation bioethanol.
— The most recent scientific results found little direct applicability to actual industrial operations.

Future research at lab scale should address those concerns. Even though pure sugars can be used to understand the fundamentals of the system, most of the work must be done using lignocellulosic biomass, so the effect of biomass-derived impurities is well understood and considered. Efforts must be done, not only increasing the furfural yields, but trying to reduce the energy requirements for its separation. In this sense, the use of organic solvents may play a critical role, as in general, the heat of evaporation is much lower than the heat of evaporation of water. A debate arises whether using low or high boiling point solvents. While low boiling point solvents may be easily separated, the purification of the furfural may be challenging. On the other side, high boiling point solvents facilitates the purification of furfural and decrease the energy required to evaporate the furfural; however, clean up and recovering of these solvents become the challenge. When choosing between monophasic and biphasic systems, it seems that monophasic systems are more advantageous for commercial scale applications. The addition of salts, is also promising, and in this sense corrosion issues are the most important. Finally, replacing homogenous catalyst by heterogenous catalysts is always a desired goal. In this direction, the catalyst should be tested in continuous reactors and for at least 100 h operation, which has been proven to be challenging as degradation products accumulate over the surface of the catalyst.

Designing new and effective methods to regenerate the catalysts will be crucial to scale any process based on heterogeneous catalysts and these steps should be included in any technoeconomic analysis performed.

References

1. F.B. Laforge, G.H. Mains, Furfural from corncobs, *Industrial and Engineering Chemistry*, 15(8) (1923) 823–829.
2. H.J. Brownlee, Furfural manufacture from Oat Hulls1: I—A study of the liquid-solid ratio, *Industrial and Engineering Chemistry*, 19(3) (1927) 422–424.
3. G. Natta, E. Beati, Processo continuo per l'estrazione del furfurolo, *La Chimica E L'industria, Ital.*, XXXIII(1951) 63–75.
4. THE FURANS. American Chemical Society Monograph Series No. 119. Signed by A. P. Dunlop and F. N. Peters. by Dunlop, A. P.; F. N. Peters: Reinhold Publishing Corporation, New York Hardcover, First Edition., Signed - Kurt Gippert Bookseller ABAA.
5. K.J. Zeitsch, *The Chemistry and Technology of Furfural and its Many By-Products*, Elsevier, The Netherlands, 2000.
6. H.E. Hoydonckx, W.M. Van Rhijn, W. Van Rhijn, D.E. De Vos, P.A. Jacobs, *Furfural and Derivatives Ullmann's Encyclopedia of Industrial Engineering*, Wiley-VCH, Weinheim, 2008.
7. Charles Kruger, Factory manager at Illovo Sugar Ltd, Private communication, 2016.
8. Illovo Sugar Ltd., Integrated annual report for the year ended 31 March 2015. https://www.illovosugarafrica.com/Overview/Integrated-Annual-Reports.
9. http://centralromana.com.do/estructura-corporativa/manufacturing/?lang =en.
10. Mohammad Yazdizadeh, Production manager at Behran Oil Company, Private communication, 2016.
11. http://www.lenzing.com/en/co-products/products/lenzing-agaustria/furfural.html, 2016.
12. http://www.tanin.si/podstrani_eng/petrochemistry/petrochemistry.php, 2016.
13. L.J. Watson, C.G. Connors, Furfural — a value adding opportunity for the Australian sugar industry, in Proceedings of the Australian Society of Sugar Cane Technology, 2008, Vol. 30, pp. 429–436.
14. L.J. Watson, Former business developer at Proserpine Co-operative Sugar Milling Association, Private communication, 2016.
15. A.T.W.M. Hendriks, G. Zeeman, Pretreatments to enhance the digestibility of lignocellulosic biomass, *Bioresource Technology*, 100(1) (2009) 10–18.
16. N. Mosier, R. Hendrickson, N. Ho, M. Sedlak, M.R. Ladisch, Optimization of pH controlled liquid hot water pretreatment of corn stover, *Bioresource Technology*, 96(18) (2005) 1986–1993.

17. A. Mandalika, T. Runge, Enabling integrated biorefineries through high-yield conversion of fractionated pentosans into furfural, *Green Chemistry*, 14(11) (2012) 3175–3184.
18. H. Liu, H. Hu, M.S. Jahan, Y. Ni, Improvement of furfural production from concentrated prehydrolysis liquor (PHL) of a kraft-based hardwood dissolving pulp production process, *Journal of Wood Chemistry and Technology*, 35(4) (2015) 260–269.
19. I. Agirrezabal-Telleria, A. Larreategui, J. Requies, M.B. Güemez, P.L. Arias, Furfural production from xylose using sulfonic ion-exchange resins (Amberlyst) and simultaneous stripping with nitrogen, *Bioresource Technology*, 102(16) (2011) 7478–7485.
20. I. Agirrezabal-Telleria, J. Requies, M.B. Güemez, P.L. Arias, Furfural production from xylose + glucose feedings and simultaneous N_2-stripping, *Green Chemistry*, 14(11) (2012) 3132–3140.
21. V. Choudhary, A.B. Pinar, S.I. Sandler, D.G. Vlachos, R.F. Lobo, Xylose isomerization to xylulose and its dehydration to furfural in aqueous media, *ACS Catalysis*, 1(12) (2011) 1724–1728.
22. V. Choudhary, S.I. Sandler, D.G. Vlachos, Conversion of xylose to furfural using Lewis and Bronsted acid catalysts in aqueous media, *ACS Catalysis*, 2(9) (2012) 2022–2028.
23. I. Agirrezabal-Telleria, C. García-Sancho, P. Maireles-Torres, P.L. Arias, Dehydration of xylose to furfural using a Lewis or Brönsted acid catalyst and N_2 stripping, *Chinese Journal of Catalysis*, 34(7) (2013) 1402–1406.
24. C. García-Sancho, I. Agirrezabal-Telleria, M.B. Güemez, P. Maireles-Torres, Dehydration of D-xylose to furfural using different supported niobia catalysts, *Applied Catalysis B: Environmental*, 152–153 (2014) 1–10.
25. I. Agirrezabal-Telleria, I. Gandarias, P.L. Arias, Production of furfural from pentosan-rich biomass: Analysis of process parameters during simultaneous furfural stripping, *Bioresource Technology*, 143 (2013) 258–264.
26. G. Marcotullio, J. Wiebren De, Technische Universiteit Delft. Method and apparatus for furfural production WO2015174840 A1, 2015.
27. J. Wiebren De, G. Marcotulio, Technische Universiteit Delft. Process for the production of furfural from pentoses and/or water soluble pentosans WO2012057625 A3, 2012.
28. J. Wiebren De, G. Marcotulio, Technische Universiteit Delft. Process for the production of furfural from pentoses and/or water soluble pentosans US9006471 B2, 2015.
29. G. Marcotullio, W. De Jong, Chloride ions enhance furfural formation from D-xylose in dilute aqueous acidic solutions, *Green Chemistry*, 12(10) (2010) 1739–1746.
30. M. Lopes, K. Dussan, J.J. Leahy, Enhancing the conversion of D-xylose into furfural at low temperatures using chloride salts as co-catalysts: Catalytic combination of AlCl3 and formic acid, *Chemical Engineering Journal*, 323 (2017) 278–286.

31. C. Liu, C.E. Wyman, The enhancement of xylose monomer and xylotriose degradation by inorganic salts in aqueous solutions at 180°C. *Carbohydrate Research*, 341 (2006) 2550–2556.

32. L. Mao, L. Zhang, N. Gao, A. Li, FeCl₃ and acetic acid co-catalyzed hydrolysis of corncob for improving furfural production and lignin removal from residue, *Bioresource Technology*, 123 (2012) 324–331.

33. J.N. Chheda, Y. Roman-Leshkov, J.A. Dumesic, Production of 5-hydroxymethylfurfural and furfural by dehydration of biomass-derived mono- and poly-saccharides, *Green Chemistry*, 9(4) (2007) 342–350.

34. R. Weingarten, J. Cho, J.Wm. Curtis Conner, G.W. Huber, Kinetics of furfural production by dehydration of xylose in a biphasic reactor with microwave heating, *Green Chemistry*, 12(8) (2010) 1423–1429.

35. H. Amiri, K. Karimi, S. Roodpeyma, Production of furans from rice straw by single-phase and biphasic systems, *Carbohydrate Research*, 345(15) (2010) 2133–2138.

36. D.-R. Hua *et al.*, Preparation of furfural and reaction kinetics of xylose dehydration to furfural in high-temperature water, *Petroleum Science*, 13(1) (2016) 167–172.

37. M. Campos, R. Mariscal, M. Ojeda, G. López, Cyclopentyl methyl ether: A green co-solvent for the selective dehydration of lignocellulosic pentoses to furfural, *Bioresource Technology*, 126 (2012) 321–327.

38. R. Xing, W. Qi, G.W. Huber, Production of furfural and carboxylic acids from waste aqueous hemicellulose solutions from the pulp and paper and cellulosic ethanol industries, *Energy & Environmental Science*, 4(6) (2011) 2193–2205.

39. E.I. Gürbüz, S.G. Wettstein, J.A. Dumesic, Conversion of hemicellulose to furfural and levulinic acid using biphasic reactors with alkylphenol solvents, *ChemSusChem*, 5(2) (2012) 383–387.

40. D.M. Alonso, S.G. Wettstein, M.A. Mellmer, E.I. Gurbuz, J.A. Dumesic, Integrated conversion of hemicellulose and cellulose from lignocellulosic biomass, *Energy & Environmental Science*, 6(1) (2012) 76–80.

41. M.A. Mellmer, C. Sener, J.M.R. Gallo, J.S. Luterbacher, D.M. Alonso, J.A. Dumesic, Solvent effects in acid-catalyzed biomass conversion reactions. *Angew. Chem. Int. Ed.* 53 (2014) 11872–11875.

42. E.I. Guerbuez, J.M.R. Gallo, D.M. Alonso, S.G. Wettstein, W.Y. Lim, J.A. Dumesic, Conversion of hemicellulose into furfural using solid acid catalysts in gamma-valerolactone, *Angewandte Chemie International Edition*, 52(4) (2013) 1270–1274.

43. D.M. Alonso *et al.*, Increasing the revenue from lignocellulosic biomass: Maximizing feedstock utilization, *Science Advances*, 3(5) (2017) e1603301.

44. L. Zhang, H. Yu, P. Wang, Y. Li, Production of furfural from xylose, xylan and corncob in gamma-valerolactone using FeCl₃•6H2O as catalyst, *Bioresource Technology*, 151 (2014) 355–360.

45. C. Burket, K.W. Hutchenson, Furfural production from biomass. E I Du Pont De Nemours And Company, US20140171664, 2014.

46. P.J. Fagan, R. Ozer, E.J. Till, Process for the production of furfural. E I Du Pont De Nemours And Company, US9181209, 2015.

47. D.R. Corbin, P.J. Fagan, S.B. Fergusson, K.W. Hutchenson, M.S. Mckinnon, R. Ozer, B. Rajagopalan, S.K. Sengupta, E.J. Till, Process for the production of furfural. E. I. Du Pont De Nemours And Company, WO2013102015A1, 2013.

48. D.R. Corbin, P.J. Fagan, S.B. Fergusson, K.W. Hutchenson, P.S. Metkar, R. Ozer, C.J. Pereira, B. Rajagopalan, S.K. Sengupta, E.J. Till, Process for the production of furfural. E I Du Pont De Nemours And Company, US9012664, 2015.

49. D.R. Corbin, P.J. Fagan, S.B. Fergusson, K.W. Hutchenson, M.S, Mckinnon, R. Ozer, B. Rajagopalan, S.K. Sengupta, E.J. Till, Process for the production of furfural. E I Du Pont De Nemours And Company, EP2797905A1, 2014.

50. D.R. Corbin, P.J. Fagan, S.B. Fergusson, K.W. Hutchenson, P.S. Metkar, R. Ozer, C.J. Pereira, B. Rajagopalan, S.K. Sengupta, E.J. Till, Process for the production of furfural. E I Du Pont De Nemours And Company, US20150152074, 2015.

51. D.R. Corbin, P.J. Fagan, S.B. Fergusson, K.W. Hutchenson, M.S. Mckinnon, R. Ozer, B. Rajagopalan, S.K. Sengupta, E.J. Till, Process for the production of furfural. E I Du Pont De Nemours And Company, CA2859898A1, 2013.

52. P.S. Metkar, E.J. Till, D.R. Corbin, C.J. Pereira, K.W. Hutchenson, S.K. Sengupta, Reactive distillation process for the production of furfural using solid acid catalysts, *Green Chemistry*, 17(3) (2015) 1453–1466.

53. C.M. Cai, T. Zhang, R. Kumar, C.E. Wyman, THF co-solvent enhances hydrocarbon fuel precursor yields from lignocellulosic biomass, *Green Chemistry*, 15(11) (2013) 3140–3145.

54. C.M. Cai, N. Nagane, R. Kumar, C.E. Wyman, Coupling metal halides with a co-solvent to produce furfural and 5-HMF at high yields directly from lignocellulosic biomass as an integrated biofuels strategy, *Green Chemistry*, 16 (2014) 3819–3829.

55. L. Zhang, H. Yu, P. Wang, Solid acids as catalysts for the conversion of D-xylose, xylan and lignocellulosics into furfural in ionic liquid, *Bioresource Technology*, 136 (2013) 515–521.

56. F. Tao, H. Song, L. Chou, Efficient process for the conversion of xylose to furfural with acidic ionic liquid, *Canadian Journal of Chemistry*, 89(1) (2011) 83–87.

57. S. Peleteiro, C.L. Da, G. Garrote, J.C. Parajó, R. Bogel-Łukasik, Simple and efficient furfural production from xylose in media containing 1-butyl-3-methylimidazolium hydrogen sulfate, *Industrial & Engineering Chemistry Research*, 54(33) (2015) 8368–8373.

58. J. Zhang, J. Zhuang, L. Lin, S. Liu, Z. Zhang, Conversion of D-xylose into furfural with mesoporous molecular sieve MCM-41 as catalyst and butanol as the extraction phase, *Biomass Bioenergy*, 39 (2012) 73–77.

59. R. Weingarten, G.A. Tompsett, W.C. Conner, G.W. Huber, Design of solid acid catalysts for aqueous-phase dehydration of carbohydrates: The role of Lewis and Bronsted acid sites, *Journal of Catalysis*, 279(1) (2011) 174–182.

60. X. Shi, Y. Wu, P. Li, H. Yi, M. Yang, G. Wang, Catalytic conversion of xylose to furfural over the solid acid SO42-/ZrO2-Al2O3/SBA-15 catalysts, *Carbohydrate Research*, 346(4) (2011) 480–487.

61. S. Lima, M. Pillinger, A.A. Valente, Dehydration of D-xylose into furfural catalysed by solid acids derived from the layered zeolite Nu-6(1), *Catalysis Communications*, 9(11–12) (2008) 2144–2148.

62. E. Lam, E. Majid, A.C.W. Leung, J.H. Chong, K.A. Mahmoud, J.H.T. Luong, Synthesis of furfural from xylose by heterogeneous and reusable nafion catalysts, *ChemSusChem*, 4(4) (2011) 535–541.

63. S.B. Kim *et al.*, Dehydration of D-xylose into furfural over H-zeolites, *Korean Journal of Chemical Engineering*, 28(3) (2011) 710–716.

64. C. García-Sancho *et al.*, Dehydration of xylose to furfural over MCM-41-supported niobium-oxide catalysts, *ChemSusChem*, 6(4) (2013) 635–642.

65. J.M.R. Gallo, D.M. Alonso, M.A. Mellmer, J.H. Yeap, H.C. Wong, J.A. Dumesic, Production of furfural from lignocellulosic biomass using beta zeolite and biomass-derived solvent, *Topics in Catalysis*, 56(18–20) (2013) 1775–1781.

66. A.S. Dias, M. Pillinger, A.A. Valente, Dehydration of xylose into furfural over micro-mesoporous sulfonic acid catalysts, *Journal of Catalysis*, 229(2) (2005) 414–423.

67. A.S. Dias, S. Lima, M. Pillinger, A.A. Valente, Modified versions of sulfated zirconia as catalysts for the conversion of xylose to furfural, *Catalysis Letters*, 114(3–4) (2007) 151–160.

Chapter 3

Renewable Chemicals, Biofuels and Resins from Furfural

Chapter 3.1

Furfuryl Alcohol and Derivatives

Pedro Maireles-Torres[*,‡] and Pedro L. Arias[†]

Universidad de Málaga, Departamento de Química Inorgánica, Cristalografía y Mineralogía (Unidad Asociada al ICP-CSIC), Facultad de Ciencias, Campus de Teatinos, 29071 Málaga, Spain

†*Department of Chemical and Environmental Engineering, Engineering School of the University of the Basque Country (UPV/EHU), Alameda Urquijo s/n, 48013 Bilbao, Spain*

3.1.1. Introduction

Among chemicals derived from biomass, furfuryl alcohol (2-furanmethanol, FOL) occupies a relevant position, because it accounts for 65% of overall furfural (FUR) production. Pure furfuryl alcohol is a clear colorless liquid, which changes to amber upon contact with the atmosphere. It has a faint odor of burning and a bitter taste, and is soluble in common organic solvents. However, although miscible with water, FOL is unstable. Several physical properties of FOL are displayed in Table 3.1.1. It may irritate skin, eyes and mucous membranes upon contact, and may be toxic by ingestion and skin contact and moderately toxic by inhalation. Its molecular structure is composed by a furanic ring and an alcohol group (Figure 3.1.1).

Furfuryl alcohol possesses a large spectrum of industrial application [1], which justify its relevance. It is used in the foundry industry (resins for high-quality cores and molds for metal casting), as

‡Corresponding author: maireles@uma.es

Table 3.1.1. Physical properties of furfuryl alcohol.

Formula	Boiling point (K)	Melting point (K)	Density (293 K) (g/mL)	Viscosity (298 K), mPa·s	Flash point (K)
$C_5H_6O_2$	443	244	1.1285	4.62	338

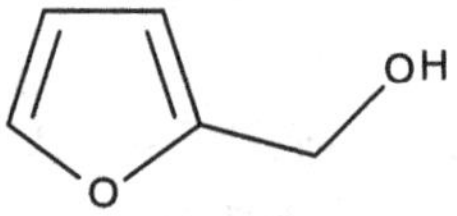

Figure 3.1.1. Structure of furfuryl alcohol.

chemical building block for drug synthesis (for instance, the antiulcer drug ranitidine), precursors for the manufacture of polyurethane foams and polyesters, reactive solvents for phenolic resins in the refractory industry, corrosion-resistant cements and mortars, paint stripper and cleaning compound formulation, wood modification and in the manufacture of fragrances, vitamin C and lysine [1, 2]. Besides, levulinic acid (LA), γ-valerolactone (GVL) and ethyl furfuryl ether (EFE), precursors of biofuels, can be also formed from FOL, although there are not yet commercial (Figure 3.1.2). It has also been demonstrated that the hydroxymethylation with formaldehyde at the 5-position, under acidic conditions, gives rise to 2,5-bis(hydroxymethyl)furan, with applications in the manufacture of polyurethane foams and polyesters. The hydrogenation of FUR, followed by oligomerization of FOL with alcohols and subsequent hydrogenation of the resulting polyethers, produces a mixture of hydrocarbons with a tetrahydrofuran skeleton, which are potential components of biofuels [3].

On the other hand, hydrogenation of furfural can yield to other chemicals, such as 2-methylfuran (MF) (from hydrogenolysis of the C–OH bond), 2-methyl tetrahydrofuran (MTHF) (hydrogenation of MF) and tetrahydrofurfuryl alcohol (THFA) (hydrogenation of the furan ring of FOL) [4]. These are the main reaction products, though minor amounts of others, such as furan and THF,

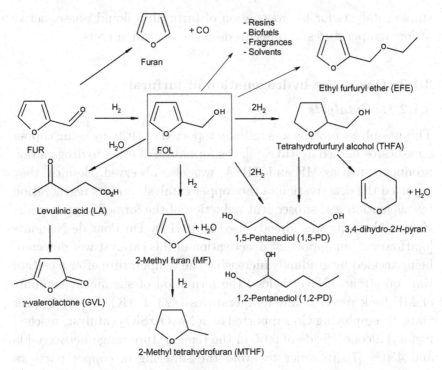

Figure 3.1.2. Furfural hydrogenation to furfuryl alcohol and other chemicals.

different pentanediols, 2-pentanone and 2-pentanol, cyclopentanone and cyclopentanol, can be also detected. However, the selectivity pattern strongly depends on both the reaction conditions and the nature of the catalyst used.

Furfuryl alcohol is prepared industrially by the catalytic hydrogenation of furfural, which can be performed in gas or liquid phase, but the former is the preferred industrial route. The major drawback of the liquid-phase process is the use of batchwise operation, which is not economically attractive for large-scale applications, due to the high operating costs associated to batch reactors and expensive equipment required for high pressure, as well as the elapsing time between successive reactions. Moreover, degradation of furfural in liquid phase is higher, thus causing a fast catalyst deactivation. In this sense, it is necessary an important scientific effort for developing stable catalysts for hydrogenation of furfural in liquid phase, active at low temperatures in order to decrease operation costs.

3.1.2. Gas phase hydrogenation of furfural

3.1.2.1. *Catalysts*

This gas-phase process was initially reported in 1929, by using copper on asbestos backed at 413 K [5], the formation of other hydrogenation products, such as MF and THFA, was also observed. Besides, they reported the deactivation of the copper catalyst, and its regeneration by calcination and subsequent reduction of the formed copper oxide. Copper chromite was already used in 1931 by Du Pont de Nemours [6]. However, an important deactivation of this catalyst was detected, being tackled by gradually increasing the temperature after a certain time on stream. Nevertheless, the formation of significant amounts of MF took place at low temperatures (423–453 K). Later, Quaker Oats, by employing Cu supported on a $Na_2O \cdot xSiO_2$ catalyst, reached furfuryl alcohol yields of 99% in the temperature range between 405 and 450 K [7]. In order to avoid the sintering of copper particles during the regeneration step, the treatment of the spent catalyst under subsequent flows of gases, comprising inert and with a low percentage of oxygen at temperatures below 1073 K was proposed. A more stable shaped catalyst, in comparison with copper chromite, composed of reduced copper and pyrogenic silica was later found, which gives rise to FOL yields higher than 98% after 53 h of operating time, at 443 K with a LHSV of $0.2\,h^{-1}$ [8].

Currently, large industrial plants for furfuryl alcohol production are based on gas-phase hydrogenation of furfural. TransFurans Chemicals is the leading chemical manufacturer of furfuryl alcohol, and its world's largest furfuryl alcohol unit in Belgium operates with an annual output of 32,500 tones [9]. Essentially, the industrial gas-phase process is conducted as follows: furfural is fed into an evaporator system formed by a packed column, a circulating pump and a heater to maintain the FUR temperature at 393 K [10]. H_2 is introduced through the bottom of the reaction column in countercurrent of liquid FUR, which is flowing downwards, thus allowing hydrogen to get saturated with the vapor pressure of FUR at 393 K. The resulting mixture of FUR and H_2 is heated at the chosen reaction temperature and put into contact with the catalyst, mainly copper chromite, conformed as pellets. The heat released during the catalytic process ($61\ kJ\cdot mol^{-1}$) is withdrawn by using oil flowing around the reactor. Reaction products are condensed and FOL is separated by distillation. Carbonaceous species are formed by unselective reactions and deposited on the catalyst surface. The catalyst is then reactivated by increasing the temperature that results in the gasification of the coke through hydrogenation. However, a temperature exists at which conversion is maintained, but selectivity to FOL decreases, and catalyst regeneration must be carried out after 7–10 weeks.

Many other catalytic systems have been proposed for this catalytic process, mainly to overcome the environmental concerns associated to the presence of chromium in copper chromite catalysts, and the experimental conditions and catalytic results of most of them have been recently summarized [11]. Table 3.1.2 gathers some representative catalytic systems, most of which corresponds to a laboratory scale. It must be emphasized the high stability of a $CuCa/SiO_2$ catalyst, which is able to maintain a FOL yield close to 100% during a time on stream as long as 80 h. This excellent catalytic performance might be attributed to the promoting effect

Table 3.1.2. Data of gas-phase hydrogenation of furfural to furfuryl alcohol with different catalysts.

Catalyst	Reaction conditions				FUR conv. (%)	FOL yield (%)	Ref.
	Space velocity (h^{-1})	$H_2/$ FUR	T (K)	Time[a] (h)			
Cu/SiO_2	0.5 (LHSV)	5	443	4	98	97	[8]
Cu/SBA-15	1.5 (WHSV)	12	443	1	92	85	[12]
Cu/MgO	4.8 (WHSV)	2.5	453	5	98	96	[13]
Cu/ZnO	0.5 (WHSV)	12	463	5	93	86	[14]
$CuCa/SiO_2$	0.33 (LHSV)	5	403	80	100	99	[15]
Ni/SiO_2	10 (WHSV)	25	493	n/a	84	31	[16]

Note: [a]Time on stream at which catalytic properties were determined.

of calcium, which prevents resinification reactions of FOL and coke deposition.

3.1.2.2. *Mechanistic studies*

As regards the active sites for the vapor phase hydrogenation of FUR to FOL, Rao *et al.* proposed a simple bimolecular surface reaction as the rate-determining step, which can account for the hydrogenation of FUR over copper chromite [17]. A relationship between Cu^+ sites and activity was found, pointing out that these sites are involved in the catalytic process, but the presence of Cu^0 sites seemed to be also required. This participation of Cu^0 and Cu^+ species in the catalytic process was corroborated later by using N_2O adsorption at 363 K and CO adsorption at 300 K, respectively, in the presence of Cu/activated carbon catalysts [18].

On the other hand, Resasco *et al.*, by using DFT calculations and DRIFTS together with a kinetic based on a Langmuir–Hinshelwood model, have studied the nature of the active species in the furfural hydrogenation over Cu/SiO_2 catalysts, at temperatures ranging between 503 and 563 K [19]. They propose that the bonding of FUR takes place on both Cu (1 1 0) and Cu (1 1 1) surfaces of metallic copper, without the participation of Cu^+ species. The heat of FUR adsorption, as deduced from the fitting of the kinetic model, was found to be significantly higher than those of FOL and MF. They concluded that adsorption of FUR takes place preferentially in a $\eta^1(O)$-aldehyde binding mode, by the lone pair of oxygen, with the FUR molecule perpendicular to the catalyst surface, while the rest of the molecule suffers a net repulsion. Then, the reaction can proceed via either an alkoxide (H addition to the C atom of carbonyl group) or a hydroxyalkyl (H attack at the O atom of carbonyl group) intermediate (Figure 3.1.3). The latter mechanism is preferred as it has a lower activation energy barrier ($32 \, kJ \cdot mol^{-1}$ compared to $45 \, kJ \cdot mol^{-1}$), which is explained by the stabilizing effect of the aromatic furan ring over the hydroxyalkyl intermediate.

Besides copper-based catalyst, Group VIII metals (Ni, Pd, Pt) exhibit hydrogenation activity at low temperatures, with high selectivity toward furfuryl alcohol, although the decarbonylation

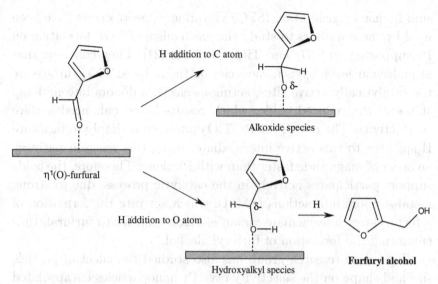

Figure 3.1.3. Comprehensive mechanism of furfural hydrogenation to furfuryl alcohol, on Cu surfaces (adapted from Ref. [15]).

process is favored at high temperatures, being MF predominantly produced. This trend has been confirmed by using platinum monolayers deposited on different supports, previously covered with a transition metal oxide [20]. Thus, it was demonstrated that the different catalytic behavior can be explained by considering the type of FUR adsorption on the catalyst surface. In this sense, the repulsion between unsaturated carbonyl compounds, like FUR (electron-rich substrate), and an electron-rich surface (MgO), weakens the reactant-active site interaction, leading to low conversion. However, a higher polarization of the catalyst surface as for TiO_2, due to its highest difference in electronegativity between Pt and Ti among the studied oxides, gave rise to a higher FUR conversion by using this support, in comparison with γ-Al_2O_3, SiO_2 and MgO. Thus, a deficiency of superficial charge favored the adsorption of furfural, and hence its conversion into FOL.

This effect of the support on the catalytic behavior of Pt-based catalysts has also been extensively studied by Somorjai's research group [21–23]. Density functional calculations (DFT) and sum frequency generation (SFG) vibrational spectroscopy have been used by these authors to study the mechanism of FOL formation on Pt supported on SiO_2 and TiO_2 catalysts [21]. They conclude that at molecular level, oxygen vacancies on the reduced TiO_2 surface are the catalytically active sites, acting as electron donors into mid-gap states of the reduced oxide, which results in an enhanced surface conductivity. The role of the Pt/TiO_2 interface is simply to facilitate H spillover to this active intermediate, being this reaction pathway an order of magnitude faster than with Pt alone. Therefore, the oxide support participates actively in the catalytic process, due to strong metal-support interactions (SMSI), which activate the formation of a furfuryl-oxy intermediate via an electron transfer to furfural, thus enhancing the formation of furfuryl alcohol.

This same research group has also studied the effect of particle size and shape on the selectivity over Pt nanoparticles encapsulated in poly(vinylpyrrolidone) and dispersed on MCF-47 mesoporous silica. They varied the metal sizes between 1.5 and 7.1 nm, with different shapes (octahedral, cube, rounded), and studied the FUR hydrogenation at 473 K, 93.3 kPa of FUR and 93 kPa of H_2 [23]. It was feasible to demonstrate that small Pt particles favor the formation of furan, whereas FOL selectivity increases from 1 to 66% as particle size goes from 1.5 to 7.1 nm and the turnover rate of FOL production is incremented until $7.6 \ 10^{-2} \, s^{-1}$, decreasing the E_a from 104 to $15 \, kJ \cdot mol^{-1}$. Regarding the particle shape, octahedral particles were found to be selective to FOL. It was concluded that two different catalytically active sites exist, whose ratio is modified with the shape and size of metal nanoparticles.

The preferential orientation of adsorbed FUR on metallic surfaces is key for the mechanism of reaction [19]. In this sense, on Cu surfaces, a strong repulsion between the furan ring and the metal surface exists, due to the overlap of the 3d band of the surface Cu atoms and the aromatic furan ring. This repulsion, as demonstrated also by Resasco *et al.* from DFT calculations [19], is absent in Group VIII metals. The adsorption of FUR molecules on a metallic surface can take place by two different modes: (i) $\eta^1(O)$-aldehyde binding mode, via either an alkoxide or hydroxyalkyl intermediate, leading

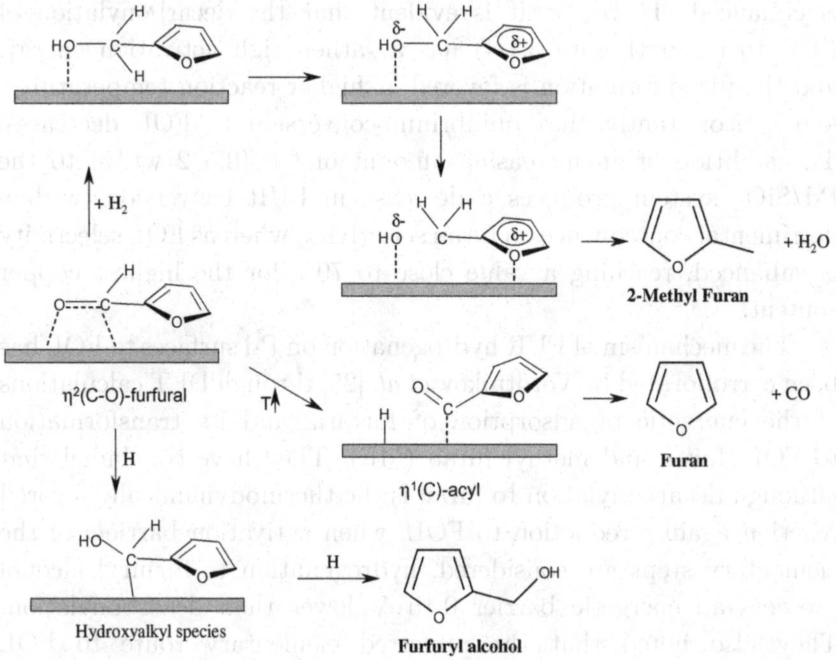

Figure 3.1.4. Comprehensive mechanism of furfural hydrogenation to furfuryl alcohol, methyl furan and furan on Pd and Ni surfaces (adapted from Refs. [16,19]).

the latter to FOL formation (Figure 3.1.3), and (ii) η^2(C–O)-furfural binding mode, forming either an hydroxyalkyl intermediate, which is subsequently hydrogenated to FOL, or an η^1(C)-acyl intermediate, whose decarbonylation gives rise to furan formation (Figure 3.1.4). Pure copper surfaces favor the η^1(O) configuration, whereas both types of binding modes have been observed on Group VIII metals.

In this context, the electronic effects associated to the nature of the active phase and the addition of promoters are very important in determining the nature of the adsorbed aldehyde species. Thus, for Pd–Cu alloys, the stability of η^2(C–O) species is decreased, thus favoring the interaction with the more electronegative atom, that is, the O of the carbonyl group, forming an intermediate similar to an η^1(O)-aldehyde [24]. Consequently, the decarbonylation rate is largely reduced over Pd–Cu catalysts, but the hydrogenation rate is enhanced. Therefore, it is evident that the decarbonylation of FUR to furan (Figure 3.1.4) has a rather high activation energy, and the furan formation is favored at higher reaction temperatures, and concomitantly the equilibrium conversion to FOL decreases. The addition of an increasing amount of Cu (0.5–2 wt.%) to the Pd/SiO$_2$ system produces a decrease in FUR conversion, with a detrimental consequence on furan selectivity, whereas FOL selectivity is enhanced, reaching a value close to 70% for the highest copper content.

The mechanism of FUR hydrogenation on Pd surfaces to FOL has been corroborated by Vorotnikov *et al.* [25] through DFT calculations of the energetic of adsorption of furfural and its transformation to FOL, furan and methyl furan (MF). They have concluded that although decarbonylation to furan is the thermodynamically favored reaction against reduction to FOL, when activation barriers of the elementary steps are considered, hydrogenation to furfuryl alcohol presents an energetic barrier 0.15 eV lower than decarbonylation. They also noted that the preferred elementary route to FOL occurs in two steps: first, hydrogenation of the carbonyl group of the adsorbed furfural, in a flat configuration, to hydroxyalkyl intermediate (barrier of 0.68 eV) and then hydrogenation of carbon bearing the hydroxyl group to adsorbed FOL (barrier of 0.80 eV). The highest energetic barrier in the transformation of furfural to furan is 0.95 eV. These authors also extended the study to the formation of MF from furfural and proposed a different route to produce MF from furfural. The DFT calculations indicated that although there exists a direct pathway involving the hydrogenation of adsorbed furfural to an alkoxy intermediate, subsequently deoxygenated and hydrogenated to MF (Figure 3.1.4), the latter route has an energy barrier (1.10 eV) higher than that of the indirect pathway from FOL. The indirect route from FOL involved, obviously first the formation of adsorbed FOL (as mentioned above has a highest energy barrier of 0.80 eV), but then this adsorbed FOL goes through a route involving dehydration and hydrogenation steps to MF, being 0.85 eV the highest activation energy of this route. However, it must be taken into account that this DFT study has been carried out with low hydrogen coverage (low H_2 pressure) and that an increase of H coverage can change the preference for a given pathway, considering the relative small differences in activation energy between them.

Therefore, the challenge in the vapor-phase hydrogenation of furfural to FOL in the presence of copper chromite, usually employed in the industrial process, or of other catalysts based on Group VIII metals, lies in the inhibition of the pathways to 2-methylfuran and furan when high conversions of furfural are attained, since it is difficult to stop the reaction at FOL.

3.1.2.3. *Catalyst deactivation*

Catalyst deactivation is the other problem that makes difficult the application of copper chromite in vapor-phase processes. This deactivation became more severe at higher FUR partial pressures, making difficult a thorough kinetic study of FUR hydrogenation [17]. Among the reasons to explain this deactivation, the formation of coke and/or poisoning by adsorption of FUR or some reaction products, or a change in the oxidation state of copper species during the catalytic process have been proposed. Recently, deactivation mechanisms of copper chromite in vapor-phase selective hydrogenation of FUR to FOL have been studied by *ex situ* and *in situ* X-ray absorption fine structure (XAFS), X-ray photoelectron spectroscopy (XPS), and Auger electron spectroscopy (AES) [26]. The comparison of AES data of fresh and spent catalysts points that the strong adsorption of species derived from FUR and FOL is the main cause for deactivation. However, loss of Cu^+ sites by hydrogenation can be discarded based on *in situ* XAFS results, which reveal that metallic copper is the active phase for this reaction and it is preserved throughout the catalytic process. Moreover, copper sintering is not responsible of deactivation due to the relatively low reaction temperature, as confirmed by *in situ* XAFS experiments. On the other hand, when the process temperature is increased until 573 K, additional covering of Cu sites by Cr species formed by decomposition of copper chromite affects significantly the catalyst activity, decreasing the selectivity to

FOL. Therefore, the development of new catalysts that cope with this problem is required.

3.1.2.4. *New gas-phase hydrogenation approaches*

Other synthetic approach for FOL production is based on the Meerwein–Ponndorf–Verley (MPV) reduction of an aldehyde, or ketone, coupled with the oxidation of a secondary alcohol, also denominated catalytic transfer hydrogenation (CTH) [27,28]. Thus, for instance, it would involve the combination of furfural hydrogenation and alcohol dehydrogenation in gas-phase conditions. This process takes advantage of the fact that external hydrogen supply is avoided, and the alcohol can be chosen in order to produce an aldehyde or ketone with industrial applications. This is the case of cyclohexanone, which is mostly used as an intermediate in the production of nylon 6 and nylon 6,6. Nagaraja *et al.* have compared the catalytic behavior of a Cu–MgO–Cr$_2$O$_3$ catalyst with a Cu–MgO coprecipitated catalyst and a commercial copper chromite [29]. XPS results reveal the presence of Cu species (Cu^0 and Cu$^+$) on the surface of Cu–MgO–Cr$_2$O$_3$ catalyst. At 473 K, with a liquid cyclohexanol and furfural feed rate of 1 mL h^{-1}, with a cyclohexanol:FUR ratio of 5:1, the highest FOL yield is obtained over the Cu–MgO–Cr$_2$O$_3$ catalyst. This better catalytic performance is attributed to the promotional effect of Cr$_2$O$_3$, a smaller Cu particle size and a high percentage of superficial Cu^0–Cu$^+$ species. This catalyst reaches the steady activity after 2 h of time-on-stream, with 85% FUR conversion and about 100% selectivity to FOL. This coupling process minimizes the formation of by-products, such as MF, THFOL and furan. Later, by using XRD and TPR techniques, this research group has corroborated this suitable promotional effect of Cr, in comparison with other promoters, which has been explained by the stabilization of active species, Cu^0–Cu$^+$, by Cr, which generates a promotion effect higher than the one between Cu and MgO [30]. This has been inferred from the higher dispersion and easier reducibility of copper oxide in the presence of the Cr promoter. However, there is not information concerning the deactivation and reutilization of this family of catalysts.

3.1.3. Liquid-phase hydrogenation of furfural

3.1.3.1. *Catalysts*

Concerning the liquid-phase hydrogenation, it was first reported by the Quaker Oats Company in 1928 [31]. Initially, a catalyst based on Ni/MgO, prepared from the corresponding nitrates by coprecipitation and subsequent nickel reduction, was used, but the process required an accurate control to avoid the formation of tetrahydrofurfuryl alcohol (THFA) by hydrogenation of the furanic ring. However, reduced copper chromite, at high H_2 pressure and temperature, has been largely used as catalyst in this process, with FOL yields higher than 90%. The FOL selectivity was further improved until 98% at 413 K and 10 MPa H_2 pressure by incorporating alkali-earth oxides (CaO, BaO) to copper chromite [32]. Later, addition of CaO to a copper chromite prepared from thermal decomposition of a complex allows to reach FOL yields close to 98% at 453 K, but at a lower H_2 pressure (2.7–3.0 MPa) [33].

From an industrial viewpoint, it has been proposed that liquid-phase hydrogenation of FUR can be performed by mixing copper chromite with FUR to form a slurry, which is continuously fed into a tubular bubble reactor, with different circulation pumps and a compressor for H_2 injection [10]. The slurry is depressurized and excess H_2 is reinjected into the reactor, whereas liquid phase and liquefied head vapors of the column are rectified to obtain pure furfuryl alcohol. Catalysts fines and high boiling polymers are discarded.

Nevertheless, the main drawback of chromium-based catalysts is the environmental concerns associated to their toxicity, and for this reason, many attempts have been made to develop new environmentally friendly catalysts. The variety of catalytic systems proposed for overcoming this drawback is very extense, and some representative examples are given in Table 3.1.3. Thus, a mixture of Cu–Al–Fe oxides and $Ca(OH)_2$ has demonstrated to be active in the liquid-phase hydrogenation of furfural to furfuryl alcohol (Table 3.1.3, entry 15), as well as heteropolyacids incorporated on a Raney nickel (Table 3.1.3, entries 8 and 9). It has been put forward that effect of the heteropolyacid is only as modifying agent,

Table 3.1.3. Data of liquid-phase hydrogenation of furfural to furfuryl alcohol by using different catalysts.

Entry	Catalyst	FUR (wt.%)	Solvent	H$_2$ (MPa)	Temp. (K)	Time (h)	Cat./FUR (wt.)	FUR conv. (%)	FOL yield (%)	Ref.
				Reaction conditions						
1	Ni-alloy	100	None	0.7	373	6	0.1	100	100	[35]
2	Ni$_{74.5}$P$_{12.1}$B$_{13.4}$	1.7	Ethanol	1.7	353	0.25	0.129	100	80	[36,37]
3	Ni$_{34.1}$Fe$_{36.0}$B$_{29.9}$	32.9	Ethanol	1	373	4	0.086	100	100	[38]
4	NiSn$_{0.2}$/SiO$_2$	5.6	Isopropanol	1	373	4	0.108	43	34	[39]
5	NiMoB/γ-Al$_2$O$_3$	24	Methanol	5	353	3	0.2	99	90	[40]
6	Mo-doped Co-B amorphous alloy	14	Ethanol	1	373	3	0.172	100	100	[41]
7	Cu–Zn/Kieselguhr	200	Water	12.4	423	2.5	0.1	95	71	[42]
8	Cu$_{3/2}$PMo$_{12}$O$_{40}$/Raney Ni	60	Ethanol	2	353	1	0.043	98	97	[43]
9	(NH$_4$)$_6$Mo$_7$O$_{24}$/Raney Ni	35	Isopropanol	2.1	333	6	0.057	99.9	98	[44]
10	Pt–Sn$_{0.3}$/SiO$_2$	5.6	Isopropanol	1	373	4	0.108	100	96	[45]
11	Ru/Zr-MOF	1.2	Water	0.5	293	4	0.862	95	95	[46]
12	Ir–ReO$_x$/SiO$_2$	10	Water	6	323	—	0.1	100	97	[47]
13	Cu–Co/SBA-15	18.6	Isopropanol	2	443	4	0.064	99	80	[48]
14	Pd–Cu/MgO	5.7	Water	0.6	383	1.3	0.08	100	99	[49]
15	Cu-Al-Fe oxides/Ca(OH)$_2$	100	None	11.8	433	0.3	0.0065	99.8	98	[50]
16	Cu–Cr oxides	41	n-Octane	6	473	4	0.082	95	78	[51]
17	Cu/Al$_2$O$_3$	0.96	Water	2	363	2	2.08	81	81	[52]
18	CuNi–MgAl mixed oxides	33	Ethanol	1	473	2	0.029	93	83	[53]
19	Cu–Zn–Cr–Zr oxides	16.8	Isopropanol	2	443	3.5	0.108	100	96	[54]

enhancing the reactivity of the carbonyl group, which facilitates its hydrogenation to FOL.

On the other hand, platinum has long been known as a furfural hydrogenation catalyst, and thus, in the pioneer work of Kaufmann *et al.* reduced platinum oxide activated with ferrous chloride was used as catalyst for the reduction of FUR, at 0.1–0.2 MPa H_2 pressure and using an alcohol as solvent, being possible to reutilize the catalyst for several runs after regeneration with O_2 [34]. Pt (and also other noble metals) supported on metal oxides are also active in furfural hydrogenation to FOL, but they tend to favor side and consecutive reactions such as hydrogenolysis of the C–O bond, ring-opening, and decarbonylation, thus decreasing the selectivity toward FOL. However, it has been shown that doping the platinum/metal oxide systems with electropositive transition metals (Sn, Fe, Ga) enhances the selectivity to furfuryl alcohol up to 80%. Thus, in the evaluation of the promoting effect of Sn on a Pt-supported silica catalyst, Merlo *et al.* found that bimetallic systems were more active than the monometallic ones. A $PtSn_{0.3}/SiO_2$ catalyst maintained a selectivity higher than 96% after three successive reaction cycles, although conversion decreases until 70%. This partial deactivation was associated to the adsorption of FUR and/or reaction products on the active sites (Table 3.1.3, entry 3). This promoting effect is explained by considering that the presence of ionic Sn species, acting as Lewis acid sites, favors the attack of hydrogen to the carbonyl group. By studying the activity as a function of the Pt/Sn ratio, they propose a synergy between the dilution effect of Pt sites, which favors the presence of η^1-(O) and η^2-(C,O) species leading to FOL, and the promoting effect of ionic Sn, yielding the highest hydrogenation rates for the highest studied Pt/Sn ratio (Pt/Sn 0.3).

Vetere *et al.* have evaluated the catalytic behavior of Pt, Rh and Ni–SiO_2, and bimetallic systems containing Sn as second metal, under the same experimental conditions (Table 3.1.3, entry 4). The results corroborate the high performance of Pt–Sn catalysts, with FOL yields of 96%, whereas for the monometallic Pt, the FUR conversion is lower than 50%. They also found that the order of activity in FUR hydrogenation to FOL is Pt > Ni > Rh. Thus, a higher expansion of the d orbital of Pt increases the repulsion with the C=C bond, and the probability of adsorption through a planar geometry (leading to hydrogenation of the furan ring) is lower, whereas the adsorption through the carbonyl group is favored.

On the other hand, by using Mo-doped Co-B amorphous alloy catalysts, at 373 K and 1 MPa of hydrogen pressure, 100% FOL yield was achieved (Table 3.1.3, entry 6). The promoting effect of the Fe and Mo dopants on the selectivity to furfuryl alcohol was also explained by taking into account the acidic character of the corresponding oxides, which favors the adsorption and activation of the carbonyl group of furfural. In this sense, a NiMoB/γ-Al$_2$O$_3$ catalyst (Mo/Ni atomic ratio of 1:7) was very effective, attaining a furfuryl alcohol yield of 91% at 353 K and 5 MPa after 3 h (Table 3.1.3, entry 5).

3.1.3.2. *Mechanistic studies*

The kinetics of the liquid-phase hydrogenation has been studied by using Pt/C catalysts in a slurry reactor, using a mixture of 2-propanol and water, at a temperature of 403–448 K and 1–2 MPa H$_2$ pressure, under which mass transfer resistances were absent [55]. The dependence of the initial rates with the hydrogen concentration, within the studied range of temperatures, is almost lineal. Above a FUR concentration of 0.13 kmol m^{-3}, the reaction is zero-order with respect to FUR, and below is 0.86. An activation energy for this reaction of 28 kJ mol^{-1} is deduced, which indirectly corroborates the absence of diffusional limitations. The experimental data were explained using Langmuir–Hinshelwood kinetics, based on a dual-site mechanism with molecular adsorption of all species. Thus, hydrogen was molecularly adsorbed on active sites different from those where FUR and FOL are adsorbed, and the reaction between adsorbed FUR and adsorbed hydrogen is proposed as rate-controlling. The activation energy of this surface reaction was found to be 60.5 kJ mol^{-1}, whereas the adsorption heats of FUR and H$_2$ are 39.8 and 21.3 kJ mol^{-1}, respectively. The reutilization studies showed that the catalyst could be reused without adverse effects.

3.1.3.3. *Catalyst deactivation*

The deactivation of metal-based catalysts in liquid-phase reactions is generally associated to sintering and leaching of copper species. For this reason, more stable noble metals have been proposed to overcome this drawback. Recently, atomic layer deposition has been revealed as a suitable technique to stabilize copper catalysts for liquid-phase catalytic reactions [56]. Thus, atomic layer deposition overcoat followed by calcination at high temperature opens the overcoating, exposing the copper underneath, but maintaining the stabilizing interaction with low coordination copper sites on the surface which prevent leaching and sintering.

3.1.3.4. *New liquid-phase hydrogenation approaches*

An interesting approach for the synthesis of FOL consists of the aqueous electrocatalytic hydrogenation of FUR using a sacrificial anode [57]. The generation of atomic hydrogen is carried out *in situ* by reduction of hydronium ions on a cathode surface using external electrons. The electrocatalytic process in a divided cell was compared with that in the undivided cell, and the effect of the use of different metals (Al, Fe, Ni, Cu and stainless steel) as cathode materials and the electrolyte solution pH on the catalytic performance and the electrochemical efficiency were evaluated. A FOL yield of 63% was reached at pH = 5.0, using a FUR concentration of 100 mM, 0.2 M NH_4Cl in 50 mL water/methanol (4:1, v/v), a current density of 600 mA/dm^2, pure nickel for cathode and anode. Under these experimental conditions, an electrochemical efficiency of 56% and a mass balance of 86%, in an undivided cell, were obtained. Similarly, Green *et al.* have employed a continuous-flow electrocatalytic membrane reactor for the reduction of an aqueous solution of FUR, where the protons are obtained from the electrolysis of water at the anode of the reactor [58]. Different catalysts have been tested as cathode, but the reduction of FUR is better accomplished on a Pd/C, with a selectivity to FOL of 54–100%, at temperatures ranging between 303 and 343 K. Besides FOL, other hydrogenated chemicals such as THFA, MF and MTHF have also been detected, changing the product selectivity as a function of the applied voltage. The current efficiency was 24–30%, being the rest involved in the production of hydrogen gas instead of FUR hydrogenation.

On the other hand, the use of supercritical CO_2 has allowed a real-time tuning of product selectivity in the hydrogenation of furfural [59]. Thus, by using two catalytic beds composed of copper chromite and Pd/C and a suitable choice of the temperature in each reactor, the selectivity to different hydrogenation products of FUR can be switched. In this sense, under a flow of CO_2 of 1 mL min^{-1}, a FUR flow of 0.05 mL min^{-1}, operating a 15 MPa of H_2 pressure, and only employing a temperature of 120°C in the first reactor (second reactor was by-passed for FOL production), a yield of 98% of FOL could be attained.

3.1.4. Biofuels from furfuryl alcohol

The hydrogenation of FUR, followed by oligomerization of FOL with alcohols and subsequent hydrogenation of the resulting polyethers, gives rise to a mixture of hydrocarbons with a tetrahydrofuran skeleton, which are potential components of biofuels. In this sense, Van Buijtenen *et al.* have filled a patent to describe the synthesis from FOL by oligomerization and subsequent hydrogenation of the C_9–C_{20} fraction yielding a mixture suitable to be used as kerosene and diesel components (Figure 3.1.5 and Table 3.1.4), whereas the

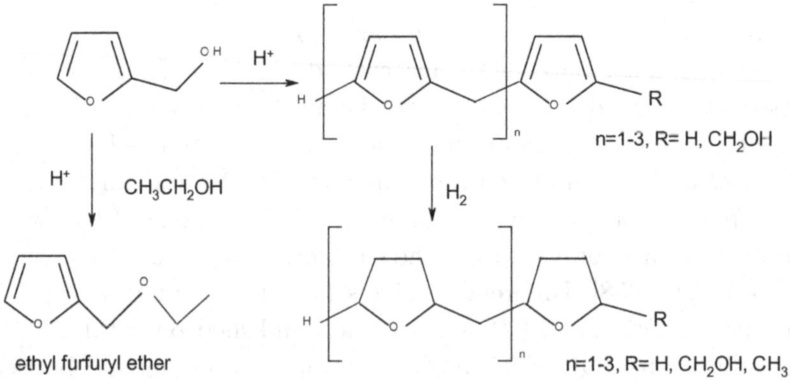

Figure 3.1.5. Formation of ethyl furfuryl ether and saturated oligomers from FOL.

Table 3.1.4. Properties of ethyl furfuryl ether (EFE) as a fuel [61].

Property	Base fuel	5 vol.% EFE in base fuel	10 vol.% EFE in base fuel
Density (g/mL)	746.5	759.9	771.6
RON	96.1	98.0	98.4
MON	85.1	85.3	85.1
Washed gum (mg/100 mL)	<1	97.6	98.0
Vapor pressure (kPa)	67.7	62.8	59.4
Induction (min)	1378	237	101

higher fraction could be fed into a catalytic cracking or hydrocracking unit, after separation by distillation, and also converted to diesel, kerosene and gasoline fraction [60].

The synthesis of the C_9–C_{20} oligomers can be performed under batch conditions or in a continuous mode. The latter process involves pumping a FOL aqueous solution (35 wt.%), containing H_2SO_4 (0.01 wt.%), through a glass column packed with glass beads. By using a residence time of 2.1 h, at 363 K, a FOL conversion of 76 mol% is attained, with an oligomer yield of 41 mol%. On the other hand, under batch conditions, after optimization of experimental variables such as residence time, reaction temperature and acid concentration, the best catalytic performance (95 mol% conversion and 52 mol% oligomers yield) was achieved at 348 K, a residence time of 24 h, a H_2SO_4 concentration of 0.001 M, in the presence of toluene as extracting solvent. The oligomerization of FOL can also be carried out under heterogeneous conditions, reaching a similar conversion and a slightly lower yield (52 mol%), by using a commercial Amberlyst-15 as catalyst (0.002 wt.%), at the same residence time and temperature. Unreacted FOL is removed from the oligomers by washing with water. The subsequent hydrogenation of the C_9–C_{20} carbon–carbon coupled oligomers is performed with a Ni/Al_2O_3 catalyst at 373 K. The hydrogenated product is distilled to isolate the diesel fraction, which can be blended (10 vol.%) with a base fuel. The determination of key parameters in the European diesel specification EN 590 has revealed that the C_9–C_{20} fraction obtained from FOL can be used for blending.

Recently, Haan *et al.* have proposed the use of alkylfurfuryl ether, prepared by reacting FOL with a low molecular weight alkyl alcohol, in the presence of an acidic zeolite catalyst at temperatures ranging from 323 to 473 K, for the preparation of gasoline composition (0.1–30 wt.% of alkylfurfuryl ether) [62]. For instance, ethyl furfuryl ether can be synthesized by mixing 120 g ethanol, 110 g FOL (molar ratio EtOH/FOL = 2.5) and 10 g acidic ZSM-5 (Si/Al = 30), for 2.5 h at 398 K, under stirring. The distillation gives a fraction boiling between 416 and 430 K, containing 77.2 wt.% ethyl furfuryl ether and 16 wt.% FOL, which blended with 95 vol.% of a gasoline increases the research octane number (RON) in two points, until 96.

3.1.5. Conclusions and future prospects

In this chapter, different aspects related to the production of furfuryl alcohol from furfural hydrogenation, have been dealt with and it is evident that there exists a large spectrum of catalysts which have been proposed as alternative to the use of copper chromite-based industrial catalysts. However, to tackle the comparison of their catalytic behavior is difficult since experimental conditions are very varied. Nevertheless, it is necessary to undertake the use of furfural coming directly from biomass treatment, not pure furfural, paying special attention to the development of environmentally friendly catalysts. Key issues to be addressed would be the catalytic stability and selectivity toward FOL, and even if catalyst is deactivated, regeneration processes should be able to recover initial catalytic performance. Moreover, experimental conditions might avoid high hydrogen pressures and reaction temperatures, and here electrochemical hydrogenation or catalytic hydrogenation transfer appear as valuable alternatives. Finally, theoretical tools coupled to *in operando* studies could allow get insights into fundamental aspects at molecular level for designing more active, selective and sustainable catalytic systems.

References

1. H.E. Hoydonckx, W.M. Van Rhijn, W.M. Van Rhijn, D.E. De Vos, P.A. Jacobs, *Furfural and Derivatives*, Wiley-VCH Verlag GmbH & Co., KGaA, Weinheim, 2012.

2. R.H. Kottke, *Kirk-Othmer Encyclopedia of Chemical Technology, Volume 12*, John Wiley Sons, New York, 1998.

3. B.J. Van, J.-P. Lange, R.J. Price, WO20110767, 2011.

4. H.Y. Zheng, Y.L. Zhu, B.T. Teng, Z.Q. Bai, C.H. Zhang, H.W. Xiang, Y.W. Li, Towards understanding the reaction pathway in vapour phase hydrogenation of furfural to 2-methylfuran, *Journal of Molecular Catalysis A: Chemical*, 246 (2006) 18–23.

5. E. Ricard, H.M. Guinot, Process for the manufacture of furfuryl alcohol and methylfurane, US1739919, 1929.

6. W. Lazier, Process for hydrogenating furfural, US2077422, 1937.

7. S. Swadesh, Catalytic production of furfuryl alcohol and catalyst therefor, US2754304, 1956.

8. M. Bankmann, J. Ohmer, T. Tacke, EUPatent 0669163 A1, 1995.

9. http://www.transfurans.be.

10. K.J. Zeitsch, *The Chemistry and Technology of Furfural and its Many By-Products, Volume 13*, Elsevier, Netherlands, 2001, pp. 338–339.

11. R. Mariscal, P. Maireles-Torres, M. Ojeda, I. Sádaba, M. López Granados, Furfural: A renewable and versatile platform molecule for the synthesis of chemicals and fuels, *Energy & Environmental Science*, 9 (2016) 1144–1189.

12. D. Vargas-Hernández, J.M. Rubio-Caballero, J. Santamaría-González, R. Moreno-Tost, J.M. Mérida-Robles, M.A. Pérez-Cruz, A. Jiménez-López, R. Hernández-Huesca, P. Maireles-Torres, Furfuryl alcohol from furfural hydrogenation over copper supported on SBA-15 silica catalysts, *Journal of Molecular Catalysis A: Chemical*, 383–384 (2014) 106–113.

13. B.M. Nagaraja, V. Siva Kumar, V. Shasikala, A.H. Padmasri, B. Sreedhar, B. David Raju, K.S. Rama Rao, A highly efficient Cu/MgO catalyst for vapour phase hydrogenation of furfural to furfuryl alcohol, *Catalysis Communications*, 4 (2003) 287–293.

14. C.P. Jiménez-Gómez, J.A. Cecilia, D. Durán-Martín, R. Moreno-Tost, J. Santamaría-González, J. Mérida-Robles, R. Mariscal, P. Maireles-Torres, Gas-phase hydrogenation of furfural to furfuryl alcohol over Cu/ZnO catalysts, *Journal of Catalysis*, 336 (2016) 107–115.

15. J. Wu, Y.M. Shen, C.H. Liu, H.B. Wang, C. Geng, Z.X. Zhang, Vapor phase hydrogenation of furfural to furfuryl alcohol over environmentally friendly Cu-Ca/SiO$_2$ catalyst, *Catalysis Communications*, 6 (2005) 633–637.

16. S. Sitthisa, W. An, D.E. Resasco, Selective conversion of furfural to methylfuran over silica-supported NiFe bimetallic catalysts, *Journal of Catalysis*, 284 (2011) 90–101.

17. R. Rao, A. Dandekar, R.T.K. Baker, M.A. Vannice, Properties of copper chromite catalysts in hydrogenation reactions, *Journal of Catalysis*, 171 (1997) 406–419.

18. R.S. Rao, R.T.K. Baker, M.A. Vannice, Furfural hydrogenation over carbon-supported copper, *Catalysis Letters*, 60 (1999) 51–57.

19. S. Sitthisa, T. Sooknoi, Y.G. Ma, P.B. Balbuena, D.E. Resasco, Kinetics and mechanism of hydrogenation of furfural on Cu/SiO$_2$ catalysts, *Journal of Catalysis*, 277 (2011) 1–13.

20. J. Kijeński, P. Winiarek, T. Paryjczak, A. Lewicki, A. Mikolajska, Platinum deposited on monolayer supports in selective hydrogenation of furfural to furfuryl alcohol, *Applied Catalysis A: General*, 233 (2002) 171–182.

21. L.R. Baker, G. Kennedy, M. Van Spronsen, A. Hervier, X. Cai, S. Chen, L.W. Wang, G.A. Somorjai, Furfuraldehyde hydrogenation on titanium oxide-supported platinum nanoparticles studied by sum frequency generation vibrational spectroscopy: Acid-base catalysis explains the molecular origin of strong metal-support interactions, *Journal of the American Chemical Society*, 134 (2012) 14208–14216.

22. K. An, N. Musselwhite, G. Kennedy, V. V. Pushkarev, L. Robert Baker, G.A. Somorjai, Preparation of mesoporous oxides and their support effects on Pt nanoparticle catalysts in catalytic hydrogenation of furfural, *Journal of Colloid and Interface Science*, 392 (2013) 122–128.

23. V. V Pushkarev, N. Musselwhite, K.J. An, S. Alayoglu, G.A. Somorjai, High structure sensitivity of vapor-phase furfural decarbonylation/ hydrogenation reaction network as a function of size and shape of Pt nanoparticles, *Nano Letters*, 12 (2012) 5196–5201.

24. S. Sitthisa, T. Pham, T. Prasomsri, T. Sooknoi, R.G. Mallinson, D.E. Resasco, Conversion of furfural and 2-methylpentanal on Pd/SiO$_2$ catalysts, *Journal of Catalysis*, 280 (2011) 17–27.

25. V. Vorotnikov, G. Mpourmpakis, D.G. Vlachos, DFT study of furfural conversion to furan, furfuryl alcohol, and 2-methylfuran on Pd(111), *ACS Catalysis*, 2 (2012) 2496–2504.

26. D. Liu, D. Zemlyanov, T. Wu, R.J. Lobo-Lapidus, J.A. Dumesic, J.T. Miller, C.L. Marshall, Deactivation mechanistic studies of copper chromite catalyst for selective hydrogenation of 2-furfuraldehyde, *Journal of Catalysis*, 299 (2013) 336–345.

27. M.A. Aramendía, V. Borau, C. Jiménez, J.M. Marinas, J.R. Ruiz, F.J. Urbano, Influence of the preparation method on the structural and surface properties of various magnesium oxides and their catalytic activity in the Meerwein-Ponndorf-Verley reaction, *Applied Catalysis A: General*, 244 (2003) 207–215.

28. A. Corma, M.E. Domine, S. Valencia, Water-resistant solid Lewis acid catalysts: Meerwein-Ponndorf-Verley and Oppenauer reactions catalyzed by tin-beta zeolite, *Journal of Catalysis*, 215 (2003) 294–304.

29. B.M. Nagaraja, A.H. Padmasri, B.D. Raju, K.S.R. Rao, B. David Raju, K.S. Rama Rao, B.D. Raju, K.S.R. Rao, Vapor phase selective hydrogenation of furfural to furfuryl alcohol over Cu-MgO coprecipitated catalysts, *Journal of Molecular Catalysis A: Chemical*, 265 (2007) 90–97.

30. B.M. Nagaraja, A.H. Padmasri, B.D. Raju, K.S.R. Rao, Production of hydrogen through the coupling of dehydrogenation and hydrogenation for the synthesis of cyclohexanone and furfuryl alcohol over different promoters supported on Cu-MgO catalysts, *International Journal of Hydrogen Energy*, 36 (2011) 3417–3425.

31. F.N. Peters, US1906973, 1933.

32. H. Adkins, R. Connor, Method of hydrogenation of furfural to furfuryl alcohol, US2094975, 1937.

33. L. Frainier, H. Fineberg, Preparation of furfuryl alcohol from furfural, US4302397, 1981.

34. W.E. Kaufmann, R. Adams, The use of platinum oxide as a catalyst in the reduction of organic compounds. IV. Reduction of furfural and its derivatives, *Journal of the American Chemical Society*, 45 (1923) 3029–3044.

35. P. Mastagli, Method for the catalytic hydrogenation of furfural, US2763666, 1956.

36. S.P. Lee, Y.W. Chen, Selective hydrogenation of furfural on Ni-P, Ni-B, and Ni-P-B ultrafine materials, *Industrial & Engineering Chemistry Research*, 38 (1999) 2548–2556.

37. S.P. Lee, Y.W. Chen, Effects of preparation on the catalytic properties of Ni-P-B ultrafine materials, *Industrial & Engineering Chemistry Research*, 40 (2001) 1495–1499.

38. H.X. Li, H.S. Luo, L. Zhuang, W.L. Dai, M.H. Qiao, Liquid phase hydrogenation of furfural to furfuryl alcohol over the Fe-promoted Ni-B amorphous alloy catalysts, *Journal of Molecular Catalysis A: Chemical*, 203 (2003) 267–275.

39. V. Vetere, A.B. Merlo, J.F. Ruggera, M.L. Casella, Transition metal-based bimetallic catalysts for the chemoselective hydrogenation of furfuraldehyde, *Journal of the Brazilian Chemical Society*, 21 (2010) 914–920.

40. S. Wei, H. Cui, J. Wang, S. Zhuo, W. Yi, L. Wang, Z. Li, Preparation and activity evaluation of NiMoB/γ-Al$_2$O$_3$ catalyst by liquid-phase furfural hydrogenation, *Particuology*, 9 (2011) 69–74.

41. X.F. Chen, H.X. Li, H.S. Luo, M.H. Qiao, Liquid phase hydrogenation of furfural to furfuryl alcohol over Mo-doped Co-B amorphous alloy catalysts, *Applied Catalysis A: General*, 233 (2002) 13–20.

42. G.G. De Witt, Catalytic Hydrogenation of Furfural, US2077409 A, 1937.

43. L. Baijun, L. Lianhai, W. Bingchun, C. Tianxi, K. Iwatani, Liquid phase selective hydrogenation of furfural on Raney nickel modified by impregnation of salts of heteropolyacids,, *Applied Catalysis A: General*, 171 (1998) 117–122.

44. W.R. De Thomas, E.V. Hort, Catalyst comprising Raney nickel with adsorbed molybdenum compound, US4153578, 1979.

45. A.B. Merlo, V. Vetere, J.F. Ruggera, M.L. Casella, Bimetallic PtSn catalyst for the selective hydrogenation of furfural to furfuryl alcohol in liquid-phase, *Catalysis Communications*, 10 (2009) 1665–1669.

46. Q. Yuan, D. Zhang, L. Van Haandel, F. Ye, T. Xue, E.J.. M. Hensen, Selective liquid phase hydrogenation of furfural to furfuryl alcohol by Ru/Zr-MOFs, *Journal of Molecular Catalysis A: Chemical*, 406 (2015) 58–64.

47. S. Liu, Y. Amada, M. Tamura, Y. Nakagawa, K. Tomishige, One-pot selective conversion of furfural into 1,5-pentanediol over a Pd-added Ir-ReO$_x$/SiO$_2$ bifunctional catalyst, *Green Chemistry*, 16 (2014) 617–626.

48. S. Srivastava, P. Mohanty, J.K. Parikh, A.K. Dalai, S.S. Amritphale, A.K. Khare, Cr-free Co-Cu/SBA-15 catalysts for hydrogenation of biomass-derived α-, β-unsaturated aldehyde to alcohol, *Cuihua Xuebao/Chinese Journal of Catalysis*, 36 (2015) 933–942.

49. K. Fulajtárova, T. Soták, M. Hronec, I. Vávra, E. Dobročka, M. Omastová, Aqueous phase hydrogenation of furfural to furfuryl alcohol over Pd-Cu catalysts, *Applied Catalysis A: General*, 502 (2015) 78–85.

50. B. Miya, Method of producing copper-iron-aluminum catalysts, US4252689, 1981.

51. K. Yan, A. Chen, Efficient hydrogenation of biomass-derived furfural and levulinic acid on the facilely synthesized noble-metal-free Cu-Cr catalyst, *Energy*, 58 (2013) 357–363.

52. M. Lesiak, M. Binczarski, S. Karski, W. Maniukiewicz, J. Rogowski, E. Szubiakiewicz, J. Berlowska, P. Dziugan, I. Witońska, Hydrogenation of furfural over Pd-Cu/Al₂O₃ catalysts. The role of interaction between palladium and copper on determining catalytic properties, *Journal of Molecular Catalysis A: Chemical*, 395 (2014) 337–348.

53. C. Xu, L. Zheng, J. Liu, Z. Huang, Furfural Hydrogenation on Nickel-promoted Cu-containing Catalysts Prepared from Hydrotalcite-Like Precursors, *Chinese Journal of Chemistry*, 29 (2011) 691–697.

54. R. V. Sharma, U. Das, R. Sammynaiken, A.K. Dalai, Liquid phase chemoselective catalytic hydrogenation of furfural to furfuryl alcohol, *Applied Catalysis A: General*, 454 (2013) 127–136.

55. P.D. Vaidya, V. V Mahajani, kinetics of liquid-phase hydrogenation of furfuraldehyde to furfuryl alcohol over a Pt/C catalyst, *Industrial & Engineering Chemistry Research*, 42 (2003) 3881–3885.

56. B.J. O'Neill, D.H.K. Jackson, A.J. Crisci, C.A. Farberow, F. Shi, A.C. Alba-Rubio, J. Lu, P.J. Dietrich, X. Gu, C.L. Marshall, P.C. Stair, J.W. Elam, J.T. Miller, F.H. Ribeiro, P.M. Voyles, J. Greeley, M. Mavrikakis, S.L. Scott, T.F. Kuech, J.A. Dumesic, Stabilization of copper catalysts for liquid-phase reactions by atomic layer deposition, *Angewandte Chemie International Edition*, 52 (2013) 13808–13812.

57. Z.L. Li, S. Kelkar, C.H. Lam, K. Luczek, J.E. Jackson, D.J. Miller, C.M. Saffron, Aqueous electrocatalytic hydrogenation of furfural using a sacrificial anode, *Electrochimica Acta*, 64 (2012) 87–93.

58. S.K. Green, J. Lee, H.J. Kim, G.A. Tompsett, W.B. Kim, G.W. Huber, The electrocatalytic hydrogenation of furanic compounds in a continuous electrocatalytic membrane reactor, *Green Chemistry*, 15 (2013) 1869–1879.

59. J.G. Stevens, R.A. Bourne, M. V. Twigg, M. Poliakoff, Real-time product switching using a twin catalyst system for the hydrogenation of furfural in supercritical CO₂, *Angewandte Chemie International Edition*, 49 (2010) 8856–8859.

60. J. Van Buijtenen, J.P. Lange, R.J. Price, Process for preparing a hydrocarbon or mixture of hydrocarbons, US20110173877, 2011.

61. J.P. Lange, E. Van Der Heide, J. Van Buijtenen, R. Price, Furfural-A promising platform for lignocellulosic biofuels, *ChemSusChem*, 5 (2012) 150–166.

62. J.-P. Haan, R.J. Lange, Gasoline composition and process for the preparation of alkylfurfuryl ether, US8372164 B2, 2013.

Chapter 3.2

Tetrahydrofurfuryl Alcohol and Derivatives

Pedro Maireles-Torres*,‡ and Pedro L. Arias†

*Universidad de Málaga, Departamento de Química Inorgánica,
Cristalografía y Mineralogía (Unidad Asociada al ICP-CSIC),
Facultad de Ciencias, Campus de Teatinos,
29071 Málaga, Spain
†Department of Chemical and Environmental Engineering,
Engineering School of the University of the Basque
Country (UPV/EHU), Alameda Urquijo s/n,
48013 Bilbao, Spain

3.2.1. Introduction

Besides furfuryl alcohol, furfural hydrogenation can lead to other more hydrogenated chemicals like tetrahydrofurfuryl alcohol (THFA). This is a clear colorless liquid, with mild odor, completely water-miscible and biodegradable, whose main physical properties are gathered in Table 3.2.1. The molecular structure is formed by a tetrahydrofuranic ring and an alcohol group (Figure 3.2.1).

THFA possesses many agricultural and industrial applications, in the fields of coating, dyes and printing inks, epoxy curing agent, and formulations for cleaning (electronic cleaning, paint stripping, biocides and pesticides, and pharmaceutics. However, its main application is the synthesis of 3,4-dihydro-2H-pyran (dihydropyran), which is used as reagent for protecting alcohols in organic synthesis [1], as well as in the preparation of relevant agrochemicals and

---

‡Corresponding author: maireles@uma.es

Table 3.2.1.   Physical properties of tetrahydrofurfuryl alcohol.

Formula	Boiling point (K)	Melting point (°C)	Density (293 K) (g/mL)	Viscosity (298 K), (MPa's)	Vapor pressure (293 K) (kPa)	Flash point (K)
$C_5H_{10}O_2$	451	<193	1.051	5.49	0.03	357

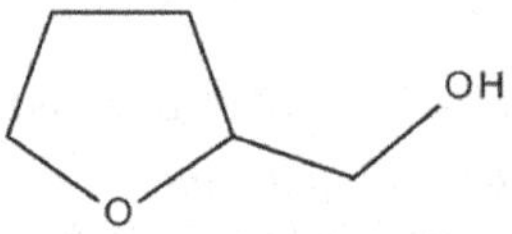

Figure 3.2.1.   Molecular structure of tetrahydrofurfuryl alcohol.

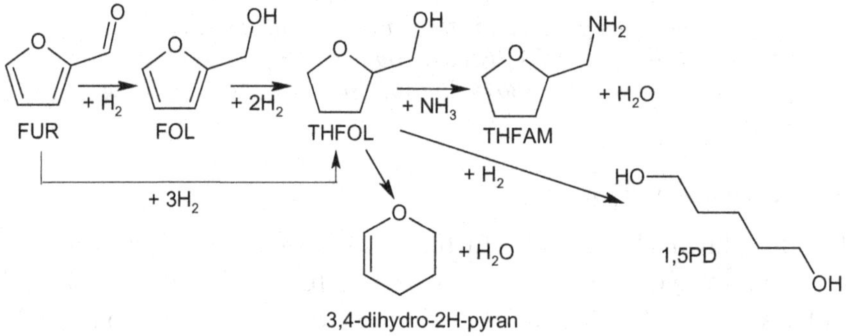

Figure 3.2.2.   Furfural hydrogenation pathways to THFA and its main derivatives.

pharmaceuticals due to its double bond that can react with H_2, H_2O, Cl_2, alcohols, glycols, and organic acids [2, 3] (Figure 3.2.2). Moreover, THFA has been proposed as a new generation biofuel or as a fuel additive, based on the similarity of some of its physico-chemical properties to those of kerosene (for instance, an octane number of 83) [4, 5]. It also finds applications as reactive diluent for epoxy resins, being a low cost, biodegradable solvent for many of the curatives and catalysts for epoxy formulations. In this sense, THFA will allow to accelerate the cure of bisphenol, resins with either aliphatic or aromatic amine curatives.

On the other hand, etherification of THFA with different alcohols or acetals gives rise to oxygenated compounds, which have also been proposed as diesel additives due to the significant decrease of particle emissions [6]. In this context, the most suitable properties have been attained with tetrahydrofurfuryl *tert*-butyl ether and ditetrahydrofurfuryl polyacetal. These condensation processes are carried out in the presence of Amberlyst-15 as an acid catalyst, under batch conditions, by mixing THFA and isobutene or dioxane, respectively, at temperatures lower than 373 K. Moreover, bis(tetrahydrofurfuryl) ether has been proposed as a non-volatile solvent in poly(3,4-propylenedioxythiophene)-based supercapacitors, thus affording a higher resistance than the neat ionic liquid electrolyte [7].

3.2.2. Production of THFA

THFA is industrially produced by furfuryl alcohol hydrogenation, using a supported Ni catalyst under moderate reaction temperatures (323–373 K), under gas or liquid-phase conditions [8]. However, direct hydrogenation of furfural has also been demonstrated to be feasible for THFA synthesis. Many different catalytic systems have been proposed, and the most relevant results reported in the literature, both for direct hydrogenation of furfural or for hydrogenation of FOL, in liquid and gas phase conditions, are summarized in Table 3.2.2. Gas-phase hydrogenation processes in a fixed-bed reactor are more suitable, compared to the liquid-phase, in terms of scaling-up (productivity), handling and the prevention of leaching. However, most of papers dealing with the hydrogenation to THFA are carried out in liquid phase.

As regards the catalytic results, Merat *et al.* evaluated the behavior of Pd-, Rh-, Ru- and Ni-supported catalysts, and their mixtures with a Cu-supported catalyst, in the liquid-phase hydrogenation of both FUR and FOL to THFA (Table 3.2.2, entries 4 and 5). The study of FUR hydrogenation demonstrated that mixtures of Ni- and Cu-supported catalysts gave rise to the best catalytic results. The reaction rates, under the most suitable experimental conditions, were 3.3–4.2 10^{-2} mmol THFA obtained s^{-1} g_{cat}^{-1}. However, the catalytic

Table 3.2.2. Catalysts for liquid-phase hydrogenation of FUR (or FOL) to tetrahydrofurfuryl alcohol.

Entry	Catalyst	FUR (FOL) (wt.%)	Solvent	H_2 (MPa)	Temp. (K)	Time (h)	Cat./FUR (wt.)	Conv. (%)	Yield (%)	Ref.
			Reaction conditions							
1	$NiCrO_4$	87	Water	0.7	413	5	0.075	100	70	[9]
2	Ru/MgO	49	Ethanol	15	383	n/a	0.042	100	78	[10]
3	Ni–Cr	100	—	4.6	308	2.25	0.1	100	90	[11]
4	Supported Ni/Cu	100	—	4	403	3	0.026	100	97	[12]
5	RuO_2+Cu	59	Methanol	5	393	1.5	0.069	100	86	[12]
6	Pd/MFI	11.5	Isopropanol	3.4	493	5	0.086	84	83	[13]
7	CuNi/CNT	12.8	Ethanol	4	403	10	0.172	100	90	[14]
8	CuNi/MgAlO	2.95	Ethanol	4	423	3	0.104	100	95	[15]
9	Pd–Ir–ReO_x/SiO_2	10	Water	6	323	2	0.1	100	78	[16]
10	Ni/Ba–Al_2O_3	5	Water	4	413	4	0.4	99	99	[17]
11	Supported Ni	(100)	None	4	403	3.8	0.03	99	96	[12]
12	SnPd/TiO_2	(0.8)	Methanol	1	298	1	0.1	100	52	[18]
13	NiPd/SiO_2	(5)	Water	8	313	2	0.21	99	96	[19]
14	RuO_2	(59)	Methanol	5	393	1.6	0.04	99	89	[12]
15	Ru/hectorite	(1.6)	Methanol	2	313	1	0.22	100	99	[20]
16	Ru/MnO_x	(10)	Water	3	333	12	0.05	91	91	[21]

mixture could not be recycled by heating under a N_2 flow, at 423 K for 1 h. In the case of FOL hydrogenation, the best catalytic performance (THFA yields as high as 98–99%) was attained with a Ni (59 wt.%) supported on a silica–alumina catalyst, without solvent and under mild operating conditions ($T = 403$ K, $P(H_2) = 4$ MPa). Moreover, this catalyst exhibits a long lifetime.

Biradar *et al.* have reported the direct hydrogenation of FUR to THFA, in liquid phase, in the presence of a Pd/Si–MFI catalyst (Table 3.2.2, entry 6). Full conversion could be reached by recycling the crude of the first hydrogenation assay over the same catalyst; moreover, this catalyst maintained its catalytic activity after five runs, with a slight decrease in the FUR conversion due to the physical loss of catalyst during withdrawing of samples between runs. By using bimetallic Cu–Ni catalysts (Cu:Ni mole ratio of 1:1) supported on carbon nanotubes (CNT), a THFA yield of 90.3% was attained (Table 3.2.2, entry 7). This was explained by considering the synergistic interactions between Ni and Cu, together with the favorable structure and properties of CNT. In this context, CuNi alloy nanocatalysts, supported on calcined MgAl hydrotalcites, with different Cu/Ni molar ratio, have also been proposed for furfural hydrogenation (Table 3.2.2, entry 8). With CuNi/MgAlO catalysts (CuNi and Cu_1Ni_3), a THFA yield of 93–95% was attained, whereas for higher Ni contents, the strong metal-support interaction diminished the reduction degree of active metal species, resulting in low hydrogenation activity. By evaluating the catalytic behavior after reduction at different temperatures, it was demonstrated that the hydrogenation of FOL to THFA is the rate-determining step, as deduced from the lower TOF values obtained for the hydrogenation of the C=C bond in the furan ring. TOF values are higher for bimetallic catalysts, in comparison to monometallic copper and nickel based catalysts. The existence of a volcano-shaped variation of the selectivity to THFA with the average metal particle size could reveal that furfural hydrogenation to THFA is structure-sensitive, thus corroborating results obtained by Nakagawa *et al.* with Ni/SiO_2 catalysts in the gas-phase hydrogenation of FUR [22]. Moreover, basicity must be controlled to avoid secondary reactions leading to pentanediols, associated with an excessive basic concentration. Finally, this work also confirms the role of the solvent in modifying the selectivity, since by using methanol instead of ethanol, the formation of FOL is favored. The good stability of CuNi alloy nanocatalysts is demonstrated during consecutive catalytic runs.

In the case of FOL hydrogenation, a positive effect of the addition of tin to a 5%Pd/TiO$_2$ catalysts has also been reported (Table 3.2.2, entry 12). Without tin, the catalyst was very selective to 2-methylfuran, but the presence of tin shifted the selectivity toward the formation of ring-saturated molecules, such as THFA and methyltetrahydrofuran. Moreover, it was found an increase of THFA yield by a suitable choice of solvent, with methanol and ethanol as most efficient ones, when a Sn:Pd weight ratio of 1:1 was used, being disfavored the hydrogenolysis process. This excellent catalytic behavior is observed at mild experimental conditions, and it is explained by assuming, as inferred from XPS data, that Pd is occluded within SnO$_2$. The best catalytic performance was achieved with Ru supported on hectorite, with a 99% THFA yield, at 313 K (Table 3.2.2, entry 15), but more drastic experimental conditions are required to attain similar catalytic performance when supported on MnO$_x$ (Table 3.2.2, entry 16).

There are not many works devoted to the gas-phase hydrogenation of furfural to tetrahydrofurfuryl alcohol. However, THFA yields higher than 94% were obtained with a Ni/SiO$_2$ catalyst, at 413 K after 30 min of time-on-stream, using a H$_2$:FUR molar ratio of 36 [22]. It was demonstrated that the reaction rate of the second hydrogenation step, that is, the conversion of FOL to THFA, is higher than that corresponding to the former hydrogenation of FUR to FOL, and adsorption of FUR on the catalyst surface is stronger than FOL, thus inhibiting the formation of THFA. This finding is in contrast to the results obtained in liquid phase, where the reaction rate of hydrogenation of FOL to THFA was lower than that of FUR to FOL.

On the other hand, Ordomsky *et al.* have combined the dehydration and hydrogenation processes involved in xylose conversion into hydrogenated products in a single biphasic (water–organic solvent) reactor [23]. By using Amberlyst-15, as dehydration catalyst (aqueous phase), and a hydrophobic Ru/C catalyst (organic phase) for the hydrogenation reaction, and organic solvents of different polarity, these authors have observed that hydrophobic solvents disfavored the hydrogenation of xylose to xylitol due to the decreasing solubility of xylose as the hydrophobicity of the organic solvent increases; thus, hydrogenation products, mainly THFA, are formed. The furfural hydrogenation takes place both at the interface and in the organic phase. Besides THFA, other reaction products such as GVL, LA and 1,2-pentanediol, and minor amounts of cyclopentanone, 1-hydroxy-4-pentanone and 1,4-pentanedione, were also formed. These products come from FOL hydration (to LA) or hydrogenation processes. The highest selectivity to THFA (50%) was reached at 408 K, for a xylose conversion of 32%, with a H_2 pressure of 2.5 MPa, by using a biphasic water–cyclohexane system. This research group has also developed a multilevel rotating foam biphasic reactor based on the use of Ru impregnated over carbon foam, as hydrogenation catalyst, and a mordenite coated over Al foam in the aqueous phase, as dehydration catalyst [24]. This system minimizes the contribution of the interface to the overall catalytic activity, and the interaction between both catalysts. However, the THFA yield was lower. The benefits of this system lie in the easy catalyst recuperation and reutilization, but it is necessary to ameliorate the catalyst–reactant contact.

On the other hand, the asymmetric hydrogenation of FOL to (S)-(+)-THFA, can be accomplished by using biopolymer-metal complex, silica-supported alginic acid–amino-Pt complex, at 303 K and 0.1 MPa H_2 pressure, with ethanol as solvent. An 88% THFA yield and 98% of the optical yield after 15 h of reaction time can be attained [25]. Moreover, the catalyst can be reused for several catalytic runs, maintaining the optical catalytic activity.

The use of supercritical CO_2 conditions is another interesting method for FUR hydrogenation, which allows switching the product selectivity in real-time in the presence of a twin catalyst system [26]. A THFA yield of 96% could be attained by using two catalytic beds formed by copper chromite and Pd/C, working at 393 and 473 K, respectively, with a flow of CO_2 of 1 mL min^{-1} (pumphead at $-10°C$

and 5.8 MPa), F(FUR) of 0.05 mL min^{-1}, and operating at a H$_2$ pressure of 15 MPa.

3.2.3. Reaction mechanisms

Concerning the reaction mechanism of liquid-phase hydrogenation, by using a Ru/TiO$_2$ catalyst, a single-site Langmuir–Hinshelwood model was proposed to best fit the data [27]. This kinetic study concluded that hydrogen is molecularly adsorbed (non-dissociative adsorption) on the catalyst surface, being the surface reaction between adsorbed H$_2$ and FOL the rate-controlling step. A competitive adsorption of TFHA with both reactants (H$_2$ and FOL) was also observed, as inferred from the reaction rate decrease with the presence of THFA. Similar conclusions were attained by Tomishige *et al.* for the gas-phase hydrogenation of FOL to THFA on Ni/SiO$_2$ catalysts [22]. The rate-determining step is the attack of adsorbed hydrogen species to the furan ring, thus producing THFA without the previous formation of free dihydrofurfuryl alcohol. This latter step is the only one that is sensitive to the catalyst structure, inasmuch as smaller Ni particles give rise to high TOF values: the hydroxyl group of the FOL molecule outside the furan ring favors the adsorption onto edges or corners of the metal particles. In this work, the authors studied the direct gas-phase hydrogenation of FUR to THFA and concluded that complete hydrogenation proceeds in two consecutive steps: the first step consists of the hydrogenation of FUR to FOL (with the existence of a strongly adsorbed FUR molecule with an η^2(C,O)-type configuration on the nickel metal surface, as previously mentioned in Section 2.1). A strongly adsorbed FUR molecule reacts with two adsorbed H atoms and then the hydrogenation of FOL to THFA occurs. Nevertheless, FUR hydrogenation is favored over FOL hydrogenation because FUR molecules are strongly adsorbed on the nickel surface.

3.2.4. Reusability studies

With respect to reusability of the catalyst, extensive research is lacking in most of papers, but Merat *et al.* found that the best

Ni catalyst could be used for 30 catalytic runs in the liquid-phase hydrogenation of FOL, although the reaction time had to be increased from run-to-run to compensate the deactivation and to maintain a THFA yield greater than 96% [12]. When a Ru/TiO_2 catalyst was tested, it slightly deactivated but it could be recycled by washing with 2-propanol to remove by-products present on the active sites responsible of the deactivation [27], and no leaching of Ru was observed. Biradar *et al.* corroborated the recyclability and stability of a Pd/MFI catalyst, even when FUR is used as the feedstock [13]. Recently, by using an alkaline earth metal modified Ni/Al_2O_3, it has been feasible to maintain the catalytic activity at least four catalytic cycles, in the liquid phase hydrogenation of FUR to THFA. The reaction, as expected, proceeds through furfuryl alcohol as an intermediate. A THFA yield close to 100% was attained at 413 K, with a $Ni/Ba–Al_2O_3$ (10 wt.% Ni and 10 mol.% Ba) catalyst (Table 3.2.2, entry 10). It was observed that the total selectivity to THFA and FOL was almost the same during the course of the catalytic process, and after 4 h, FOL was not detected. The resinification of FUR, one of the causes of catalyst deactivation and favored by the presence of acid sites, was attenuated by the addition of alkaline earth metals. The presence of the modifier influences the interaction between Ni and support, limiting the formation of $NiAl_2O_4$ species with no hydrogenation activity.

3.2.5. Conclusions and future prospects

The catalytic results concerning the hydrogenation of FUR and FOL to THFA demonstrate that most of the studies have been carried out in liquid phase, where higher THFA yields have been attained. However, it is necessary to develop gas-phase processes working with low H_2/substrate molar ratios, preferentially using FUR aqueous solutions as feedstock and highly selective catalysts, in order to avoid time-consuming separation steps of THFA. Moreover, mechanistic studies are required to get insights about the different kinetics observed in liquid-and gas-phase hydrogenation of FUR to THFA.

In this sense, the role of the active sites nature must be clarified to design more active, selective and stable catalysts working under softer experimental conditions.

References

1. S. Sato, J. Igarashi, Y. Yamada, Stable vapor-phase conversion of tetrahydrofurfuryl alcohol into 3,4-2H-dihydropyran, *Applied Catalysis A: General*, 453 (2013) 213–218.
2. N. Künzle, US20110087, 2011.
3. A.C. Ott, M.F. Murray, R.L. Pederson, The Reaction of Dihydropyran with Steroidal Alcohols. Utility in the Syntheses of Testosterone Acyl Esters, *Journal of the American Chemical Society*, 74 (1952) 1239–1241.
4. S. Bayan, E. Beati, Furfural and its derivatives as motor fuels, *Chimica e Industria*, 23 (1941) 432-434.
5. C. Stamigna, D. Chiaretti, E. Chiaretti, P.P. Prosini, Oil and furfural recovery from Brassica carinata, *Biomass Bioenergy*, 39 (2012) 478–483.
6. T. Lacôme, X. Montagne, B. Delfort, F. Paille, Diesel fuel compositions containing oxygenated compounds derived from tetrahydrofurfuryl alcohol, US6537336 B2, 2003.
7. J.D. Stenger-Smith, L. Baldwin, A. Chafin, P.A. Goodman, Synthesis and Characterization of bis(Tetrahydrofurfuryl) Ether, *ChemistryOpen*, 5 (2016) 297–300.
8. H.E. Hoydonckx, W.M. Van Rhijn, W.M. Van Rhijn, D.E. De Vos, P.A. Jacobs, *Furfural and Derivatives*, Wiley-VCH Verlag GmbH & Co. KGaA, Weinheim, 2012.
9. G.D. Graves, Reduction of furfural to tetrahydrofurfuryl alcohol, US1794453A, 1931.
10. B.W. Howk, Process for the hydrogenation of furfural, US2487054A, 1949.
11. H. Priickner, Process for producing tetrahydrofurfuryl alcohol, US2071704, 1937.
12. N. Merat, C. Godawa, A. Gaset, High selective production of tetrahydrofurfuryl alcohol: Catalytic hydrogenation of furfural and furfuryl alcohol, *Journal of Chemical Technology & Biotechnology*, 48 (1990) 145–159.
13. N.S. Biradar, A.M. Hengne, S.N. Birajdar, P.S. Niphadkar, P.N. Joshi, C. V Rode, Single-pot formation of THFAL via catalytic hydrogenation of FFR over Pd/MFI catalyst, *ACS Sustainable Chemistry & Engineering*, 2 (2014) 272–281.
14. L. Liu, H. Lou, M. Chen, Selective hydrogenation of furfural to tetrahydrofurfuryl alcohol over Ni/CNTs and bimetallic Cu-Ni/CNTs catalysts, *International Journal of Hydrogen Energy*, 41 (2016) 14721–14731.
15. J. Wu, G. Gao, J. Li, P. Sun, X. Long, F. Li, Efficient and versatile CuNi alloy nanocatalysts for the highly selective hydrogenation of furfural, *Applied Catalysis B: Environmental*, 203 (2017) 227–236.

16. S. Liu, Y. Amada, M. Tamura, Y. Nakagawa, K. Tomishige, One-pot selective conversion of furfural into 1,5-pentanediol over a Pd-added Ir-ReO$_x$/SiO$_2$ bifunctional catalyst, *Green Chemistry*, 16 (2014) 617–626.

17. Y. Yang, J. Ma, X. Jia, Z. Du, Y. Duan, J. Xu, Aqueous phase hydrogenation of furfural to tetrahydrofurfuryl alcohol on alkaline earth modified Ni/Al$_2$O$_3$, *RSC Advances*, 6 (2016) 51221–51228.

18. G.M. King, S. Iqbal, P.J. Miedziak, G.L. Brett, S.A. Kondrat, B.R. Yeo, X. Liu, J.K. Edwards, D.J. Morgan, D.K. Knight, G.J. Hutchings, An Investigation of the Effect of the Addition of Tin to 5 %Pd/TiO$_2$ for the Hydrogenation of Furfuryl Alcohol, *ChemCatChem*, 7 (2015) 2122–2129.

19. Y. Nakagawa, K. Tomishige, Total hydrogenation of furan derivatives over silica-supported Ni-Pd alloy catalyst, *Catalysis Communications*, 12 (2010) 154–156.

20. F.A. Khan, A. Vallat, G. Süss-Fink, Highly selective low-temperature hydrogenation of furfuryl alcohol to tetrahydrofurfuryl alcohol catalysed by hectorite-supported ruthenium nanoparticles, *Catalysis Communications*, 12 (2011) 1428–1431.

21. B. Zhang, Y. Zhu, G. Ding, H. Zheng, Y. Li, Selective conversion of furfuryl alcohol to 1,2-pentanediol over a Ru/MnO$_x$ catalyst in aqueous phase, *Green Chemistry*, 14 (2012) 3402–3409.

22. Y. Nakagawa, H. Nakazawa, H. Watanabe, K. Tomishige, Total hydrogenation of furfural over a Silica-supported Nickel catalyst prepared by the reduction of a Nickel Nitrate Precursor, *ChemCatChem*, 4 (2012) 1791–1797.

23. V.V Ordomsky, J.C. Schouten, J. Van Der Schaaf, T.A. Nijhuis, Biphasic single-reactor process for dehydration of xylose and hydrogenation of produced furfural, *Applied Catalysis A: General*, 451 (2013) 6–13.

24. V.V Ordomsky, J.C. Schouten, J. van der Schaaf, T.A. Nijhuis, Multilevel rotating foam biphasic reactor for combination of processes in biomass transformation, *Chemical Engineering Journal*, 231 (2013) 12–17.

25. W.L. Wei, H.Y. Zhu, C.L. Zhao, M.Y. Huang, Y.Y. Jiang, symmetric hydrogenation of furfuryl alcohol catalyzed by a biopolymer-metal complex, silica-supported alginic acid-amino acid-Pt complex, *Reactive and Functional Polymers*, 59 (2004) 33–39.

26. J.G. Stevens, R.A. Bourne, M.V. Twigg, M. Poliakoff, Real-time product switching using a twin catalyst system for the hydrogenation of furfural in supercritical CO$_2$, *Angewandte Chemie International Edition*, 49 (2010) 8856–8859.

27. M.A. Tike, V.V. Mahajani, Kinetics of liquid-phase hydrogenation of furfuryl alcohol to tetrahydrofurfuryl alcohol over a Ru/TiO$_2$ Catalyst, *Industrial & Engineering Chemistry Research*, 46 (2007) 3275–3282.

Chapter 3.3

Catalytic Transformations of Furfural and its Derived Compounds into Pentanediols

Sibao Liu, Masazumi Tamura, Yoshinao Nakagawa
and Keiichi Tomishige*
*Department of Applied Chemistry,
School of Engineering, Tohoku University,
6-6-07 Aoba, Aramaki, Aoba-ku, Sendai 980–8579, Japan*

3.3.1. General introduction

Furfural was regarded as one of the most important platform molecules in biorefinery, which has been manufactured on an industrial scale by hydration + dehydration of the hemicellulose part of agricultural waste and forest residues [1, 2]. Tetrahydrofurfuryl alcohol (THFA) and furfuryl alcohol (FFA) can be produced from furfural by total hydrogenation and selective hydrogenation, respectively. The catalytic synthesis of them has been introduced in detail in Chapters 3.1 and 3.2. They can also serve as renewable building blocks. The catalytic transformations of these renewable furanic compounds into valuable chemicals and fuels have attracted much attention in recent years [3–8]. The hydrogenation of the C=C bonds and/or hydrogenolysis of the C–O bond in a furan ring is considered to be the most effective method for the synthesis of useful pentanediols from these furanic compounds (Scheme 3.3.1). The

*Corresponding author: tomi@erec.che.tohoku.ac.jp

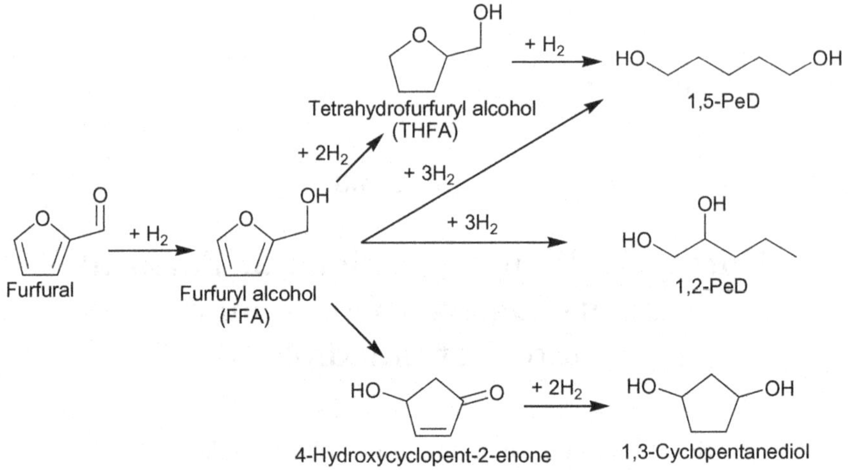

Scheme 3.3.1. Conversion of furfural and its derivatives into pentanediols (PeDs).

pentanediols (PeDs) include linear pentanediols (1,5-PeD and 1,2-PeD) and cyclic pentanediol (1,3-cyclopentanediol). These diols can be used as monomers of polyesters and polyurethanes. In addition, if these diols are transformed to diamines, the diamines can be used as the monomer of polyamides. Here, in this chapter, we introduce the catalytic conversion of furfural and its derived compounds (THFA and FFA) into pentanediols as target products.

3.3.2. Conversion of furfural derived THFA into 1,5-PeD

There are two different methods (indirect and direct routes) for the production of 1,5-PeD from THFA. The first one is the three-step route, which was first described in 1946 [9]. The three steps are rearrangement of THFA to dihydropyran, hydrolysis of dihydropyran to δ-hydroxyvaleraldehyde and hydrogenation of δ-hydroxyvaleraldehyde to 1,5-PeD. This route yields only 70% 1,5-PeD and requires the isolation and purification of the intermediates. Another route is the direct selective C–O hydrogenolysis of THFA into 1,5-PeD by using heterogeneous catalysts in a single

Table 3.3.1. Conversion of THFA into 1,5-PeD over different catalysts.

Entry	Catalyst	Reactor	Solvent	T (K)	P_{H2} (MPa)	Yield (%)		Ref.
						1,5-PeD	1,2-PeD	
1	Rh–ReO$_x$/SiO$_2$	Batch	H$_2$O	393	8	77	0	[10]
2	Rh–MoO$_x$/SiO$_2$	Batch	H$_2$O	373	8	85	0	[11]
3	Rh–ReO$_x$/C	Batch	H$_2$O	373	8	94	0	[15]
4	Ir–ReO$_x$/SiO$_2$	Batch	H$_2$O	373	8	83	0	[16]
5	Ir–MoO$_x$/SiO$_2$	Flow[a]	H$_2$O	393	6	52	0	[19]
6	Ir–VO$_x$/SiO$_2$	Flow[a]	H$_2$O	393	6	57	0	[20]
7	Rh/MCM-41	Batch	scCO$_2$ (14 MPa)	353	4	73	0	[22]

Note: [a]THFA aqueous solution flow rate 0.04 mL min^{-1}, H$_2$ flow rate 60 mL min^{-1}.

step. This route was first reported in 2009 using Rh-ReO$_x$/SiO$_2$ catalysts. Metallic Rh or Ir catalysts modified with a partially reduced metal oxide species such as ReO$_x$, MoO$_x$, WO$_x$ or VO$_x$ supported on SiO$_2$ or carbon materials were reported as effective catalysts for this reaction (Table 3.3.1) [10–20]. Both, batchwise operation and continuous operation, have been explored. In this reaction, water was used as a solvent. The highest yield of 1,5-PeD could reach 94% by using Rh–ReO$_x$/C as catalyst at 373 K, 8 MPa H$_2$, and 24 h of reaction in batch reactor [15]. Ir–ReO$_x$/SiO$_2$ showed higher TOF than Rh–ReO$_x$/SiO$_2$ and the selectivity was comparable [16].

Various characterization techniques including XRD, H$_2$-TPR, TEM, XAFS, XPS, CO adsorption and FT-IR of CO adsorption were used for clarifying the structures of these modified Rh or Ir catalysts [10–21]. All of these catalysts showed a similar structure: noble metal particles are partially covered with low-valent metal oxide species via direct bonds between noble metal and additive metal atoms, while the structures of the low-valent metal oxide species may be different: MoO$_x$ has a monomeric structure on Rh [13], while ReO$_x$ forms clusters on Rh [14] or Ir [21]. Two-dimensional ReO$_x$ clusters were formed on Rh metal, while three-dimensional ones were formed on Ir metal. Monometallic catalysts (Rh, Ir, Re and Mo) showed much lower activity compared to these modified Rh or Ir catalysts, indicating the strong synergetic effect of the noble-metal particles

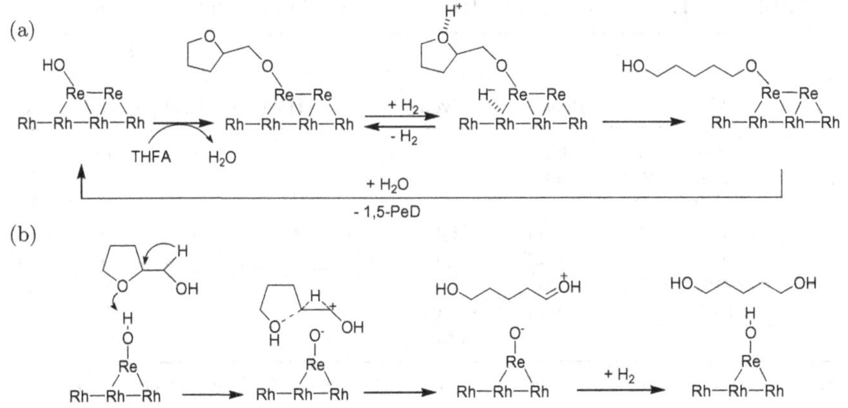

Scheme 3.3.2. Proposed mechanisms of THFA hydrogenolysis: (a) regioselective hydride attack mechanism [12] and (b) concerted acid mediated mechanism [18].

and low-valent metal oxide species. The catalytically active site may be the interface between the noble-metal particle and low-valent metal oxide species.

Two different mechanisms have been proposed for the hydrogenolysis of THFA (Scheme 3.3.2). A regioselective hydride attack mechanism has been proposed on the basis of the reactivity of related substrates, kinetic studies, and the reaction of THFA with D_2 [10–16]. First, THFA is initially chemisorbed on the surface of the metal oxide species (ReO_x, MoO_x) at the $-CH_2OH$ group to form terminal alkoxide. Then hydrogen is heterolytically activated on Rh or Ir metal at the interface with metal oxide species to produce a hydride and proton. After that, the hydride species on the Rh or Ir metal attacks the 2-position of the alkoxide to break the C–O bond via an S_N2 reaction. Finally, the hydrolysis of the reduced alkoxide releases the product. On the other hand, Dumesic group proposed a concerted acid mediated mechanism for THFA hydrogenolysis over Rh–ReO_x/C [18]. Acidic Re–OH formed on Rh metal particles transfers a proton to the tetrahydrofuran ring. Then, the α-hydrogen atom in the $-CH_2OH$ group is concertedly migrates to the β-position (2-position of the tetrahydrofuran ring). This step breaks the O–C bond between the 1- and 2-positions to open the ring. Later, hydrogenation of the formed protonated aldehyde over Rh metal gives 1,5-PeD. One of the differences between those mechanisms is the source of hydrogen atom at 2-position in 1,5-PeD. One hydrogen atom is incorporated to this position from H_2 in hydride attack mechanism while the hydrogen atom at 2-position in 1,5-PeD is from THFA molecule in the concerted acid mediated mechanism. The deuterium label study using $Ir-ReO_x/SiO_2$-catalyzed THFA hydrogenolysis with D_2 showed the incorporation of D atom from D_2 [16].

The stability of the catalysts is very important for the practical biomass conversion. In the batch reaction mode, the recycling $Ir-ReO_x/SiO_2$ needs regeneration step by calcination to regain similar level of activity to fresh catalyst while $Rh-ReO_x/SiO_2$ and $Rh-MoO_x/SiO_2$ can be used as recovered. Both $Ir-ReO_x/SiO_2$ and $Rh-MoO_x/SiO_2$ showed better stability compared to $Rh-ReO_x/SiO_2$, probably due to the leaching of Re from $Rh-ReO_x/SiO_2$ [10, 11, 16]. In the flow reaction mode, $Ir-MoO_x/SiO_2$ and $Ir-VO_x/SiO_2$ showed medium stability [19,20]. The conversion of THFA dropped to the half of the initial value while the selectivity of 1,5-PeD was kept after 30 h reaction. The leaching of Mo and V during the reaction could be responsible for the decrease of THFA conversion. Although the leaching of partially reduced metals (ReO_x, MoO_x or VO_x) could be the reason for all the catalysts' deactivation, further characterizations of used catalysts is still necessary to analyze the structure change during the recovery and/or regeneration step.

Apart from bimetallic catalysts, Rh/MCM-41 was also effective for this reaction in supercritical carbon dioxide (scCO$_2$) media without any other co-solvent [22]. ScCO$_2$ is a well-established environmental benign reaction medium and the use of scCO$_2$ overcomes the low solubility of hydrogen in aqueous media. It was hypothesized that both presence of Rh_2O_3 and Rh^0, which was confirmed by XRD, was very necessary for achieving high selectivity towards 1,5-PeD. With optimal composition of Rh_2O_3 and Rh^0, the selectivity of 1,5-PeD attained 91.2% at 80% conversion of THFA at 353 K under 14 MPa CO_2 and 4 MPa H_2 pressure. However, the stability of the catalyst was not provided.

It should be noted that M–M′O_x-based catalysts (M = Rh or Ir, M′ = Re or Mo) showed wide applicability to various reactions such as the hydrogenolysis of glycerol [21, 24–33], the hydrogenolysis of tetrahydropyran-2-methanol and 2,5-bis(hydroxymethyl)tetrahydrofuran [15, 34–39], the hydrogenolysis of erythritol [40], the hydrogenation of unsaturated aldehyde to unsaturated alcohol [41, 42], the hydrogenation of the carboxyl groups [43–47], the conversion of sugar alcohols, cellulose, and hemicellulose to alkanes or monoalcohols [48–53], dehydration of fructose to 5-hydroxymethylfurfural [34], and so on.

3.3.3. Conversion of furfural derived FFA into PeDs

In the 1930s, Adkins and Conner reported the first hydrogenolysis of FFA over $CuCr_2O.1,2$-PeD and 1,5-PeD with yields of 40% and 30% were achieved from neat FFA at 448 K and 10–15 MPa H_2, respectively [54, 55]. However, the toxicity of chromium-containing catalyst and high pressure hydrogen could limit the practical application of this system. To overcome these problems, recently, Chen *et al.* have developed two different effective Cr-free Cu-based catalysts, $Cu–Mg_3AlO_{4.5}$ [56] and $Cu–Al_2O_3$ [57], for conversion of FFA into PeDs by using ethanol as solvent under mild reaction conditions (413 K, 6 MPa H_2). Both of the catalysts were prepared by coprecipitation method. The supports include basic layered double oxides ($Mg_3AlO_{4.5}$) and acidic Al_2O_3. $Cu–Mg_3AlO_{4.5}$ showed higher conversion of FFA and higher selectivity towards PeDs, probably due to the basic support suppressing polymerization of FFA. The maximum yield of PeDs attained up to 80% (51.2% 1,2-PeD and 28.8% 1,5-PeD) over 10 wt.% $Cu–Mg_3AlO_{4.5}$. The selectivity of 1,2-PeD is higher than that of 1,5-PeD. The size of active Cu particle is very important to attain high PeDs yield. In the case of $Cu–Mg_3AlO_{4.5}$, the catalyst with 1.7 nm Cu particles at the average showed the highest TOF while in the case of $Cu–Al_2O_3$, Cu particles with the average size of 1.9–2.4 nm gave the highest TOF. The proposed reaction mechanisms over both of the catalysts were similar, regardless of basic and acidic supports. Initially, the hydroxymethyl moiety of FFA is adsorbed on and interacted with the surface basic sites of $Mg_3AlO_{4.5}$ support or the Lewis acid sites of the Al_2O_3 support to form an alcoholate species. Subsequently, the $C^2=C^3$ or $C^4=C^5$ bond in the furan ring is partially hydrogenated to form semi-hydrogenated species, most likely via hydrogenation of C^2 or C^5 atoms. This induces the ring opening reaction by decreasing the barrier to C–O cleavage. Finally, the ring-opened species are rapidly hydrogenated to the target products 1,2-PeD or 1,5-PeD. The higher yield of 1,2-PeD could be attributed to the steric hindrance of the –CH_2OH group that makes the partial hydrogenation of $C^4=C^5$ bond to form the intermediate that generates 1,2-PeD easier. Both of the catalysts could be readily recycled over several repeated runs without significant loss in either the FFA conversion or the PeDs selectivities. The $Cu-Mg_3AlO_{4.5}$ catalyst showed higher stability than the $Cu-Al_2O_3$ catalyst. The slight decrease in activity and selectivity of the $Cu-Al_2O_3$ catalyst may be attributed to the aggregation of Cu particles during each run.

Noble metal catalysts were also explored for opening the furan ring of FFA to produce PeDs (Table 3.3.2). The catalysts included Pt-based catalysts and Ru-based catalysts on basic support or combination of alkali. 1,2-PeD was the dominant product in this reaction, while 1,5-PeD was rarely formed. The main byproduct is THFA. Koch *et al.* of Symrise company demonstrated the conversion of FFA into 1,2-PeD over PtO_2/Al_2O_3 catalyst in ethanol solvent,

Table 3.3.2. Conversion of FFA into PeDs over different catalysts.

Entry	Catalyst	Solvent	T (K)	P_{H2} (MPa)	Yield (%) 1,5-PeD	1,2-PeD	THFA	Ref.
1	$CuCr_2O_4$	neat	448	10–15	30	40	n.r.	[54]
2	$Cu-Mg_3AlO_{4.5}$	EtOH	413	6	28.8	51.2	1.8	[56]
3	$Cu-Al_2O_3$	EtOH	413	6	19.0	41.7	2.3	[57]
4	PtO_2/Al_2O_3	EtOH	273–278	0.1	1	80	15	[58]
5	Pt/hydrotalcite	2-PrOH	423	3	7	80	10	[59]
6	Ru/MnO_x	H_2O	423	1.5	n.r.	42.1	37	[60]
7	Ru/Al_2O_3 + Na_2CO_3	H_2O	473	10	n.r.	32	57	[61]

Note: n.r.: not reported.

in their patent application [58]. The reaction condition used here is very mild. The reaction temperature is below room temperature (273–278 K) and the hydrogen pressure is very low (0.1 MPa). The maximum yield of 1,2-PeD could reach 80% (accompany with 15% yield of THFA). Mizugaki *et al.* achieved 80% yield of 1,2-PeD from the hydrogenolysis of FFA at 423 K and under 3 MPa H_2 using a hydrotalcite-supported Pt catalyst [59]. In addition, 10% yield of THFA and 7% yield of 1,5-PeD were formed. Ru-based catalysts gave lower yield of 1,2-PeD compared to Pt-based catalysts. Zhang *et al.* reported Ru/MnO_x catalyst for conversion of FFA into 1,2-PeD [60]. The catalyst was prepared by incipient wetness impregnation method. The MnO_x support turned into basic $Mn(OH)_2$ in aqueous phase reaction. Both low hydrogen pressure and high reaction temperature favor the generation of 1,2-PeD. The highest yield of 1,2-PeD was 42.1% at 423 K under 1.5 MPa H_2 after 6 h reaction. Meanwhile, about 37% yield of THFA was obtained. The recyclability of the Ru/MnO_x catalyst was feasible without regeneration. Claus *et al.* studied Ru/Al_2O_3 and Ru/C catalysts in conversion of FFA to 1,2-PeD in aqueous phase [61, 62]. The selectivity towards 1,2-PeD over Ru/Al_2O_3 is higher than that over Ru/C, due to the smaller amount of acid sites of Al_2O_3 support compared to carbon materials that reduced the formation of polymer from FFA. The addition of homogenous basic additive, Na_2CO_3, enhanced the selectivity of 1,2-PeD by suppressing the undesired polymerization of FFA. Only 32% yield of 1,2-PeD was attained at 473 K, 10 MPa H_2 pressure over Ru/Al_2O_3 catalyst in Na_2CO_3 solution and a high yield (57%) of THFA was obtained.

During the production of 1,2-PeD from FFA over noble metal catalysts, the key step is the hydrogenolysis of C^5–O^1 bond of FFA. THFA couldn't be converted over these catalysts, suggesting that THFA is the parallel product of 1,2-PeD in the conversion of FFA. It was proposed that 1-hydroxy-2-pentanone is an intermediate for the production of 1,2-PeD. This intermediate was actually observed when the reaction was conducted with a flow system over PtO_2/Al_2O_3 catalyst [58] or the reaction proceeded over Ru/MnO_x catalyst [60] and Pt/hydrotalcite [59] at the beginning stage. The presence

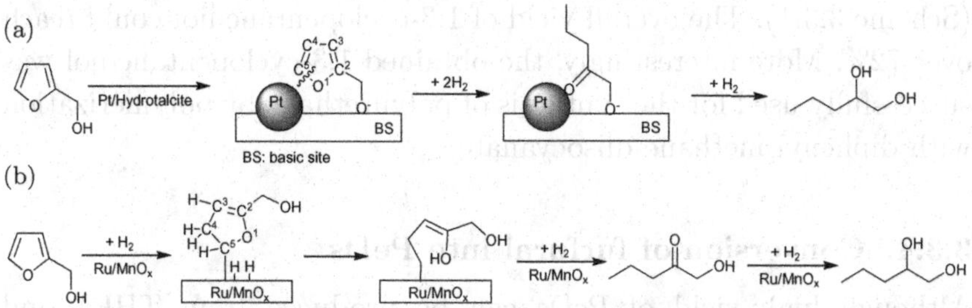

Scheme 3.3.3. Mechanisms of hydrogenolysis of FFA: (a) over the Pt/hydrotalcite catalyst [59], (b) over the Ru/MnO_x catalyst [60].

of the $C^2=C^3$ bond of furan ring can weaken the C^5-O^1 bond [58]. Consequently, to obtain high yield of 1,2-PeD, suppressing the hydrogenation of FFA to THFA is the key. In the reaction mechanism, the proposed adsorption mode and the role of base support are different over Pt/hydrotalcite and Ru/MnO_x catalysts (Scheme 3.3.3). In the case of Pt/hydrotalcite catalyst [59], the proposed adsorption site was located at the interface between Pt particles and the basic sites of the support. The hydroxymethyl moiety and furan ring of FFA were adsorbed on the surface basic sites of hydrotalcite and Pt particles, respectively. The role of basic site is to remove the proton of FFA, which gives alcoholate species. In the case of Ru/MnO_x catalyst [60], it was proposed that FFA was adsorbed at the C^5 position on the Ru surface before the conversion to 1,2-PeD. The role of the basic MnO_x is to stabilize the partial hydrogenation intermediate species by forming an adduct with Mn ion that suppressed the generation of THFA. Nevertheless, the reaction mechanism for the hydrogenolysis of FFA into 1,2-PeD remains controversial, especially for the ring opening of FFA and the role of base. Consequently, this requires further clarification.

Besides synthesis of linear pentanediols from FFA, there is another interesting cyclic pentanediol, 1,3-cyclopentanediol, which can be produced from FFA through a two-step method [63]. At first, FFA was rearranged to 4-hydroxycyclopent-2-enone in aqueous phase catalyzed by base catalysts (MgAl-hydrotalcite) and then the hydrogenation of 4-hydroxycyclopent-2-enone in tetrahydrofuran solvent over metal catalysts (Ru/C or Raney Ni) gave the product

(Scheme 3.3.1). The overall yield of 1,3-cyclopentanediol could reach over 72%. More interestingly, the obtained 1,3-cyclopentanediol was successfully used for the synthesis of polyurethane by polymerization with diphenyl-methane-diisocyanate.

3.3.4. Conversion of furfural into PeDs

Although high yield of PeDs can be produced from THFA and FFA, the productions of THFA and FFA from furfural need additional hydrogenation and purification steps. Consequently, direct conversion of furfural into PeDs would be more promising with respect to the industrial implementation. Pt-based catalysts on basic supports [59, 64, 65] and noble metal modified Ir–ReO$_x$/SiO$_2$ catalysts were effective for this reaction [66, 67] (Table 3.3.3). Pt-based catalysts gave high yield of 1,2-PeD while noble metal modified Ir–ReO$_x$/SiO$_2$ catalysts gave high yield of 1,5-PeD. The reaction paths were different: for Pt-based catalysts, production of PeDs proceeds through the partial hydrogenation of furfural to FFA and ring opening of FFA while for the noble metal modified Ir–ReO$_x$/SiO$_2$ catalysts, 1,5-PeD was formed via total hydrogenation of furfural into THFA and hydrogenolysis of formed THFA.

Adams *et al.* pioneered this reaction process and reported the first direct conversion of furfural into PeDs in ethanol solvent over PtO$_2$ with FeCl$_2$ and 1,2-PeD and 1,5-PeD with the yield of 27% and

Table 3.3.3.   Conversion of furfural into PeDs over different catalysts.

Entry	Catalyst	Solvent	T (K)	P_{H2} (MPa)	1,5-PeD	1,2-PeD	THFA	Ref.
					Yield (%)			
1	PtO$_2$ + FeCl$_2$	EtOH	n.r.	0.1	11	27	48	[64]
2	Li–Pt/Co$_2$AlO$_4$	EtOH	413	1.5	34.9	16.2	31.3	[65]
3	Pt/hydrotalcite	2-PrOH	423	3	8	73	14	[59]
4	Cu–Mg$_3$AlO$_{4.5}$	EtOH	413	6	25.5	45.2	3.1	[56]
5	Pd–Ir–ReO$_x$/SiO$_2$	H$_2$O	313[a]–373[b]	6	78.2	1.2	6.4	[66]
6	Rh–Ir–ReO$_x$/SiO$_2$	H$_2$O	313[a]–373[b]	6	71.4	1.4	4.4	[67]

Notes: [a]Low temperature step.
[b]High temperature step.
n.r.: Not reported.

11% were achieved under 0.1–0.2 MPa H_2 pressure, respectively [64]. However, the reaction temperature was not given.

Xu *et al.* conducted direct hydrogenolysis of furfural in ethanol solvent by using Li-promoted Pt/Co_2AlO_4 catalyst with the aim of production of 1,5-PeD [65]. The catalyst was prepared by coprecipitation method. The maximum yield of 1,5-PeD was 34.9% together with 16.2% yield of 1,2-PeD at 413 K under 1.5 MPa H_2 pressure after 24 h reaction. The time course of furfural conversion over Pt/Co_2AlO_4 catalyst showed that the reaction sequence proceeds stepwise via FFA as the intermediate and the reaction route via THFA to 1,5-PeD is negligible. It was proposed that Co^{3+} cations of the support are mainly responsible for the adsorption of C=C bonds and the breaking the C–O bond of furan ring, while Pt works for the following hydrogenation. The addition of Li increased the basicity of the catalyst and partly suppresses the undesirable reaction of FFA to 2-methyltetrahydrofuran. The catalyst's reusability was also tested. Although the conversion of furfural and yield of 1,5-PeD were almost constant after three cycles, the distribution of other compounds especially 1,2-PeD and 1,4-PeD changed significantly. Further characterizations of the used catalyst need to be studied.

Mizugaki *et al.* performed hydrogenolysis of furfural to 1,2-PeD over hydrotalcite-supported Pt catalysts [59]. The catalyst was prepared by impregnation of hydrotalcite with an aqueous solution of H_2PtCl_6 and subsequent reduction with KBH_4. A notably high yield (73%) of 1,2-PeD was obtained from furfural at 150°C under 3 MPa H_2 pressure. The time course of the reaction and reactions with different furan compounds shows that 1,2-PeD is produced via formation of 1-hydroxy-2-pentanone from FFA. The reaction mechanism was similar to the conversion of FFA, which was described in the last section. The catalyst can be reused without any loss of activity and selectivity by simple separation.

Chen *et al.* attempted the direct conversion of furfural to PeDs in ethanol solvent using a $Cu–Mg_3AlO_{4.5}$ catalyst under mild reaction conditions (413 K, 6 MPa H_2) and attained moderate yields of 1,2-PeD (45.2%) and 1,5-PeD (25.5%) [56].

The one-pot two-step conversion of furfural into 1,5-PeD with high yield in aqueous phase by using noble metal modified Ir–ReO$_x$/SiO$_2$ catalysts has been attempted [66, 67]. In the previous researches, Ir–ReO$_x$/SiO$_2$ catalyst showed excellent performance in the hydrogenation of C=O bond of unsaturated aldehydes such as furfural to FFA [40, 41] and hydrogenolysis of THFA into 1,5-PeD [16], however, this catalyst was inert in the hydrogenation of C=C bonds of furan ring. Consequently, the addition of another noble metal (Pd, Rh, Pt, Ru) to Ir–ReO$_x$/SiO$_2$ catalyst was tested with the aim of improving the activity of hydrogenation of furan ring to produce THFA and maintaining activity of the C–O hydrogenolysis of THFA to 1,5-PeD on the basis of the studies of total hydrogenation of furfural to THFA [68, 69]. It is found that Pd- or Rh-modified Ir–ReO$_x$/SiO$_2$ catalyst was effective for this reaction [66, 67]. Two-step reaction at different temperature that combines hydrogenation of furfural to THFA and hydrogenolysis of formed THFA into 1,5-PeD was applied in the reduction of furfural. The hydrogenation of furfural to THFA was conducted at low temperature ($\leq$ 323 K) and the hydrogenolysis of the formed THFA was later accomplished in the second step by increasing the temperature to $\geq$ 373 K. The amount of added metal and the reaction conditions were optimized for Pd–Ir–ReO$_x$/SiO$_2$ and Rh–Ir–ReO$_x$/SiO$_2$ catalysts. The optimum loading amount of Rh and Pd was 0.66 wt.% for both catalysts. The highest yield of 1,5-PeD from diluted furfural (furfural:water = 1:9, weight ratio) was 78.4% and 71.4% over Rh–Ir–ReO$_x$/SiO$_2$ and Pd–Ir–ReO$_x$/SiO$_2$, respectively. Even from highly concentrated furfural (furfural:water = 1:1, weight ratio), the highest yield of 1,5-PeD reached 71.1% and 61.5%, respectively. The Rh–Ir–ReO$_x$/SiO$_2$ catalyst was superior in view of the reaction time as well as the highest yield of 1,5-PeD.

The time courses of furfural reaction over Rh– and Pd–Ir–ReO$_x$/SiO$_2$ catalysts were compared. For the total hydrogenation of furfural to THFA at low temperature step, Pd–Ir–ReO$_x$/SiO$_2$ has higher activity while Rh–Ir–ReO$_x$/SiO$_2$ was more active in the hydrogenolysis of THFA into 1,5-PeD at high temperature step. Low temperature step was very important for the formation of THFA intermediate since furfural or FFA tended to polymerize or hydrolyze in water solvent over these catalysts with the acidic property at high temperature. Enough reaction time in the low temperature step is necessary to obtain good 1,5-PeD yield over Rh–Ir–ReO$_x$/SiO$_2$ in the two-step reaction of furfural.

The characterization results of TPR, XRD, XANES, EXAFS, STEM-EDX and FT-IR of CO adsorption indicated that Pd–Ir–ReO$_x$/SiO$_2$ catalyst had the structure of mixture of Pd–ReO$_x$ particles and Ir–ReO$_x$ particles, which were separately supported on silica, while Rh–Ir–ReO$_x$/SiO$_2$ catalyst showed the structure of Rh–Ir alloy partially covered with ReO$_x$ species. In the case of Pd–Ir–ReO$_x$/SiO$_2$ catalyst, ReO$_x$-promoted Pd particles catalyzed total hydrogenation of furfural into THFA at low temperature step and Ir particles partially covered with ReO$_x$ species was responsible for the hydrogenolysis of THFA into 1,5-PeD at high temperature step. In the case of Rh–Ir–ReO$_x$/SiO$_2$ catalyst, both Ir–Rh alloy formation and ReO$_x$ modification of the alloy particles are responsible for the good activities in hydrogenation of furfural to THFA and hydrogenolysis of formed THFA to 1,5-PeD. The reaction mechanism for the hydrogenolysis of THFA to 1,5-PeD was the same as that mentioned previously in Section 3.3.2.

The Rh–Ir–ReO$_x$/SiO$_2$ catalyst has worse stability than the Pd–Ir–ReO$_x$/SiO$_2$ catalyst, though it has higher activity. The deactivation is probably due to the leaching and sintering of the metal particles. The aggregation of metal particles also occurred in the case of Pd–Ir–ReO$_x$/SiO$_2$ catalyst. Since this one-pot reaction consists of two steps, the stabilities of the catalysts in the hydrogenation step are still unclear. The insufficient hydrogenation of furfural may affect the hydrogenolysis step. Furthermore, thorough characterizations of used catalysts are still needed to figure out the reason for the deactivation.

3.3.5. Conclusions and future prospects

In this chapter, synthesis of valuable pentanediols (PeDs) from furfural and its derived compounds (THFA and FFA) was critically reviewed. These pentanediols can be used as monomers of polyesters, polyurethanes and even polyamides. The hydrogenolysis of THFA gave very high yield of 1,5-PeD in water solvent. This reaction proceeded over bi-metallic catalysts where partially reduced metal oxide species (ReO_x, MoO_x, WO_x or VO_x) modified Rh or Ir metal particles supported on SiO_2 or carbon materials. The selective ring opening of FFA provided 1,2-PeD as a dominant product. Production of 1,5-PeD was also possible. Cu-based catalysts and Ru- and Pt-based catalysts with basic support or combined with base additive were effective for this reaction. Suppressing both the hydrogenation of FFA to THFA and polymerization of FFA is the key to obtain high yield of 1,2-PeD. 1,3-Cyclopentanediol can be synthesized from FFA by rearrangement and subsequent hydrogenation in two separated steps.

Direct conversion of furfural combining hydrogenation and hydrogenolysis reactions is feasible for the production of PeDs. Pt-based catalysts with basic support were active in converting furfural into 1,2-PeD and 1,5-PeD via selective hydrogenation furfural to FFA and opening-ring of FFA in one step. However, Pd- and Rh-Ir-ReO_x/SiO_2 catalysts converted furfural to 1,5-PeD involving total hydrogenation of furfural to THFA and hydrogenolysis of THFA in two steps. The total hydrogenation reaction should be conducted at low temperature step since the instability of furfural or FFA in the presence of these catalysts at high temperature in water solvent.

The production of PeDs from THFA and FFA is extensively explored, however, direct conversion of furfural to PeDs is still in its infancy. It appears that the direct conversion of furfural has been dominated so far by the use of noble-metal catalysts. Therefore, it is urgent to develop many other efficient and durable catalysts without containing precious metals for this reaction with respect to the industrial implementation in the near future. The C–O hydrogenolysis mechanism of furanic compounds is still under debate, therefore future works with the aid of advanced spectroscopic tools are necessary to understand the reaction mechanisms and these will give insights into designing highly efficient multifunctional catalysts.

References

1. J.J. Bozell, G.R. Petersen, Technology development for the production of biobased products from biorefinery carbohydrates — the US Department of Energy's "Top 10" revisited, *Green Chemistry*, 12 (2010) 539–554.
2. K. Zeitsch, *The Chemistry and Technology of Furfural and its Many By-Products*, Elsevier Science, Amsterdam, 2000.
3. Y. Nakagawa, M. Tamura, K. Tomishige, Catalytic reduction of biomass-derived furanic compounds with hydrogen, *ACS Catalysis*, 3 (2013) 2655–2668.
4. J.P. Lange, E. van der Heide, J. van Buijtenen, R. Price, Furfural- a promising platform for lignocellulosic biofuels, *The Chemical Record*, 5 (2012) 150–166.
5. R. Mariscal, P. Maireles-Torres, M. Ojeda, I. Sadaba, M. Lopez Granados, Furfural: a renewable and versatile platform molecule for the synthesis of chemicals and fuels, *Energy & Environmental Science*, 9 (2016) 1144–1189.
6. Y. Nakagawa, K. Tomishige, Catalyst development for the hydrogenolysis of biomass-derived chemicals to value-added ones, *Catalysis Surveys from Asia*, 15 (2011) 111–116.
7. Y. Nakagawa, K. Tomishige, Production of 1,5-pentanediol from biomass via furfural and tetrahydrofurfuryl alcohol, *Catalysis Today*, 195 (2012) 136–143.
8. Y. Nakagawa, M. Tamura, K. Tomishige, Catalytic conversions of furfural to pentanediols, *Catalysis Surveys from Asia*, 19 (2015) 249–256.
9. L.E. Schniepp, H.H. Geller, Preparation of dihydropyran, δ-hydroxyvaler-aldehyde and 1,5-pentanediol form tetrahydrofurfuryl alcohol, *Journal of the American Chemical Society*, 68 (1946) 1646–1648.
10. S. Koso, I. Furikado, A. Shimao, T. Miyazawa, K. Kunimori, K. Tomishige, Chemoselective hydrogenolysis of tetrahydrofurfuryl alcohol to 1,5-pentanediol, *Chemical Communications*, (2009) 2035–2037.
11. S. Koso, N. Ueda, Y. Shinmi, K. Okumura, T. Kizuka, K. Tomishige, Promoting effect of Mo on the hydrogenolysis of tetrahydrofurfuryl alcohol to 1,5-pentanediol over Rh/SiO_2, *Journal of Catalysis*, 267 (2009) 89–92.
12. S. Koso, Y. Nakagawa, K. Tomishige, Mechanism of the hydrogenolysis of ethers over silica-supported rhodium catalyst modified with rhenium oxide, *Journal of Catalysis*, 280 (2011) 221–229.
13. S. Koso, H. Watanabe, K. Okumura, Y. Nakagawa, K. Tomishige, Comparative study of Rh-MoO_x and Rh-ReO_x supported on SiO_2 for the hydrogenolysis of ethers and polyols, *Applied Catalysis B*, 111–112 (2012) 27–37.
14. S. Koso, H. Watanabe, K. Okumura, Y. Nakagawa, K. Tomishige, Stable low-valence ReOx cluster attached on Rh metal particles formed by hydrogen reduction and its formation mechanism, *The Journal of Physical Chemistry C*, 116 (2012) 3079–3090.
15. K. Chen, S. Koso, T. Kubota, Y. Nakagawa, K. Tomishige, Chemoselective hydrogenolysis of tetrahydropyran-2-methanol to 1,6-hexanediol over

rhenium-modified carbon-supported rhodium catalysts, *ChemCatChem*, 2 (2010) 547–555.

16. K. Chen, K. Mori, H. Watanabe, Y. Nakagawa, K. Tomishige, C-O bond hydrogenolysis of cyclic ethers with OH groups over rhenium-modified supported iridium catalysts, *Journal of Catalysis*, 294 (2012) 171–183.

17. K. Tomishige, M. Tamura, Y. Nakagawa, Role of Re species and acid cocatalyst on Ir-ReO$_x$/SiO$_2$ in the C-O hydrogenolysis of biomass-derived substrate, *The Chemical Record*, 14 (2014) 1041–1054.

18. M. Chia, Y.J.P. Torres, D. Hibbitts, Q. Tan, H.N. Pham, A.K. Datye, M. Neurock, R.J. Davis, J.A. Dumesic, Selective hydrogenolysis of polyols and cyclic ethers over bifunctional surface sites on rhodium-rhenium catalysts, *Journal of the American Chemical Society*, 133 (2011) 12675–12689.

19. Z. Wang, B. Pholjaroen, M. Li, W. Dong, N. Li, A. Wang, X. Wang, Y. Cong, T. Zhang, Chemoselective hydrogenolysis of tetrahydrofurfuryl alcohol to 1,5-pentanediol over Ir-MoO$_x$/SiO$_2$ catalyst, *Journal of Energy Chemistry*, 23 (2014) 427–434.

20. B. Pholjaroen, N. Li, Y. Huang, L. Li, A. Wang, T. Zhang, Selective hydrogenolysis of tetrahydrofurfuryl alcohol to 1,5-pentanediol over vanadium modified Ir/SiO$_2$ catalyst, *Catalysis Today*, 245 (2015) 93–99.

21. Y. Amada, H. Watanabe, M. Tamura, Y. Nakagawa, K. Okumura, K. Tomishige, Structure of ReO$_x$ clusters attached on the Ir metal surface in Ir-ReO$_x$/SiO$_2$ for the hydrogenolysis reaction, *The Journal of Physical Chemistry C*, 116 (2012) 23503–23514.

22. M. Chatterjee, H. Kawanami, T. Ishizaka, M. Sato, T. Suzuki, A. Suzuki, An attempt to achieve the direct hydrogenolysis of tetrahydrofurfuryl alcohol in supercritical carbon dioxide, *Catalysis Science & Technology*, 1 (2011) 1466–1471.

23. Y. Nakagawa, Y. Shinmi, S. Koso, K. Tomishige, Direct hydrogenolysis of glycerol in-to 1,3-propanediol over rhenium- modified iridium catalyst, *Journal of Catalysis*, 272 (2010) 191–194.

24. Y. Amada, Y. Shinmi, S. Koso, T. Kubota, Y. Nakagawa, K. Tomishige, Reaction mechanism of the glycerol hydrogenolysis to 1,3-propanediol over Ir-ReO$_x$/SiO$_2$ catalyst, *Applied Catalysis B*, 105 (2011) 117–127.

25. Y. Nakagawa, X. Ning, Y. Amada, K. Tomishige, Solid acid co-catalyst for the hydro-genolysis of glycerol to 1,3-propanediol over Ir-ReO$_x$/SiO$_2$, *Applied Catalysis A*, 433–434 (2012) 128–134.

26. Y. Nakagawa, K. Mori, K. Chen, Y. Amada, M. Tamura, K. Tomishige, Hydrogenolysis of C-O bond over Re-modified Ir catalyst in alkane solvent, *Applied Catalysis A*, 468 (2013) 418–425.

27. C. Deng, X. Duan, J. Zhou, X. Zhou, W. Yuan, S.L. Scott, Ir-Re alloy as a highly active catalyst for the hydrogenolysis of glycerol to 1,3-propanediol, *Catalysis Science & Technology*, 5 (2015) 1540–1547.

28. C. Deng, L. Leng, X. Duan, J. Zhou, X. Zhou, W. Yuan, Support effect on the bimetallic structure of Ir-Re catalysts and their performances in glycerol hydrogenolysis, *Journal of Molecular Catalysis A*, 41 (2015) 81–88.

29. W. Luo, Y. Liu, L. Gong, H. Du, T. Wang, Y. Ding, Selective hydrogenolysis of glycerol to 1,3-propanediol over egg-shell type Ir-ReO$_x$ catalysts, *RSC Advances*, 6 (2016) 13600–13608.

30. W. Luo, Y. Liu, L. Gong, H. Du, M. Jiang, Y. Ding, The influence of impregnation se-quence on glycerol hydrogenolysis over iridium-rhenium catalyst, *Reaction Kinetics, Mechanisms and Catalysis*, 118 (2016) 481–496.

31. M. Tamura, Y. Amada, S. Liu, Z. Yuan, Y. Nakagawa, K. Tomishige, Promoting effect of Ru on Ir-ReO$_x$/SiO$_2$ catalyst in hydrogenolysis of glycerol, *Journal of Molecular Catalysis A*, 388–389 (2014), 177–187.

32. A. Shimao, S. Koso, N. Ueda, Y. Shinmi, I. Furikado, K. Tomishige, Promoting effect of Re addition to Rh/SiO$_2$ on glycerol hydrogenolysis, *Chemistry Letters*, 38 (2009) 540–541.

33. Y. Shinmi, S. Koso, T. Kubota, Y. Nakagawa, K. Tomishige, Modification of Rh/SiO$_2$ catalyst for the hydrogenolysis of glycerol in water, *Applied Catalysis B*, 94 (2010) 318–326.

34. M. Chia, B.J. O'Neill, R. Alamillo, P.J. Dietrich, F.H. Ribeiro, J.T. Miller, J.A. Dumesic, Bimetallic RhRe/C catalysts for the production of biomass-derived chemicals, *Journal of Catalysis*, 308 (2013) 226–236.

35. A.C. Alba-Rubio, C. Sener, S.H. Hakim, T.M. Gostanian, J.A. Dumesic, Synthesis of supported RhMo and PtMo bimetallic catalysts by controlled surface reactions, *ChemCatChem*, 7 (2015) 3881–3886.

36. S.H. Hakim, C. Sener, A.C. Alba-Rubio, T.M. Gostanian, B.J. O'Neill, F.H. Ribeiro, J.T. Miller, J.A. Dumesic, Synthesis of supported bimetallic nanoparticles with controlled size and composition distributions for active site elucidation, *Journal of Catalysis*, 328 (2015) 75–90.

37. T. Buntara, S. Noel, P.H. Phua, I. Melian-Cabrera, J. G. de Vries, H.J. Heeres, Caprolactam from renewable resources: Catalytic conversion of 5-hydroxymethylfurfural into caprolactone, *Angewandte Chemie International Edition*, 50 (2011) 7083–7087.

38. T. Buntara, S. Noel, P.H. Phua, I. Melian-Cabrera, J.G. de Vries, H.J. Heeres, From 5-hydroxymethylfurfural (HMF) to polymer precursors: Catalyst screening studies on the conversion of 1,2,6-hexanetriol to 1,6-hexanediol, *Topics in Catalysis*, 55 (2012) 612–619.

39. T. Buntara, I. Melian-Cabrera, Q. Tan, J.L.G. Fierro, M. Neurock, J.G. de Vries, H.J. Heeres, Catalyst studies on the ring opening of tetrahydrofuran-dimethanol to 1,2,6-hexanetriol, *Catalysis Today*, 210 (2013) 106–116.

40. Y. Amada, H. Watanabe, Y. Hirai, Y. Kajikawa, Y. Nakagawa, K. Tomishige, Production of biobutanediols by the hydrogenolysis of erythritol, *ChemSusChem*, 5 (2012) 1991–1999.

41. M. Tamura, K. Tokonami, Y. Nakagawa, K. Tomishige, Rapid synthesis of unsaturated alcohol in mild conditions by highly selective hydrogenation, *Chemical Communications*, 49 (2013) 7034–7036.

42. M. Tamura, K. Tokonami, Y. Nakagawa, K. Tomishige, Selective hydrogenation of crotonaldehyde to crotyl alcohol over metal oxide modified Ir catalysts and the mechanistic insight, *ACS Catalysis*, 6 (2016) 3600–3609.

43. M. Tamura, R. Tamura, Y. Takeda, Y. Nakagawa, K. Tomishige, Insight into the mechanism of hydrogenation of amino acids to amino alcohols catalyzed by a heterogeneous MoO_x-modified Rh catalyst, *Chemistry — A European Journal*, 21 (2015) 3097–3107.

44. Y. Nakagawa, R. Tamura, M. Tamura, K. Tomishige, Combination of supported bimetallic rhodium-molybdenum catalyst and cerium oxide for hydrogenation of amide, *Science and Technology of Advanced Materials*, 16 (2015) 014901 (7 pp).

45. M. Tamura, R. Tamura, Y. Takeda, Y. Nakagawa, K. Tomishige, Catalytic hydrogena-tion of amino acids to amino alcohols with complete retention of configuration, *Chemical Communications*, 50 (2014) 6656–6659.

46. M. Li, G. Li, N. Li, A. Wang, W. Dong, X. Wang, Y. Cong, Aqueous phase hydrogenation of levulinic acid to 1,4-pentanediol, *Chemical Communications*, 50 (2014) 1414–1416.

47. Z. Wang, G. Li, X. Liu, Y. Huang, A. Wang, W. Chu, X. Wang, N. Li, Aqueous phase hydrogenation of acetic acid to ethanol over Ir-MoO_x/SiO$_2$ catalyst, *Catalysis Communications*, 43 (2014) 38–41.

48. K. Chen, M. Tamura, Z. Yuan, Y. Nakagawa, K. Tomishige, One-pot conversion of sugar and sugar polyols to *n*-alkanes without C-C dissociation over the Ir-ReO$_x$/SiO$_2$ catalyst combined with H-ZSM-5, *ChemSusChem*, 6 (2013) 613–621.

49. S. Liu, M. Tamura, Y. Nakagawa, K. Tomishige, One-pot conversion of cellulose into *n*-hexane over the Ir-ReO$_x$/SiO$_2$ catalyst combined with HZSM-5, *ACS Sustainable Chemistry & Engineering*, 2(7) (2014) 1819–1827.

50. S. Liu, Y. Okuyama, M. Tamura, Y. Nakagawa, A. Imai, K. Tomishige, Production of renewable hexanols from mechanocatalytically depolymerized cellulose by using Ir-ReO$_x$/SiO$_2$ catalyst, *ChemSusChem*, 8 (2015) 628–635.

51. Y. Nakagawa, S. Liu, M. Tamura, K. Tomishige, Catalytic total hydrodeoxygenation of biomass-derived polyfunctionalized substrates to alkanes, *ChemSusChem*, 8 (2014) 1114–1132.

52. S. Liu, Y. Okuyama, M. Tamura, Y. Nakagawa, A. Imai, K. Tomishige, Selective trans-formation of hemicellulose (xylan) into *n*-pentane, pentanols or xylitol over a rhenium-modified iridium catalyst combined with acids, *Green Chemistry*, 18 (2016) 165–175.

53. S. Liu, Y. Okuyama, M. Tamura, Y. Nakagawa, A. Imai, K. Tomishige, Catalytic conversion of sorbitol to gasoline-ranged products without external hydrogen over Pt-modified Ir-ReO$_x$/SiO$_2$, *Catalysis Today*, 269 (2016) 122–131.

54. H. Adkins, R. Connor, The catalytic hydrogenation of organic compounds over copper chromite, *Journal of the American Chemical Society*, 53 (1931) 1091–1095.

55. R. Connor, H. Adkins, Hydrogenolysis of oxygenated organic compounds, *Journal of the American Chemical Society*, 54 (1932) 4678–4690.

56. H. Liu, Z. Huang, F. Zhao, F. Cui, X. Li, C. Xia, J. Chen, Efficient hydrogenolysis of biomass-derived furfuryl alcohol to 1,2- and 1,5-pentanediols over a non-precious Cu-Mg$_3$AlO$_{4.5}$ bifunctional catalyst, *Catalysis Science & Technology*, 6 (2016) 668–671.

57. H. Liu, Z. Huang, H. Kang, C. Xia, J. Chen, Selective hydrogenolysis of biomass-derived furfuryl alcohol into 1,2- and 1,5-pentanediol over highly dispersed Cu-Al$_2$O$_3$ catalysts, *Chinese Journal of Catalysis*, 37(5) (2016) 700–710.

58. O. Koch, A. Köckritz, M. Kant, A. Matin, A. Schöning, U. Armbruster, M. Baroszek, S. Evert, B. Lange, R. Bienert, Method for producing 1,2-pentanediol, US Patent No. US8921617 (2014).

59. T. Mizugaki, T. Yamakawa, Y. Nagatsu, Z. Maeno, T. Mitsudome, K. Jitsukawa, K. Kaneda, Direct transformation of furfural to 1,2-pentanediol using a hydrotalcite-supported platinum nanoparticle catalyst, *ACS Sustainable Chemistry & Engineering*, 2 (2014) 2243–2247.

60. B. Zhang, Y. Zhu, G. Ding, H. Zheng, Y. Li, Selective conversion of furfuryl alcohol to 1,2-pentanediol over a Ru/MnO$_x$ catalyst in aqueous phase, *Green Chemistry*, 14 (2012) 3402–3409.

61. D. Götz, M. Lucas, P. Claus, C-O bond hydrogenolysis vs. C=C group hydrogenation of furfuryl alcohol: towards sustainable synthesis of 1,2-pentanediol, *Reaction Chemistry & Engineering*, 1 (2016) 161–164.

62. M. Omeis, M. Neumann, V. Brehme, C. Theis, D. Wolf, P. Claus, M. Lucas, R. Eckert, Hydrogenolysis of furfuryl alcohol to 1,2-pentanediol, US Patent Application No. 2014/0243562 (2014).

63. G. Li, N. Li, M. Zheng, S. Li, A. Wang, Y. Cong, X. Wang, T. Zhang, Industrially scalable and cost-effective synthesis of 1,3-cyclopentanediol with furfuryl alcohol from lignocellulose, *Green Chemistry*, 18 (2016) 3607–3613.

64. W.E. Kaufmann, R. Adams, The use of platinum oxide as a catalyst in the reduction of organic compounds. IV. Reduction of furfural and its derivatives, *Journal of the American Chemical Society*, 45 (1923) 3029–3044.

65. W. Xu, H. Wang, X. Liu, J. Ren, Y. Wang, G. Lu, Direct catalytic conversion of furfural to 1,5-pentanediol by hydrogenolysis of the furan ring under mild conditions over Pt/Co$_2$AlO$_4$ catalyst, *Chemical Communications*, 47 (2011) 3924–3926.

66. S. Liu, Y. Amada, M. Tamura, Y. Nakagawa, K. Tomishige, One-pot selective conversion of furfural into 1,5-pentanediol over Pd-added Ir-ReO$_x$/SiO$_2$ bifunctional catalyst, *Green Chemistry*, 16 (2014) 617–626.

67. S. Liu, Y. Amada, M. Tamura, Y. Nakagawa, K. Tomishige, Performance and characterization of rhenium-modified Rh-Ir alloy catalyst for one-pot conversion of furfural into 1,5-pentanediol, *Catalysis Science & Technology*, 4 (2014) 2535–2549.

68. Y. Nakagawa, K. Tomishige, Total hydrogenation of furan derivatives over silica-supported Ni-Pd alloy catalyst, *Catalysis Communications*, 12 (2010) 154–156.

69. Y. Nakagawa, K. Takada, M. Tamura, K. Tomishige, Total hydrogenation of furfural and 5-hydroxymethylfurfural over supported Pd-Ir alloy catalyst, *ACS Catalysis*, 4 (2014) 2718–2726.

Chapter 3.4

2-Methyl Furan and Derived Biofuels

Manuel López Granados*, Inaki Gandarias[†], Iker Obregón[†]
and Pedro L. Arias[†,‡]
*Institute of Catalysis and Petrochemistry (CSIC),
Sustainable Chemistry and Energy Group,
C/Marie Curie 2, Campus de Cantoblanco, 28049 Madrid, Spain

[†]Department of Chemical and Environmental Engineering,
Faculty of Engineering, University of the Basque Country
(UPV/EHU), Alameda Urquijo Street s/n, 48013 Bilbao, Spain

3.4.1. Introduction

2-Methyl furan (MF), also known as sylvan, is a flammable, water-insoluble liquid with a typical odor. It is used as solvent and as feedstock for the production of antimalarial drugs (chloroquine), methyltetrahydrofuran (MTHF), chrysanthemate pesticides, perfume intermediates, nitrogen and sulfur heterocycles, and functionally substituted aliphatic compounds. Moreover, a new route for the production of liquid fuels with high alkanes and low oxygenates contents from MF has been recently proposed [1, 2].

As it is shown in Table 3.4.1, MF itself presents interesting properties as alternative fuel in the gasoline range. MF has an octane number of 103 [6] and a thermal efficiency higher than gasoline and 2,5-dimethylfuran (2,5-DMF), due to its fast burning rate and notable better knock suppression ability [7]. This potential of MF as biofuel has been corroborated by Lange et al. [8], by analyzing

---

[‡]Corresponding author: pedroluis.arias@ehu.es

Table 3.4.1. Properties of fuels (adapted from Refs. [3–5]).

	RON 95 E10 Gasoline (EN51626-1)	Ethanol	MF	MTHF
Boiling temperature (°C)	36–190	78	64	78
Vapor pressure at 20°C (kPa)	—	5.8	13.9	13.6
Lower heating value (MJ/L)	30.8	21.1	27.6	28.2

a simplified footprint by considering only the investment costs associated to the production of H_2 required for the hydrogenation steps and the net CO_2 emissions related to the upgrade process from furfural (FUR). The capital index of MF is approximately the half of those of MTHF and trimer hydrocarbons, and similar to that of ethyl furfuryl ether (EFE), and the overall CO_2 footprint represents only less than 25% of the well-to-wheel CO_2 emissions of fossil transportation fuels (estimated at around $84\,g\ MJ^{-1}$). MF falls in the gasoline boiling range, with a blending research octane numbers (BRON) close to 140, higher than that of ethanol. In a road trip, with vehicles using a 10 vol.% MF added to a gasoline, for a combined total distance of 90,000 km, regulated vehicle emissions fulfill the EURO 4/5 compliance, whereas the loss of fuel economy was approximately 1%, and no detrimental impact on engine-oil degradation and engine wear were observed. Finally, it has to be taken into account that a pressurized vehicle tank system may be required, in order to avoid the excessive evaporation under high ambient temperatures, and to account for the MF excellent plastic solvent properties [3].

3.4.2. Production processes

MF is currently produced as a by-product of the synthesis of furfuryl alcohol (FOL) from FUR [9]. However, attempts have been made to produce it directly from FUR or from FOL (see scheme of Figure 3.4.1). In Table 3.4.2, the results and operating conditions of a selection of the results are reported.

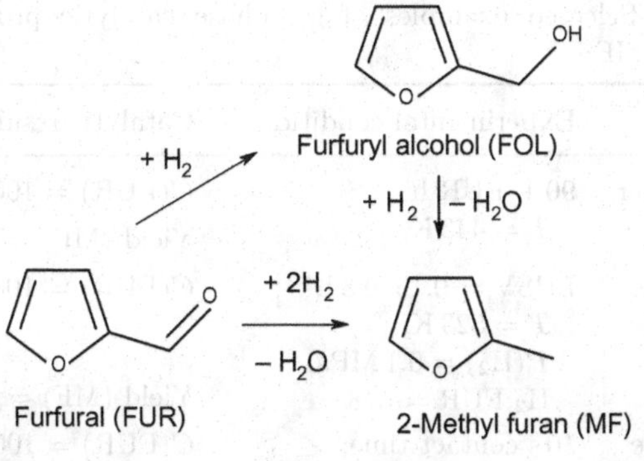

Figure 3.4.1. Furfural hydrogenation to 2-methyl furan.

3.4.2.1. *Gas phase processes*

The earliest work on MF production goes back to the first half of the 20th century and are described in two patents of Guinot [10, 11]. A procedure for the manufacture of FOL and MF via gas-phase hydrogenation of FUR was disclosed, with MF yields up to 40% (see Table 3.4.1) using a reduced copper oxide catalyst on different backing substances (asbestos, pumice stone, silica, and kaolin). It was possible to increase the MF yield until 100% by feeding FOL instead of FUR, at 443 K . Although there was no information regarding the stability of the catalysts, the authors claimed that no resins or by-products were detected since the large amount of heat evolved in the hydrogenation process was rapidly carried away.

Several works report the use of copper chromite [12–15], a well-known hydrogenating catalyst that can, however, generate toxicity problems. For instance, Bremner *et al.* using copper chromite catalysts achieved a MF yield of 95% at 523 K with a H_2:FUR ratio between 5 and 8 [12]. They reported a regular change from the predominance of MF to that of FOL in the hydrogenation of FUR with an increase in the amount of alkali added to a copper chromite catalyst, but deactivation of catalysts was always observed. The MF yield increased up to 99.5% when the copper chromite catalyst was dispersed on activated charcoal [13], at slightly lower working

Table 3.4.2. Selected examples of gas-phase catalytic processes for FUR conversion into MF.

Catalyst	Experimental conditions	Catalytic results	Ref.
Supported copper	90 L FUR h^{-1}; $T = 443$ K	C(FUR) = 100% Yield (MF) = 40%	[11]
Copper chromite	LHSV = 0.15–0.3 h^{-1}; $T = 523$ K; $P(H_2) = 0.1$ MPa; H_2:FUR = 5–8	C(FUR) = 100% Yield (MF) = 95%	[12]
Copper chromite on activated charcoal	10 s contact time; $T = 498$ K; $P(H_2) = 0.1$ MPa; H_2:FUR = 15	C(FUR) = 100% Yield (MF) = 95%	[13]
Reduced CuO–Cr$_2$O$_3$– MnO$_2$–BaCrO$_4$	F(FUR+H$_2$) = 2.5 mL min^{-1}; $T = 448$ K	Yield (MF) = 50%	[15]
Cu–Fe/SiO$_2$			[16]
Ni–Fe/SiO$_2$	$W/F = 0.1$ h; $T = 523$ K; H_2/feed ratio = 25;	Yield (MF) = 80% (from FOL) Yield (MF) < 40% (from FUR)	[17]
CuNiMgAlO$_y$ (activation H$_2$@ 573 K)	GHSV = 4000 h^{-1}; $T = 493$ K; H_2:FUR = 10	C(FUR) = 87% Yield (MF) = 43.5%	[18]
Cu–CeO$_2x$ (activation H$_2$ @ 573 K)	WHSV = 1.5 h^{-1}; $T = 463$ K; H_2:FUR = 7	C(FUR) = 88% Yield (MF) = 45%	[19]

temperatures, *i.e.*, 498 K. Nonetheless, the catalyst deactivation problem persisted and, after 7 days, the catalyst was completely deactivated and they found temperature variation of 493–503 K between the center of the catalytic bed and the reactor wall.

There are also other studies reporting the use of Cu-based catalysts different to Cu chromite. Lessard *et al.* reached a MF yield of 98% in the vapor-phase hydrogenation/hydrogenolysis of FUR, [16] in the presence of a Cu–Fe/SiO$_2$ catalyst, using toluene as diluting agent for FUR. The catalytic activity dropped after 20 h of operation, and it could be only partially recovered after the catalyst regeneration. When Cu was substituted by Ni (Ni–Fe/SiO$_2$), the MF yield decreased to 40%, accompanied by an increase in the FOL yield (31%) [17]. The effect of the type of metal active sites on the reaction mechanism is discussed in Section 3.4.2.5. Xu *et al.* demonstrated that a CuNiMgAlO$_y$ catalyst, prepared from a hydrotalcite precursor, after activation at 573 K under H$_2$, could be reused for 36 h without deactivation, with a FUR conversion close to 85% and a selectivity toward MF of 47% [18]. Similar values (FUR conversion of 85% and 45% MF yield) have recently been reported using Cu–CeO$_2$ catalysts [19]. However, the catalyst deactivated and the selectivity towards MF decreased with Time on Stream (TOS), which provoked a rise of the FOL yield. This deactivation was ascribed to the formation of carbonaceous deposits on the hydrogenating sites, and by the oxidation of the active sites by the water generated in the hydrogenolysis of FOL to MF.

The research works summarized above reflect that the stability of the catalyst is the main drawback of vapor phase processes for MF formation. Three major mechanisms have been suggested to cause this deactivation: (i) the formation of coke, (ii) metal sintering/leaching [20], and (iii) support migration over the active sites [21]. The stability of copper chromite catalysts could be significantly improved by the atomic layer deposition (ALD) of Al$_2$O$_3$ overcoatings [22]. Both the sintering of Cu and the coke formation decreased when the metal nanoparticles where covered with thin alumina layers. Moreover, ALD of Al$_2$O$_3$ also decreased the activation energy of FUR hydrogenation.

In general, high temperatures are required for the gas-phase synthesis of MF using Cu based catalyst. Both the stability of the catalyst and the formation of many different by-products, which can be formed by hydrogenation of MF (2-pentanone, 2-MTHF, 1- and 2-pentanol) or by decarbonylation of FUR and FOL (furan), limit the commercial implementation of this process.

3.4.2.2. *Liquid-phase reactions*

Liquid-phase hydrogenation reactions have also been explored in order to overcome the deactivation problems associated with the use of higher temperatures in the gas-phase process. The most relevant

Table 3.4.3. Selected examples of liquid-phase catalytic processes for FUR conversion into MF.

Catalyst	Experimental conditions	Catalytic results	Ref.
1% Ru–4% Pd/TiO$_2$	15 mL octane; 1 g FUR; 0.1 g cat.; $P = 0.3$ MPa; $T = 293$ K; $t = 120$ min	C(FUR) $= 39.3\%$ Yield (MF) $= 20.2\%$	[24]
5% Pd/C	WHSP $= 1.7$ h^{-1}; 0.2 M FUR in ethyl acetate; $P = 5$ MPa; $T = 363$ K	C(FUR) $= 100\%$ Yield (MF) $= 75\%$	[26]
Ru/Co$_3$O$_4$	3 mL THF, 0.017 g FUR; 0.1 g cat; $P = 1$ MPa; $T = 443$ K; $t = 24$ h	C(FUR) $= 100\%$ Yield $= 93\%$	[27]
Cu–Co/γ-Al$_2$O$_3$	20 mL 2-propanol; 4.3 g FUR; 1 g cat; $P = 4$ MPa; $T = 493$ K; $t = 4$ h	C(FUR) $= 100\%$ Yield (MF) $= 65\%$	[28]
Cu–Fe	5 mL octane; 2.32 g FUR; $P = 9$ MPa; $T = 493$ K; $t = 14$ h	C(FUR) $= 99\%$ Yield (MF) $= 50\%$	[29]
cis-[Ru(6,6′-Cl$_2$bpy)$_2$(OH)$_2$] (CF$_3$SO$_3$)$_2$	10 ml EtOH; 0.12 g FUR; 1.0 mol.% cat.; $P = 5.1$ MPa; $T = 303$ K; $t = 4$ h	C(FUR) $= 100\%$ Yield (MF) $= 20\%$	[30]

results are summarized in Table 3.4.3. The reaction of FUR into MF is usually carried out in organic solvents, because water catalyzes the transformation of FOL into cyclopentanone [23]. Aldosari *et al.* studied the effect of different solvents at room temperature and 0.3 MPa of H$_2$ pressure using Pd:Ru/TiO$_2$ catalyst prepared by impregnation [24]. Methanol and 1,2-dichloroethane promoted the formation of acetals and polymeric species, while the selective production of MF was promoted using non-polar solvents such as toluene or octane. The addition of Ru to the Pd/TiO$_2$ catalyst decreased the conversion but increased the selectivity towards MF by decreasing side reactions. On the contrary, when the reaction started from FOL, MF was only selectively obtained when 1,2-dichloroethane was used as solvent [25]. Under the operating conditions used, metal leaching was not detected, but no information was provided regarding the reusability of the catalysts.

The continuous flow hydroconversion of FUR was studied over different catalysts with novel or base metals as hydrogenating active sites [26]. Catalysts containing noble metals gave significantly higher FUR conversions as compared to those with base metals. The highest yield of MF (75%) was obtained with the 5% Pd/C catalyst. Nonetheless, the selectivity of the processes rapidly changed with TOS, and at only 120 min FOL was the main product. Authors suggested that Pd(II) species were the active sites of the hydrogenolysis reaction of FOL to MF. They ascribed the deactivation of the catalyst to the gradual reduction of the Pd(II) species to Pd(0) under the operating conditions used. With a Ru/Co_3O_4 catalyst and THF as solvent, a remarkable 93% MF yield was reported [27]. Although the catalyst/FUR ratio was as high as 6, the relevance of the work arises from the direct use of the FUR obtained after the hydrolysis of biomass.

When base metals are utilized, harsher conditions are required. Srivastava et al. tested Cu–Co catalysts over different supports (SiO_2, H-ZSM-5 and γ-Al_2O_3) at 220°C, 4 MPa H_2 pressure and 4 h reaction time [28]. They reported a significant MF yield of 65% with the Cu–Co/Al_2O_3 catalyst. The catalyst suffered deactivation due to coke formation, but after calcinations it could be recycled without significant activity lost. The hydrogenation of highly concentrated FUR in n-octane (41 wt.%) over a Cu–Fe catalyst yielded FOL as the main reaction product at temperatures lower that 473 K [29]. Nonetheless, a significant 51% MF yield was attained at 493 K and 9 MPa of H_2 pressure. Although the catalyst was not pre-reduced, all the spent catalysts showed the presence of Cu^0, which evidenced the in situ reduction of the copper species under the operating conditions utilized.

Gowda et al. have proposed the use of ruthenium(II) bis(diimine) complexes for the hydrogenation of FUR and FOL, and they found that the catalytic activity is related to the Lewis acidity of Ru^{2+} centers and the steric hindrance, which can lead to a determined coordination of the FUR and FOL molecules [30]. The MF selectivity increased from 20% to 25% by using FOL as feedstock at lower conversions (90%).

In general, most of the reported activity tests results in batch reactors lack from reusability studies to assess the stability of the catalysts. Among the different solvents studied, those that are used to obtain FUR from hemicellulose will present a higher applicability.

3.4.2.3. *Unconventional reaction systems*

Catalytic transfer hydrogenation (CTH), coupling the dehydrogenation of an alcohol and the hydrogenation of FUR has also been employed for synthesizing MF from FUR without the presence of molecular hydrogen. The main results are summarized in Table 3.4.4. Zhu's research group has extensively studied this coupling process by using different alcohols in gas phase [31–33]. This process

Table 3.4.4.   Selected examples of for FUR conversion into MF using CTH.

Catalyst	Experimental conditions	Catalytic results	Ref.
Cu–Zn–Al (gas-phase)	FUR:BDO molar ratio = 1:1; LHSV (FUR+BDO) = $0.7\,h^{-1}$; H_2:(FUR+BDO) molar ratio = 13:1; residence time = 4.7 s; $T = 498\,K$	C(FUR) = 100% Yield (MF) = 93%	[32]
Cu/Mn/Si (gas-phase)	FUR:CHN molar ratio = 0.3:1; LHSV (FUR+CHN) = $0.49\,h^{-1}$; H_2:CHN molar ratio = 10:1; $T = 552\,K$	C(FUR) = 100% Yield (MF) = 94%	[33]
Mg/Fe/O (gas-phase)	FUR:methanol molar ratio 10:1; residence time 1.1 s; $T = 653\,K$	C(FUR) = 100% Yield (MF) = 83%	[34]
2 wt.% Pd/Fe$_2$O$_3$ (liquid-phase)	0.4 M FUR in 2-propanol (40 mL); $P = 1.5\,MPa(N_2)$; 0.5 g cat; $T = 453\,K$; $t = 7.5\,h$	C(FUR) = 100%; Yield (MF+MTHF) = 13%	[35]
5% Ru/C (liquid-phase)	1 wt.% FUR in 2-propanol (24 mL); $P = 2\,MPa(N_2)$; 0.1 g cat; $T = 433\,K$; $t = 15\,h$	C(FUR) = 95%; Yield (MF) = 64%	[36]
10% Cu–3% Pd/ZrO$_2$ (l–v equilibrium)	0.096 g FUR in 2-propanol (14 mL); $P =$ autogenerated; 0.120 g cat; $T = 493\,K$; $t = 4\,h$	C(FUR) = 95%; Yield (MF+MTHF) = 83.9%	[37]

helps to overcome the limitations of the conventional individual alcohol dehydrogenation and FUR hydrogenation, mainly due to the difficulty to control the temperature over the process, the poor hydrogen utilization and conversion constrained by thermodynamic equilibrium. In this sense, the evaluation of the reduction of FUR in the presence of 1,4-butanediol, which underwent cyclization to supply the hydrogen for the reduction process, was studied over a Cu/Zn catalyst, reaching a MF yield of 96.5% and a 99.4% yield of γ-butyrolactone. When this process was carried out by supplying H_2 gas, the MF yield was lower (88.6%). The same research group also studied a series of Cu–Zn–Al catalysts obtained by co-precipitation of the corresponding nitrates, calcined at different temperatures and subsequently reduced at 543 K, in a tubular fixed-bed reactor [32]. A mixture of 1,4-butanediol (BDO) and FUR in hydrogen was fed into the reactor. The calcination temperature was optimized, finding that the optimum catalytic activity and selectivities to γ-butyrolactone and MF are achieved after calcination at 623 K , with a MF yield close to 93%. These authors conclude that the two coupled reactions take place on different active sites, but they do not provide any information about the nature of the real active sites and their role in reaction. In a similar study of Cu/Mn, Cu/Si, and three Cu/Mn/Si catalysts, but using cyclohexanol as alcohol, they found that the interactions between the catalyst components, as well as the loading of the active phase, affect the acidity, surface area and copper dispersion [33]. They observe that manganese exerts a positive effect on the catalytic performance, mainly associated to the interface interaction between Cu and MnO, and the optimum molar composition was Cu:Mn:Si = 1:1.12:1.13. Using a magnesium–iron catalyst and methanol as the hydrogen donor molecule, MF was only the main reaction product when significantly high reaction temperatures (*i.e.*, 653 K) were utilized [34]. The catalyst suffered deactivation due to a decrease in the surface area and the formation of methanol/formaldehyde oligomers onto the catalyst surface.

Liquid-phase CTH processes have also been studied, and 2-propanol is the most selected hydrogen donor molecule. Sholz *et al.*, tested the activity of Fe_2O_3-supported Cu, Ni and Pd catalysts, with

2-propanol as hydrogen donor [35]. They found that the strong metal-support interaction of Pd/Fe_2O_3 favors its higher catalytic activity. However, only moderate MF yields (13%) were reported starting from FUR. The MF yield increased by employing lower substrate concentrations. Panagiotopoulou *et al.* reported a significant MF yield of 50% using 2-propanol as the hydrogen donor molecule, and a Ru/C catalyst that was activated by reduction in H_2 at 300°C followed by oxidation for 3 h at 130°C in O_2/He flow [36]. The yield rose to 61% when FOL was used as the reactant. The authors highlighted the role of RuO_x species on the activity of the catalyst. In fact, the deactivation of the catalyst was related to the partial reduction of the Ru/C catalyst under the operating conditions used. The regeneration was successful after applying the same thermal treatment as of the activation. In 2-propanol and using non-catalytic amounts of a Cu–Pd/ZrO_2 catalyst (1.25 g of catalyst/g of FUR), a combined yield of MF+MTHF of 83.9% was obtained [37]. The reactor was not pressurized with N_2, and therefore the reaction took place in a vapor–liquid equilibrium. Interestingly, the catalyst could be recycled five times without significant deactivation.

Another interesting approach to enhance the MF selectivity is to carry out the reaction under H_2 stripping conditions in order to remove the produced MF from the reaction medium, and avoid the consecutive hydrogenation to MTHF [38]. However, the reactive distillation requires a strict control of the operating conditions. Thus, by using a commercial CuCrBa catalyst, at 443 K , WHSV of 0.2 g_{FUR} g_{cat}^{-1} h^{-1}, H_2-to-FUR ratio of 71 and 45 NL_{H2}/h, a MF yield of 31% was attained, whereas it was only 16.6% in the absence of H_2 stripping.

Supercritical CO_2 has also been used in a two-step process for the synthesis of MF from FUR, combining copper chromite (first reactor) and Pd (5%) supported on activated carbon (second reactor) catalysts. This process takes advantage, as already indicated in other sections, of its flexibility to vary the selectivity toward the desired hydrogenation product by modifying the temperature of each reactor [39]. Thus, a MF yield of 90% can be attained by using a CO_2 flow of 1 mL min^{-1} (pumphead at 263 K and 5.8 MPa), FUR flow of

$0.05\,mL\,min^{-1}$, 15 MPa of operating pressure, by operating at 513 K in the first reactor, and turning off the temperature of the second one.

On the other hand, the aqueous electrocatalytic hydrogenation of FUR using a sacrificial anode, as previously explained in the section devoted to FOL and derivatives, can be also used for the synthesis of MF [40]. In this process, atomic hydrogen is generated *in situ* by reduction of hydronium ions on a catalytic cathode surface using external electrons. Different metals (Al, Fe, Ni, Cu and stainless steel) were evaluated as cathode materials, and the results demonstrated the important effect on the electrocatalytic hydrogenation of FUR, where Ni and Fe gave the best results. MF formation was favored at pH = 1.0, although yields were always lower than 10%. Green *et al.* [41] have also detected the formation of MF, but with a selectivity lower than 10%, in the electrocatalytic hydrogenation of aqueous solution of FUR in a continuous membrane reactor. Protons were obtained from the electrolysis of water at the anode, whereas reduction of FUR took place at a Pd/C cathode, at temperatures between 303 and 343 K . The current efficiency was 24–30%, and the use of H_2 gas instead as a source of protons barely affected the rate of FUR conversion at a lower power input. A similar strategy has been used by Nilges and Shröder [42], attaining a full conversion of FUR with a selectivity to MF of 80% when a copper electrode was used, with 500 mM H_2O_4 and a water–acetonitrile mixture. Solid copper and electrochemically deposited copper can be used as electrode, and no electrode deterioration was detected over time. The nature of the interaction of FUR with the different types of electrodes is under study.

3.4.2.4. *Kinetics and reaction mechanisms*

Zheng *et al.* studied in a fixed-bed reactor the catalytic behavior of a commercial (Cu:Zn:Al:Ca = 59:33:6:1 atomic ratio) and a synthesized (Cu:Cr:Ni:Zn:Fe = 43:45:8:3:1 atomic ratio) catalysts in the hydrogenation of FUR and some reaction intermediates in order to elucidate the mechanism of formation of MF and its further transformations [43]. The catalyst containing Ni promoted the hydrogenation of the ring and therefore MTHF was the main reaction product. MF was not stable under the reaction conditions and it could react to MTHF, 1-pentanol, 2-pentanol or 2-pentanone. On the contrary, MTHF showed very low reactivity even at 523 K.

Concerning the kinetics and mechanisms of hydrogenation of FUR in gas phase, Sitthisa *et al.* have studied the hydrogenation/hydrodeoxygenation of FUR on silica-supported Ni and Ni–Fe bimetallic catalysts, at 0.1 MPa and 483–523 K [44]. The highest MF yield (close to 80%) was achieved over a 5 wt.% Ni–2 wt.% Fe/SiO$_2$ catalyst at 523 K, by using FOL as feedstock, whereas from FUR, the MF yield barely attained 40%, with an important contribution of C$_4$ products. It was observed that only FOL and furan were formed on the monometallic nickel catalyst, which were transformed into MF (via C–O hydrogenolysis) and C$_4$ products (via ring opening, giving rise to butanal, butanol and butane), respectively. Ni catalyzes decarbonylation over hydrogenation, mainly at high temperatures, similar to Pd catalysts [17], that leads to further ring opening hydrogenolysis of furan to produce C$_4$ acyclic compounds, whereas Fe suppresses this decarbonylation activity of Ni by forming a bimetallic Fe–Ni alloy (as inferred from XRD and TPR studies). DFT calculations suggest that differences in selectivity could be due to the stability of the η^2(C,O) surface species. The oxophilic nature of Fe strengthened the oxygen interaction with the η^2(C,O) surface, with a significant lengthening of the C1–O1 bond, which can be readily hydrogenated to FOL and then covert into MF, via hydrogenolysis, thus inhibiting the formation of an acyl species. These acyl species are found on Ni surfaces, where they decompose to produce furan and CO. In this reaction scheme, the hydrogenolysis step is proposed to occur prior to any atomic-H addition to the molecule, although this fact is not discussed.

In a recent study, DFT calculations have been again performed to investigate the adsorption of FUR, FOL, MF and furan on Pd(111) surfaces, and study the reaction barriers for their interconversion [45]. It was found that the most stable conformation was with the furan ring lying flat on the metal surface, centered on a hollow site. Thermodynamics concerns clearly favor the hydrogenation of FUR to furan and CO, with an apparent reaction barrier 14 kJ/mol higher than that of hydrogenation to FOL. However, at high hydrogen coverage, the conformation FUR becomes tilted [46]. Moreover, the selectivity to hydrogenation over decarbonylation increases with H coverage or hydrogen pressure and decreases with increasing temperature. The formation of MF occurs in two steps: hydrogenation of FUR to FOL and hydrogen-assisted dehydration, rather than dehydroxylation of FOL to MF. Other mechanistic paths involve that FOL first dehydrogenates to a methoxy intermediate, which directly deoxygenates, being MF then formed via hydrogenation, or dehydroxylation followed by hydrogenation. The authors affirm that the preference for a particular pathway will depend on the hydrogen coverage.

Shi *et al.* carried out computational studies on the mechanism of FUR hydrogenation to MF over $Mo_2C(101)$ surface [47]. Authors claimed that FUR, FOL, MF and furan preferentially adsorb on the unsaturated Mo_A sites, while H_2 dissociative adsorption prefers the surface unsaturated C_A sites. On the clean surface, FUR dissociation into F-CO is thermodynamically and kinetically favored, and therefore furan rather than MF should be the main product. Nonetheless, the computed minimum energy path showed that high H_2 partial pressure can promote MF formation and suppress furan formation, which matches the above-mentioned work from Vorotnikov *et al.* [46]. Based on these results, authors recommend a two-step protocol for experiments, with an initial pretreatment of the catalyst with H_2, and the subsequent FUR hydrogenation on the hydrogen precovered catalyst.

In summary, gas-phase processes for the production of MF from FUR require elevated temperatures (448–473 K) at which the catalyst is rapidly deactivated, and this hinders the commercial implementation of these processes [9]. With regards to liquid phase processes, much diluted furfural concentrations in organic solvents are required. This implies expensive distillation costs to separate FUR from the aqueous phase where it is obtained, and after the reaction, to separate MF from the organic solvent. The fact that many studies report higher yields of MF when the reaction starts from FOL [25, 30, 36], suggest that a two-step process (FUR → FOL and FOL → MF) could be an option to minimize secondary reactions. The effect of the two-step reaction on the stability of the catalyst should also be assessed.

3.4.3. 2-Methyl furan trimer derived biofuels

Branched alkanes in the diesel range can be obtained from 2-MF by condensation with different aldehydes and ketones. It requires a two-step process: the first step is the formation of the diesel precursor via the hydroxyalkylation/alkylation of MF with an aldehyde or ketone. In the second step, these precursors are subsequently subjected to hydrodeoxygenation (HDO) to afford the diesel range alkanes (see Figure 3.4.2).

This strategy was first demonstrated by Corma *et al.* [1, 2, 48, 49] and originally they used the trimerization of MF (see scheme I of Figure 3.4.2). In practice, the reaction initiates by the hydrolysis of MF to 4-oxopentanal (via ring opening and keto-enol tautomerism) and then the trimerization takes place by two consecutive hydroxyalkylation and alkylation of MF (also described as an aromatic substitution of the ring first with the electrophile 4-oxopentanal and then with the formed alcohol). The yield of C_{15} oxygenated trimer, 5,5-bis(sylvyl)-2-pentanone, was ca. 74% (using 12 wt.% aqueous H_2SO_4 as catalyst and 16 h at 333 K). Advantageously the diesel precursor separates from the aqueous phase what simplifies the recovery and the reusability of the catalyst [2]. The subsequent HDO step of the C_{15} diesel precursor was also demonstrated with a fixed catalytic bed reactor using a composite 3 wt.% Pt/C–TiO$_2$ catalyst, 4–5 MPa H_2 and $W/F = 1.12\,h^{-1}$ (in a continuous mode). A gradient of temperature along the reactor was recommended to prevent cracking of the trimer (393 K for the top zone of the reactor and 623 K for the middle zone) [2] and a molar yield of the diesel range fraction (C_9–C_{24}) greater than 93% was obtained [2]. The diesel fraction spontaneously separates from the aqueous phase, thus facilitating the separation of the fuel without distillation. The HDO step was successfully operated for more than 140 h [1]. This latter step ideally should render branched 6-butylundecane but side reactions

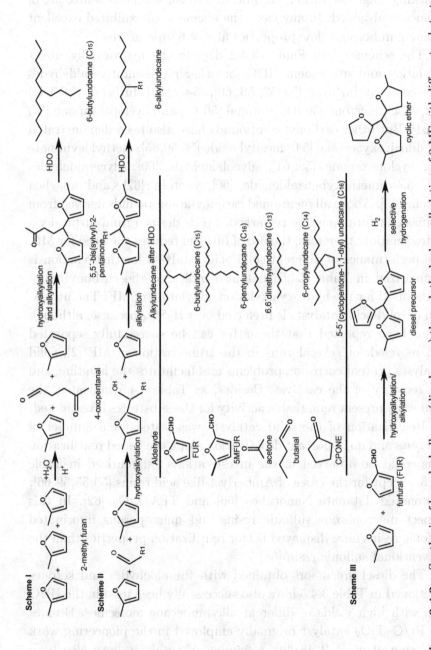

Figure 3.4.2. Synthesis of branched alkanes (diesel range) through the hydroxyalkylation of 2-MF with different substrates and subsequent hydrodeoxygenation of the intermediate oxygenates.

(cracking, oligomerization, etc.) occur and consequently a mixture of alkanes is obtained. In any case, the alkane pool exhibited excellent cetane number and flow properties at low temperature.

The scheme II in Figure 3.4.2 depicts the hydroxyalkylation/ alkylation and subsequent HDO for other representative aldehydes and ketones: furfural [50–53, 56, 60], 5-methyl furfural (MFUR) [1,2,48,49], acetone [56,64], butanal [56,64] and cyclopentanone [57, 58,61]. But other carbonyl compounds have also been demonstrated like dihydroxyacetone [54], mesityl oxide [56,59,65], methyl levulinate [50], cyclohexanone [58, 61], glycolaldehyde [60], glyceraldehydes [60], 3,4-dimethoxybenzaldehyde [60], acetoin [62], and angelica lactone [63]. Above all mentioned carbonylic compounds derived from biomass, furfural must be remarked. MF is derived from furfural via hydrogenation; therefore, the use of furfural reduces the need for MF. The performance of different solid acid catalysts for this reaction is summarized in Table 3.4.5 that also indicated the carbonyl compound used for the hydroxyalkylation/alkylation of MF. The utilization of solid acid catalysts is preferred over H_2SO_4 because, although it has been reported that the latter can be successfully separated and recycled for several runs in the trimerization of MF [2], solid catalysts prevent corrosion problems and facilitates the handling and the recovery of the catalyst. Besides, as Table 3.4.5 indicates, the solid acids present remarkable activity for the substrates investigated.

Reutilization of the solid catalysts was tested for a number of catalysts and no important deactivation along the tested reutilization runs could be observed in the investigations summarized in Table 3.4.3, except for the cases of Amberlyst-like acid resins [50,55,59,66], Protonated Titanate Nanotubes [56] and TFA/ZrO_2 [62]. In this respect macroporous sulfonic resins and macroporous fluorinated sulfonic resins have displayed better reutilization properties than the conventional sulfonic resin [58].

The diesel precursors obtained with the aldehydes and ketones mentioned in Table 3.4.5 have also successfully been used in the HDO step with high yields of different alkylundecane molecules. Besides the Pt/C–TiO_2 catalyst originally employed in the pioneering work of Corma *et al.* [1, 2, 48, 49], a number of catalysts have also been

Table 3.4.5. Hydroxyalkylation/alkylation of MF with different aldehydes and ketones to different diesel precursors.

Substrates (mol. ratio)	Catalyst	Cat. (wt.%)[a]	Temp. (K)	Time (h)	Yield (%)	Ref.
MF	H_2SO_4	12	333	16	74	[2]
MF	Amberlyst-15[b]	2.6	358	52	50	[5]
MF/MFUR (5:1)	p-TSA	1.8	323	6	93	[48]
MF/FUR (2:1)	Nafion-212	2.9	323	2	70	[50]
MF/FUR (2:1)	LF resin[c]	2.9	333	12	89	[51]
MF/FUR (2.05:1)	Pd/NbOPO$_4$	3.7	353	5	93	[52]
MF/FUR (2.2:1)	R-SO$_3$H-SiO$_2$	1.6	338	2	88	[53]
MF/HA (2:1)	Nafion-212	3.1	338	2	>65	[54]
MF/acetone (2:1)	Nafion-212	3.4	323	25	75.5	[55]
MF/butanal (2:1)	Nafion-212	3.2	323	4	89.5	[55]
MF/butanal (2:1)	PTN[d]	3.2	323	6	77	[56]
MF/CPONE (2:1)	Nafion-212	3.0	338	12	91.3	[57]

(*Continued*)

Table 3.4.5. (*Continued*)

Substrates (mol. ratio)	Catalyst	Cat. (wt.%)[a]	Temp. (K)	Time (h)	Yield (%)	Ref.
MF/CPONE (2:1)	Macroporous PS[e]	3.7	328	9	50.1	[58]
MF/CPONE (2:1)	Macroporous fluorinated PS[f]	3.7	328	9	68.5	[58]
MF/Mesityl oxide (1:1)	Nafion-212	4.2	333	2	~65%	[59]
MF/formalin (2:1)	SBA-15 supported AIL[g]	4.1	333	8	88	[60]
MF/cyclohexanone (2:1)	Nafion-212	2.9	333	6	89	[61]
MF/cyclohexanone (2:1)	Macroporous PS[e]	1.8	328	9	73	[58]
MF/cyclohexanone (2:1)	Macroporous fluorinated PS[f]	1.8	328	9	88	[58]
MF/acetoin (2.1:1)	TFA/ZrO$_2$[h]	3.7	333	2	95	[62]
MF/angelica lactone (2:1)	Nafion-212	3.7	323	1	83	[63]

Notes: [a]% with respect to total reactants.
[b]15 wt.% H_2O was incorporated to MF.
[c]Lignosulfonate-formaldehyde resin.
[d]Protonated Titanate Nanotubes.
[e]Macroporous sulfonic resin.
[f]Macroporous fluorinated sulfonic resin.
[g]AIL (acidic ionic liquid): 1-methyl-3-(4-sulfobutyl)imidazolium methanesulfonate.
[h]Zirconia-supported trifluoromethanesulfonic acid.
FUR: Furfural, 5-MFUR: 5-methyl furfural, HA: hydroxyacetone, p-TSA: p-toluenesulphonic acid.

explored in the HDO step, affording also high yields of diesel fraction liquids: Pt/Zr phosphate [50], Pd/C [54,55,63], Pd/charcoal and Ru/SiO$_2$–Al$_2$O$_3$ [5], Pd/C and Hβ [58, 62], Pd/NbOPO$_4$ [67], carbon supported Ni–W$_x$C [55], Ni supported on different acid supports (SiO$_2$/Al$_2$O$_3$ and Hβ, ZSM-5 and USY zeolites) [57,65,68], Ni–W$_2$C/SiO$_2$ and Ni–Mo$_2$C/SiO$_2$ [59], Ni/Hβ zeolite [68], Ni/H–ZSM5 zeolite [66], and Ni/ZrO$_2$–SiO$_2$ [69]. It is remarkable that the Pd/NbOPO$_4$ system can be used for both reactions as it possesses strong Brønsted acid sites for the first step and metal functionalities for the HDO reaction. Actually, the reaction was run in a one-reactor-two-step mode with a yield to diesel high alkanes of 89% [67]. Although the catalyst deactivates after four runs when conducting the HDO reaction, the activity was largely recovered by removal of the coke deposits with calcination at 773 K [67].

Stability studies were also conducted in some of the HDO catalysts and showed that Ni–W$_x$C/C was stable for 24 h on stream [55], Ni/Hβ zeolite (Si/Al = 394) and Ni/H–ZSM-5 for 24 h [66,68], 10 wt.%Ni/ZrO$_2$–SiO$_2$ for more than 110 h.

The total removal of oxygen atoms and the reduction of the double bonds in the oxygenated and unsaturated products obtained in the first step (hydroxyalkylation/alkylation) imply the utilization of large amounts of H$_2$. Thus, in the case of MF and furfural (6-butylundecane) ca. 4.17 mol of H$_2$ per mol of initial furfural is required (MF is also derived from furfural) to obtain the undecane. An alternative to the processes described above is to obtain the saturated oxygenated molecules by selective hydrogenation of the double bonds without proceeding to the hydrogenolysis of the ether C–O in the furan ring (see scheme III in Figure 3.4.2 for the case of using MF and FUR as furanic molecules for the hydroxyalkylation/alkylation step). The resulting product has been proposed as a diesel blending agent [70]. Remarkably, the cyclic ether presents excellent fuel properties: a high cetane number (60.4), low freezing point (< 233 K), high volumetric energy density (32.6 MJ/L) and good lubricity properties [70]. The five additional molecules of H$_2$ needed to obtain the corresponding 6-butylundecane are saved with this approach, which represents a 40% saving of H$_2$ (including the H$_2$

needed to produce MF from furfural). A catalytic system based on Pd nanoparticles supported on ionic liquid-modified SiO_2 was used at 333 K, 2.08 MPa H_2, and 0.5 wt.% of metal loading. For the MF/FUR case described in the Figure 3.4.2, a 92% yield of the cyclic ether was achieved after 30 h.

3.4.4. Summary and prospective

The production of MF from FUR still faces many challenges that should be confronted in order to develop an economic and technically feasible process. Gas-phase processes entail elevated temperatures (448–473 K) at which the catalyst is rapidly deactivated, and this hinders the commercial implementation of these processes [9]. Moreover, water free FUR has to be fed, which implies an energy intensive two step distillation in order to separate the FUR from the aqueous phase where it is obtained after the hydrolysis step. With regards to liquid phase processes, much diluted furfural concentrations in organic solvents are required. Again, expensive distillation costs are required to separate FUR from the aqueous phase where it is obtained, and after the reaction, to separate MF from the organic solvent. The fact that many studies report higher yields of MF when the reaction starts from FOL [25, 30, 36], suggests that a two-step process (FUR → FOL and FOL → MF) could be an option to minimize secondary reactions. The effect of this two-step reaction on the stability of the catalyst should also be assessed.

Diesel range biofuels can be readily produced by a two-step process consisting of, first, the hydroxyalkylation/alkylation of MF with an aldehyde or ketone, and second, of the hydrodeoxygenation of the resulting oxygenated and unsaturated precursors to render branched alkanes. A number of robust solid catalysts based on strong Brønsted acid sites have been identified for the first step that may replace the problematic use of H_2SO_4 as catalyst. For the second step, catalysts based on supported metal particles have also been investigated. The process has also been practiced with a wide variety of ketones and aldehydes derivable from biomass. However most of the reports so far published have been conducted with high-grade reactants and there is a lack of information regarding the robustness of the process using reactants obtained directly from biomass. In this context, it is remarkable the investigation conducted by Wang *et al.* that have technically demonstrated the production of aviation fuel from furfural and MF directly derived from corncob (MF was obtained by hydrogenation of furfural) using homogeneous inorganic acids [69]. This research must inspire future investigation needed on the robustness of solids acid catalyst when using the low-grade MF and ketones/aldehydes obtained from biomass. Another type of information lacking in the scientific literature concerns the economic and life cycle analyzes to assess on the economic viability and on the environmental impact and the savings of Greenhouse Gas Emissions of this process when confronted with the petrofuel.

References

1. A. Corma, O. De La Torre, M. Renz, N. Villandier, Production of high-quality diesel from biomass waste products, *Angewandte Chemie International Edition*, 50 (2011) 2375–2378.
2. A. Corma, O. De La Torre, M. Renz, Production of high quality diesel from cellulose and hemicellulose by the Sylvan process: Catalysts and process variables, *Energy & Environmental Science*, 5 (2012) 6328–6344.
3. M. Thewes, M. Muether, S. Pischinger, M. Budde, A. Brunn, A. Sehr, P. Adomeit, J. Klankermayer, Analysis of the impact of 2-methylfuran on mixture formation and combustion in a direct-injection spark-ignition engine, *Energy and Fuels,* 25 (2011) 5549–5561.
4. F. Hoppe, U. Burke, M. Thewes, A. Heufer, F. Kremer, S. Pischinger, Tailor-made fuels from biomass: Potentials of 2-butanone and 2-methylfuran in direct injection spark ignition engines, *Fuel*, 167 (2016) 106–117.
5. R. Christensen, E. Yanovitz, J. Ratcliff, M. McCormick, Renewable oxygenate blending effects on gasoline properties, *Energy Fuels*, 25 (2011) 4723–4733.
6. X. Ma, C. Jiang, H. Xu, H. Ding, S. Shuai, Laminar burning characteristics of 2-methylfuran and isooctane blend fuels, *Fuel*, 116 (2014) 281–291.
7. C. Wang, H. Xu, R. Daniel, A. Ghafourian, J.M. Herreros, S. Shuai, X. Ma, Combustion characteristics and emissions of 2-methylfuran compared to 2,5-dimethylfuran, gasoline and ethanol in a DISI engine, *Fuel*, 103 (2013) 200–211.
8. J.-P. Lange, E. van der Heide, J. van Buijtenen, R. Price, Furfural — a promising platform for lignocellulosic biofuels, *ChemSusChem*, 5 (2012) 150–166.

9. H.E. Hoydonckx, W.M. Van Rhijn, D.E.E. De Vos, P.A. Jacobs, Furfural and derivatives, in: *Ullmann's Encyclopedya of Industrial Chemistry*, Wiley-VCH Verlag GmbH & Co. KGaA, 2007: pp. 335–340.

10. E. Ricard, H.M. Guinot, Process for the manufacture of furfuryl alcohol and methylfurane, US1739919 A, 1929.

11. H.M. Guinot, Process for catalytically hydrogenating organic substances, US2456187 A, 1948.

12. J.G.M. Bremner, R.K.F. Keeys, The hydrogenation of furfuraldehyde to furfuryl alcohol and sylvan (2-methylfuran), *Journal of the Chemical Society*, 0, (1947) 1068–1080.

13. L.W. Burnett, I.B. Johns, R.F. Holdren, R.M. Hixon, Production of 2-methylfuran by vapor-phase hydrogenation of furfural, *Industrial Engineering Chemistry*, 40 (1948) 502–505.

14. R.F. Holdren, Manufacture of methylfuran, US2445714 A, 1948.

15. I. Ahmed, Processes for the preparation of 2-methylfuran and 2-methyltetrahydrofuran, US20060229458 A1, 2002.

16. J. Lessard, J.F. Morin, J.F. Wehrung, D. Magnin, E. Chornet, High yield conversion of residual pentoses into furfural via zeolite catalysis and catalytic hydrogenation of furfural to 2-methylfuran, *Topics in Catalysis*, 53 (2010) 1231–1234.

17. S. Sitthisa, D.E. Resasco, Hydrodeoxygenation of furfural over supported metal catalysts: A comparative study of Cu, Pd and Ni, *Catalysis Letters*, 141 (2011) 784–791.

18. C.H. Xu, L.K. Zheng, D.F. Deng, J.Y. Liu, S.Y. Liu, Effect of activation temperature on the surface copper particles and catalytic properties of Cu-Ni-Mg-Al oxides from hydrotalcite-like precursors, *Catalysis Communications*, 12 (2011) 996–999.

19. C.P. Jiménez-gómez, J.A. Cecilia, I. Márquez-rodríguez, R. Moreno-tost, J. Santamaría-gonzález, J. Mérida-robles, P. Maireles-torres, Gas-phase hydrogenation of furfural over Cu/CeO_2 catalysts, *Catalysis Today*, 279 (2017) 327–338.

20. B.J. O'Neill, D.H.K. Jackson, A.J. Crisci, C.A. Farberow, F. Shi, A.C. Alba-Rubio, J. Lu, P.J. Dietrich, X. Gu, C.L. Marshall, P.C. Stair, J.W. Elam, J.T. Miller, F.H. Ribeiro, P.M. Voyles, J. Greeley, M. Mavrikakis, S.L. Scott, T.F. Kuech, J.A. Dumesic, Stabilization of copper catalysts for liquid-phase reactions by atomic layer deposition, *Angewandte Chemie — International Edition*, 52 (2013) 13808–13812.

21. D. Liu, D. Zemlyanov, T. Wu, R.J. Lobo-Lapidus, J.A. Dumesic, J.T. Miller, C.L. Marshall, Deactivation mechanistic studies of copper chromite catalyst for selective hydrogenation of 2-furfuraldehyde, *Journal of Catalysis*, 299 (2013) 336–345.

22. H. Zhang, C. Canlas, A. Jeremy Kropf, J.W. Elam, J.A. Dumesic, C.L. Marshall, Enhancing the stability of copper chromite catalysts for the selective hydrogenation of furfural using ALD overcoating, *Journal of Catalysis*, 326 (2015) 172–181.

23. M. Hronec, K. Fulajtárová, I. Vávra, T. Soták, E. Dobrocka, M. Micusïk, Carbon supported Pd-Cu catalysts for highly selective rearrangement of furfural to cyclopentanone, *Applied Catalysis B: Environmental*, 181 (2016) 210–219.

24. O.F. Aldosari, S. Iqbal, P.J. Miedziak, G.L. Brett, D.R. Jones, X. Liu, J.K. Edwards, D.J. Morgan, D.K. Knight, G.J. Hutchings, Pd-Ru/TiO2 catalyst — an active and selective catalyst for furfural hydrogenation, *Catalysis Science & Technology*, 6 (2016) 234–242.

25. S. Iqbal, X. Liu, O.F. Aldosari, P.J. Miedziak, J.K. Edwards, G.L. Brett, A. Akram, G.M. King, T.E. Davies, D.J. Morgan, D.K. Knight, G.J. Hutchings, Conversion of furfuryl alcohol into 2-methylfuran at room temperature using Pd/TiO2 catalyst, *Catalysis Science & Technology*, 4 (2014) 2280.

26. A.J. Garcia-Olmo, A. Yepez, A.M. Balu, A.A. Romero, Y. Li, R. Luque, Insights into the activity, selectivity and stability of heterogeneous catalysts in the continuous flow hydroconversion of furfural, *Catalysis Science & Technology*, 6 (2016) 4705–4711.

27. J. Wang, X. Liu, B. Hu, G. Lu, Y. Wang, Efficient catalytic conversion of lignocellulosic biomass into renewable liquid biofuels via furan derivatives, *RSC Advances*, 4 (2014) 31101-31107.

28. S. Srivastava, G.C. Jadeja, J. Parikh, A versatile bi-metallic copper–cobalt catalyst for liquid phase hydrogenation of furfural to 2-methylfuran, *RSC Advances*, 6 (2016) 1649–1658.

29. K. Yan, A. Chen, Selective hydrogenation of furfural and levulinic acid to biofuels on the ecofriendly Cu-Fe catalyst, *Fuel*, 115 (2014) 101–108.

30. A.S. Gowda, S. Parkin, F.T. Ladipo, Hydrogenation and hydrogenolysis of furfural and furfuryl alcohol catalyzed by ruthenium(II) bis(diimine) complexes, *Applied Organometallic Chemistry*, 26 (2012) 86–93.

31. H.-Y. Zheng, J. Yang, Y.-L. Zhu, G.-W. Zhao, Synthesis of g-butyrolactone and 2-methylfuran through the coupling of dehydrogenation and hydrogenation over copper-chromite catalyst, *Reaction Kinetics and Catalysis Letters*, 82 (2004) 263–269.

32. J. Yang, H.Y. Zheng, Y.L. Zhu, G.W. Zhao, C.H. Zhang, B.T. Teng, H.W. Xiang, Y.W. Li, Effects of calcination temperature on performance of Cu-Zn-Al catalyst for synthesizing γ-butyrolactone and 2-methylfuran through the coupling of dehydrogenation and hydrogenation, *Catalysis Communications*, 5 (2004) 505–510.

33. H.Y. Zheng, Y.L. Zhu, L. Huang, Z.Y. Zeng, H.J. Wan, Y.W. Li, Study on Cu-Mn-Si catalysts for synthesis of cyclohexanone and 2-methylfuran through the coupling process, *Catalysis Communications*, 9 (2008) 342–348.

34. L. Grazia, A. Lolli, F. Folco, Y. Zhang, S. Albonetti, F. Cavani, Gas-phase cascade upgrading of furfural to 2-methylfuran using methanol as a H-transfer reactant and MgO based catalysts, *Catalysis Science & Technology*, 6 (2016) 4418-4427.

35. D. Scholz, C. Aellig, I. Hermans, Catalytic transfer hydrogenation/ hydrogenolysis for reductive upgrading of furfural and 5-(hydroxymethyl) furfural, *ChemSusChem*, 7 (2014) 268–275.

36. P. Panagiotopoulou, N. Martin, D.G. Vlachos, Effect of hydrogen donor on liquid phase catalytic transfer hydrogenation of furfural over a Ru/RuO2/C catalyst, *Applied Catalysis A: General*, 392 (2014) 223–228.

37. X. Chang, A.-F. Liu, B. Cai, J.-Y. Luo, H. Pan, Y.-B. Huang, Catalytic transfer hydrogenation of furfural to 2-methylfuran and 2-methyltetrahydrofuran over bimetallic copper-palladium catalysts, *ChemSusChem*, (2016) 1–9.

38. J.P. Lange, B.J. Van, Process for the hydrogenolysis of furfuryl derivatives, US20110184195 A1, 2011.

39. J.G. Stevens, R.A. Bourne, M. V Twigg, M. Poliakoff, Real-time product switching using a twin catalyst system for the hydrogenation of furfural in supercritical CO2, *Angewandte Chemie International Edition*, 49 (2010) 8856–8859.

40. Z.L. Li, S. Kelkar, C.H. Lam, K. Luczek, J.E. Jackson, D.J. Miller, C.M. Saffron, Aqueous electrocatalytic hydrogenation of furfural using a sacrificial anode, *Electrochimica Acta*, 64 (2012) 87–93.

41. S.K. Green, J. Lee, H.J. Kim, G.A. Tompsett, W.B. Kim, G.W. Huber, The electrocatalytic hydrogenation of furanic compounds in a continuous electrocatalytic membrane reactor, *Green Chemisty*, 15 (2013) 1869–1879.

42. B. Zhao, M. Chen, Q. Guo, Y. Fu, Electrocatalytic hydrogenation of furfural to furfuryl alcohol using platinum supported on activated carbon fibers, *Electrochimica Acta*, 135 (2014) 139–146.

43. H.Y. Zheng, Y.L. Zhu, B.T. Teng, Z.Q. Bai, C.H. Zhang, H.W. Xiang, Y.W. Li, Towards understanding the reaction pathway in vapour phase hydrogenation of furfural to 2-methylfuran, *Journal of Molecular Catalysis A: Chemical*, 246 (2006) 18–23.

44. S. Sitthisa, W. An, D.E. Resasco, Selective conversion of furfural to methylfuran over silica-supported NiFe bimetallic catalysts, *Journal of Catalysis*, 284 (2011) 90–101.

45. V. Vorotnikov, G. Mpourmpakis, D.G. Vlachos, DFT Study of Furfural Conversion to Furan, Furfuryl Alcohol, and 2-Methylfuran on Pd(111), *ACS Catalysis*, 2 (2012) 2496–2504.

46. S. Wang, V. Vorotnikov, D.G. Vlachos, Coverage-induced conformational effects on activity and selectivity: Hydrogenation and decarbonylation of furfural on Pd(111), *ACS Catalysis*, 5 (2015) 104–112.

47. Y. Shi, Y. Yang, Y.-W. Li, H. Jiao, Mechanisms of Mo2C(101)-catalyzed furfural selective hydrodeoxygenation to 2-methylfuran from computation, *ACS Catalysis*, 6 (2016) 6790–6803.

48. A. Corma, O. Delatorre, M. Renz, High-quality diesel from hexose- and pentose-derived biomass platform molecules, *ChemSusChem*, 4 (2011) 1574–1577.

49. A. Corma Canos, M. Renz, O. De La Torre, Production of liquid fuels (sylvan-liquid-fuels) from 2-methylfuran, US9199955 B2, 2012.

50. G.Y. Li, N. Li, Z.Q. Wang, C.Z. Li, A.Q. Wang, X.D. Wang, Y. Cong, T. Zhang, Synthesis of high-quality diesel with furfural and 2-methylfuran from hemicellulose, *ChemSusChem*, 5 (2012) 1958–1966.

51. S. Li, N. Li, G. Li, L. Li, A. Wang, Y. Cong, X. Wang, T. Zhang, Lignosulfonate-based acidic resin for the synthesis of renewable diesel and jet fuel range alkanes with 2-methylfuran and furfural, *Green Chemistry*, 17 (2015) 3644–3652.

52. Q. Xia, Y. Xia, J. Xi, X. Liu, Y. Zhang, Y. Guo, Y. Wang, Selective one-pot production of high-grade diesel-range alkanes from furfural and 2-methylfuran over Pd/NbOPO4, *ChemSusChem,* 10 (2017) 747–753.

53. M. Balakrishnan, E.R. Sacia, A.T. Bell, Syntheses of biodiesel precursors: Sulfonic acid catalysts for condensation of biomass-derived platform molecules, *ChemSusChem,* 7 (2014) 1078–1085.

54. G. Li, N. Li, S. Li, A. Wang, Y. Cong, X. Wang, T. Zhang, Synthesis of renewable diesel with hydroxyacetone and 2-methyl-furan, *Chemical Communications*, 49 (2013) 5727–5729.

55. G. Li, N. Li, J. Yang, A. Wang, X. Wang, Y. Cong, T. Zhang, Synthesis of renewable diesel with the 2-methylfuran, butanal and acetone derived from lignocellulose, *Bioresource Technology*, 134 (2013) 66–72.

56. S. Li, N. Li, G. Li, L. Li, A. Wang, Y. Cong, X. Wang, G. Xu, T. Zhang, Protonated titanate nanotubes as a highly active catalyst for the synthesis of renewable diesel and jet fuel range alkanes, *Applied Catalysis B: Environmental*, 170–171 (2015) 124–134.

57. G. Li, N. Li, X.X. Wang, X. Sheng, S. Li, A. Wang, Y. Cong, T. Zhang, X.X. Wang, T. Zhang, Synthesis of diesel or jet fuel range cycloalkanes with 2-methylfuran and cyclopentanone from lignocellulose, *Energy and Fuels,* 28 (2014) 5112–5118.

58. X. Zhang, Q. Deng, P. Han, J. Xu, L. Pan, L. Wang, J.-J.J. Zou, Hydrophobic mesoporous acidic resin for hydroxyalkylation/alkylation of 2-methylfuran and ketone to high-density biofuel, *AIChE Journal*, 63 (2017) 680–688.

59. S. Li, N. Li, G. Li, A. Wang, Y. Cong, X. Wang, T. Zhang, Synthesis of diesel range alkanes with 2-methylfuran and mesityl oxide from lignocellulose, *Catalysis Today*, 234 (2014) 91–99.

60. H. Li, S. Saravanamurugan, S. Yang, A. Riisager, Catalytic alkylation of 2-methylfuran with formalin using supported acidic ionic liquids, *ACS Sustainable Chemisty & Engineering*, 3 (2015) 3274–3280.

61. Q. Deng, P. Han, J. Xu, J.-J.J. Zou, L. Wang, X. Zhang, Highly controllable and selective hydroxyalkylation/alkylation of 2-methylfuran with cyclohexanone for synthesis of high-density biofuel, *Chemical Engineering Science*, 138 (2015) 239–243.

62. C. Zhu, T. Shen, D. Liu, J. Wu, Y. Chen, L. Wang, K. Guo, H.-J.H. Ying, P. Ouyang, J. Klein, H.-J.H. Ying, J.-J. Xie, X.-J. He, H.-J.H. Ying, Production of liquid hydrocarbon fuels with acetoin and platform molecules derived from lignocellulose, *Green Chemistry*, 18 (2016) 2165–2174.

63. W. Wang, N. Li, S. Li, G. Li, F. Chen, X. Sheng, A. Wang, X. Wang, Y. Cong, T. Zhang, Synthesis of renewable diesel with 2-methylfuran and angelica lactone derived from carbohydrates, *Green Chemistry*, 18 (2016) 1218–1223.

64. G. Li, N. Li, J. Yang, A. Wang, X. Wang, Y. Cong, T. Zhang, Synthesis of renewable diesel with the 2-methylfuran, butanal and acetone derived from lignocellulose, *Bioresource Technology*, 134 (2013) 66–72.

65. S. Li, N. Li, W. Wang, L. Li, A. Wang, X. Wang, T. Zhang, Synthesis of jet fuel range branched cycloalkanes with mesityl oxide and 2-methylfuran from lignocellulose, *Scientific Reports*, 6 (2016) 32379-32386.

66. S. Li, N. Li, G. Li, L. Li, A. Wang, Y. Cong, X. Wang, T. Zhang, Lignosulfonate-based acidic resin for the synthesis of renewable diesel and jet fuel range alkanes with 2-methylfuran and furfural, *Green Chemistry*, 17 (2015) 3644–3652.

67. Q. Xia, Y. Xia, J. Xi, X. Liu, Y. Zhang, Y. Guo, Y. Wang, Selective one-pot production of high-grade diesel-range alkanes from furfural and 2-methylfuran over Pd/NbOPO4, *ChemSusChem*, 10 (2017) 747–753.

68. G. Li, N. Li, J. Yang, L. Li, A. Wang, X. Wang, Y. Cong, T. Zhang, Synthesis of renewable diesel range alkanes by hydrodeoxygenation of furans over Ni/Hβ under mild conditions, *Green Chemistry*, 16 (2014) 594–599.

69. T. Wang, K. Li, Q. Liu, Q. Zhang, S. Qiu, J. Long, L. Chen, L. Ma, Aviation fuel synthesis by catalytic conversion of biomass hydrolysate in aqueous phase, *Applied Energy*, 136 (2014) 775–780.

70. M. Balakrishnan, E.R. Sacia, A.T. Bell, Selective hydrogenation of furan-containing condensation products as a source of biomass-derived diesel additives, *ChemSusChem*, 7 (2014) 2796–2800.

Chapter 3.5

2-Methyl Tetrahydrofuran (MTHF) and its Use as Biofuel

Iker Obregón, Inaki Gandarias and Pedro L. Arias*

*Department of Chemical and Environmental Engineering,
Faculty of Engineering, University of the Basque Country
(UPV/EHU), Alameda Urquijo Street s/n, 48013 Bilbao, Spain*

3.5.1. Introduction

Despite the fact that 2-methyltetrahydrofuran (MTHF) production from levulinic acid (LA) was reported as early as 1947 [1], special attention has been paid to this chemical, MTHF, only in view of recent research such as the formulation of the *P-Series Fuels* [2,3]. These fuels are composed of ethanol, *pentanes plus* (hydrocarbons from natural gas with more than four carbon atoms), and MTHF to reduce the vapor pressure of the blend [4]. These formulations contain up to 70% renewable chemicals, considering that both ethanol and MTHF can be derived from biomass, and can be used alone or mixed with gasoline in any proportion [4].

Aside from the use of renewable chemicals, further benefits of these blends concern the high production process efficiency, between 1.75 and 2.25 kJ produced/kJ spent [3]. Besides, the U.S. Environmental Protection Agency (U.S. EPA) reported an important reduction in the ozone formation potential of the exhaust gases

---

*Corresponding author: pedroluis.arias@ehu.es

when using the P-Series Fuels compared to those of currently used conventional fuels [2, 3, 5].

However, these fuels can only be used in the so-called *flexible-fuel engines* due to the slightly corrosive nature of ethanol. In addition, the energy density of ethanol is lower than that of gasoline resulting in lower fuel economy. Furthermore, the addition of hydrophilic compounds, such as ethanol, to gasoline increases water solubility and, hence, the risk of phase separation and removal of ethanol from the mixture [6, 7].

These drawbacks can be overcome by using pure MTHF as gasoline additive. Despite the fact that MTHF has a lower energy density than conventional gasoline, its higher specific gravity results in a similar mileage. Unlike ethanol, MTHF is a hydrophobic chemical which greatly facilitates its implementation on the currently used fuel infrastructures. Owing to the mentioned properties current engines can run on gasoline-MTHF blends containing up to 70 vol.% MTHF without engine performance decrease [8, 9]. Nevertheless, MTHF produces peroxides in contact with air which could be a security issue for its storage and transportation if the necessary safety measures are not taken [10, 11].

Apart from its use as fuel, MTHF has a number of potential uses such as general solvent [4] being a green substitute of tetrahydrofuran (THF) [12] with more suitable physicochemical properties [13], a reagent in biphasic reactions [14] and as a substitute for the increasingly regulated chlorinated solvents. In addition, the use of MTHF as solvent is increasing in the pharmaceutical industry [13, 15–18]. Its use in such a delicate sector as pharmaceutical is possible because MTHF possess no mutagenicity nor genotoxicity characteristics and the human permitted daily exposure limit is 6.2 mg/day [13]. The biggest market potential, however, is expected to be as a fuel with a projected market larger than 4500 metric tons per year [19].

MTHF production can be accomplished by hydrogenation of platform molecules derived either form pentoses (FUR) or hexose (LA), as can be seen in Figure 3.5.1.

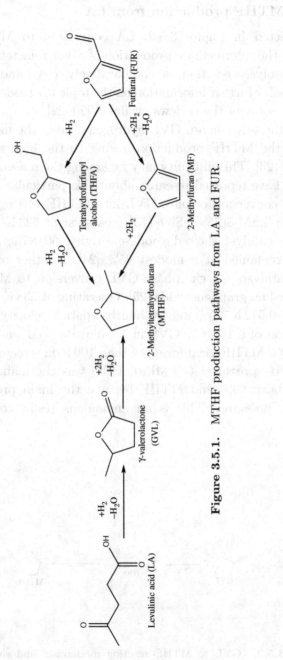

Figure 3.5.1. MTHF production pathways from LA and FUR.

3.5.2. MTHF production from LA

As depicted in Figure 3.5.2, LA conversion to MTHF proceeds through the intermediate production of γ-valerolactone (GVL). This hydrogenolysis reaction is comparatively easy and well studied, therefore for further information on the topic the reader is encouraged to check some of the reviews available [20–22].

On the other hand, GVL hydrogenation is the most demanding step in the MTHF production, owing to the high stability of the GVL [11,23]. This difficulty may be among the reasons why very few authors have reported research about this particular reaction.

The conversion of pure GVL into MTHF was reported using a Pt(0.7%)/ZSM-5(25%)–SiO$_2$(75%) catalyst at 523 K under 4.5 MPa H$_2$. The catalyst showed good selectivity, 60–85%, but GVL conversion remained at a modest 25% [24]. Another report on noble metal catalysts for gas phase GVL conversion to MTHF used Ru supported on graphene oxide [25]. Operating at 538 K under 2.5 MPa H$_2$ and 0.512 h^{-1} of weight hourly spatial velocity (WHSV) full conversion of a 10 wt.% GVL in 1,4-dioxane feed was achieved with up to 70% MTHF yield for more than 100 h on stream. Interestingly, for low H$_2$ pressures (<1 MPa), PDO was the main product with yields above 80% and MTHF became the main product only for high H$_2$ pressures. This is an anomalous result considering that

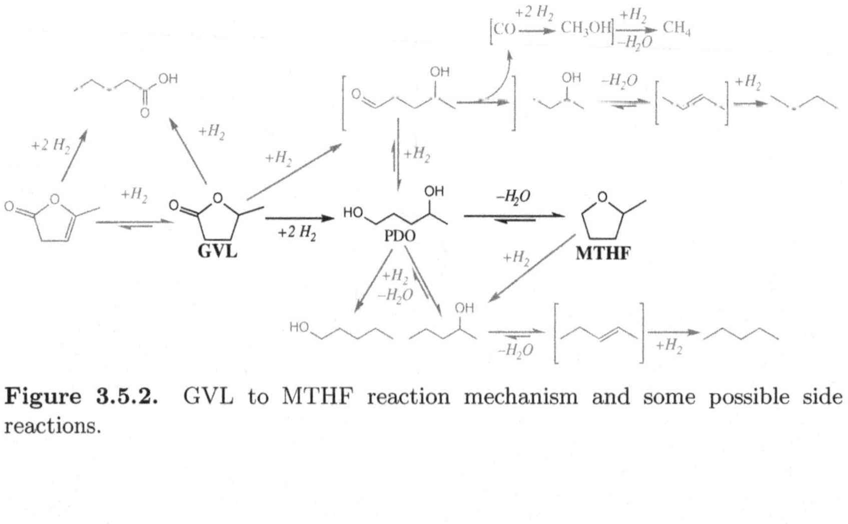

Figure 3.5.2. GVL to MTHF reaction mechanism and some possible side reactions.

the conversion of PDO to MTHF is an acid-catalyzed dehydration and the mechanism should not involve hydrogen. Besides, using a commercial Ru(5%)/C catalyst the reaction network leading from GVL to MTHF was thoroughly investigated [26]. Using reaction intermediate products (such as 2-butanol or MTHF) as substrates under different reaction atmospheres helped to elucidate the reaction mechanism of the intermediate and side reactions.

The use of non-noble metals was also reported for this reaction. A $CuCr_2O_3$ catalyst, which showed to be selective for the GVL to PDO reaction with up to 83% yields, also favored the dehydration of PDO to MTHF at 543–563 K under 20 MPa H_2 [1]. A Cu(30%)/ZrO_2 catalyst facilitated the reaction of a 6 wt.% GVL solution in ethanol at 473–513 K under 6 MPa H_2. This catalyst was found to be especially interesting due to the fact that, depending on the reduction temperature, its selectivity could be tailored towards the production of PDO (up to 96% yield reducing at 973 K and reacting at 473 K) or to MTHF (up to 91% yield reducing at 673 K and reacting at 513 K) [17].

Despite the fact that the following articles selectively produce PDO instead of MTHF, they will also be described in this section because PDO is the intermediate between GVL and MTHF as shown in Figure 3.5.2. First, a series of M–MoO_x/SiO_2 catalysts, where M was a noble metal (Pt, Rh, Ru, Pd, Ir), were tested in a continuous set-up for the conversion of a 10 wt.% LA aqueous solution under 6 MPa H_2 at 353 K [27]. The most active metal was Rh, and its activity was attributed to the special synergy existing between this noble metal and the oxophilic promoter (MoO_x). Further evidence supporting this statement was provided by carrying out experiments with a physical mixture of Rh/SiO_2 and MoO_x/SiO_2, which showed activity for GVL production from LA but not for its further conversion to MTHF.

A similar noble metal and oxophilic promoter screening was carried out testing several catalysts, prepared by impregnation of Pt over oxophilic supports, finding the best promoter to be MoO_x. Then, different noble metals impregnated over MoO_x were tested and found that Pt was the most active one, reaching up to 73% PDO yield in 6 h [28]. The activity of the catalyst was improved when both the noble metal and the oxophilic promoter were impregnated over inorganic matrices. Different supports were tested and it was found that a hydroxyapatite (HAP) provided the highest activity, reaching 93% PDO yield in 5 h at 403 K under 5 MPa H_2. Considering that the impregnated Pt/HAP was only active to produce GVL from an aqueous 4 wt.% LA solutions, the presence of an oxophilic promoter, and its interaction with the noble metal, seems to be necessary in aqueous phase to activate the stable GVL to produce PDO.

Finally, PDO conversion to MTHF is an acid catalyzed dehydration and, hence, it can be accomplished by heating PDO in the presence of acids. When mineral acids were employed high temperatures (above 523 K) were required, however, the use of ion exchange resins (such as perfluorinated *Nafion-H*) reported up to 90% MTHF yields at conditions as mild as 408 K [7,29]. This process is thermodynamically favored (ΔG [523 K] = −73 kJ/mol) [7], yet the reaction is reversible [30]. The equilibrium constant for a 1 mol/L PDO solution in liquid water at 573 K from a long-term reaction process was determined to be:

$$Kc[573 \, \text{K}] = \frac{[\text{MTHF}][\text{H}_2\text{O}]}{[\text{PDO}]} = 132 \pm 23$$

3.5.2.1. *Direct production of MTHF from LA*

The production of MTHF was first reported as a by-product of the LA hydrogenation over a $CuCr_2O_3$ catalyst [1]. More recently, the direct conversion to MTHF was provided using Cu/SiO_2 catalysts in gas phase, at 538 K under 2.5 MPa H_2 pressure with a feed made of a 10 wt.% LA solution in 1,4-dioxane and a H_2/LA molar ratio of 80 [31]. It was observed that the selectivity towards MTHF increased with the increase in the metal load (0.1% yield for 5 wt.% Cu *vs.* 64% yield for 80 wt.% Cu) of the catalyst, consistently with previous publications [32,33]. The process was further improved by promoting the catalyst with 8 wt.% of Ni. This catalyst Cu(72%)–Ni(8%)/SiO_2 provided stable 89% MTHF yield for more than 300 h on stream.

The Ni–Cu promotion effect was further studied using 2-propanol as solvent in a batch system. The effect of the Ni–Cu ratio was investigated for a 35% total metal load over γ-Al_2O_3 support [34]. In this paper Cu was found to be the most selective metal towards MTHF but with low activity. Ni, on the other hand, was significantly more active but also produced up to 34% by-product yields. The optimum Ni–Cu ratio was 2, which achieved 56% MTHF yield at 523 K under 7 MPa H_2 after 5 h reaction time and 62% at 503 K after 24 h. The different activity of this catalyst was associated to the formation of a Ni–Cu alloy which was detected by XRD and TPR analysis.

Using noble and transition metal based catalysts the effect of the hydrogen source on the LA to MTHF reaction was investigated using catalytic transfer hydrogenation (CTH) and H_2 hydrogenation reaction conditions [35]. In this paper, the high hydrogen availability required for the GVL conversion was confirmed by the negligible MTHF yields achieved under N_2 atmosphere using good hydrogen donor molecules such as 2-propanol at 523 K. When changing the atmosphere to H_2, on the other hand, significant MTHF yields were achieved after 5 h reaction time and a nice trend relating the CTH activity of the catalyst-solvent system to the MTHF yield was observed. These results suggest that both the CTH and the H_2 hydrogenation mechanisms play an important and synergetic role on the reaction even under high H_2 pressures, allowing for up to 80% MTHF yields using the a bimetallic Ni(23%)–Cu(12%)/Al_2O_3 catalyst after 20 h reaction time.

A graphene oxide supported Ru catalyst was reported to enable significant MTHF yields from a 10 wt.% LA in 1,4-dioxane solution [25]. After a variable screening, the best operation conditions produced a stable 48% MTHF yield for over 100 h on stream operating at 538 K under 2.5 MPa H_2 pressure and 0.512 h^{-1} WHSV. Interestingly, in this reaction set-up, MTHF yield strongly depended on the H_2 pressure. For pressures below 1 MPa up to 90% GVL yields were obtained. Besides, for WHSV values above 10 h^{-1} GVL was produced with >80% yields. These two facts corroborate the high hydrogen availability required for the hydrogenation of GVL and the fact that this step (GVL conversion) is the rate limiting step of the overall reaction.

Nevertheless, CTH reaction using formic acid as the hydrogen source for the LA to MTHF reaction was recently reported [36]. Under microwave irradiation a commercial Pd(5%)/C catalyst enabled up to 72% MTHF yields operating at 423 K from a 1:3 by volume LA solution in formic acid. When the reaction was set-up in a continuous flow reactor, however, the yields notably decreased. Maximum 45–48% MTHF yields were achieved using Cu-based catalysts (supported on Al_2O_3 or SiO_2), which underwent deactivation after 10–30 min reaction time.

Using noble metal based catalysts, Pd(5%)–Re(5%)/C, a procedure for the direct production of MTHF from LA at 494–515 K under 10 MPa H_2 was patented [37]. The system allowed total LA conversion, fed as 60 vol.% solution in 1,4-dioxane or as pure LA, with up to 90% selectivity. Aqueous phase reaction was also effective over Pt–Mo based catalysts supported on acidic supports. Pt(3.9%)–Mo(0.13%)/H-β was found to be the most active catalyst, achieving up to 86% MTHF yield from a 4 wt.% LA aqueous solution at 403 K under 5 MPa H_2 in 24 h [38]. The activity of the studied catalyst series was nicely related to the activity of the catalysts supports to promote the dehydration of PDO to MTHF. The authors suggest that the hydrophobic nature of the H-β zeolite surface allows for the equilibrium reaction to generate MTHF even in the presence of water. Considering this statement, it is possible that water competes with the reactants for adsorption on the surface active sites.

Finally, solvent free LA hydrogenation to MTHF was also carried out using a commercial Ru(5%)/C catalyst with up to 61% yield. The one-pot three-step reaction consisted on (i) LA hydrogenation to GVL at 463 K under 1.2 MPa H_2 for 45 min followed by (ii) evaporation of the produced water, catalyst separation, washing and drying and (iii) GVL hydrogenation to MTHF at 463 K under 10 MPa (loaded at room temperature) for 4 h [26].

As a summary, Table 3.5.1 shows the highlights of the reported MTHF production from either LA or GVL using heterogeneous catalysts.

Table 3.5.1.   Literature overview of MTHF and PDO production from LA or GVL using heterogeneous catalysts.

Entry	Catalyst	Feed	T(K)	P(MPa)	Reactor type	Y(%)		Ref.
1	$CuCr_2O_3$	100% GVL	543–563	20	Batch	83	PDO	[1]
2	$Rh-MoO_x/SiO_2$	10% LA in H_2O	353	6	FBR	70	PDO	[27]
3	$Pt-MoO_x/HAP$	4% LA in H_2O	403	5	Batch	93	PDO	[28]
4	Ru/GO	10% GVL in dioxane	538	2.5	FBR	70	MTHF	[25]
5	Cu/ZrO_2	6% GVL in ethanol	513	6	Batch	91	MTHF	[17]
6	Ru/GO	10% LA in dioxane	538	2.5	FBR	48	MTHF	[25]
7	Pd–Re/C	60 % LA in dioxane	494–515	10	FBR	89	MTHF	[37]
8	$Cu-Ni/SiO_2$	10 % LA in dioxane	538	2.5	FBR	89	MTHF	[31]
9	Pd/C	24% LA in formic acid	423	Autogen.	MW	72	MTHF	[36]
10	Cu/SiO_2	24% LA in formic acid	423	Autogen.	FBR	48	MTHF	[36]
11	Ru/C[a]	100% LA	463	10	Batch	61	MTHF	[26]
12	$Pt-MoO_x/H-\beta$	4% LA in H_2O	403	5	Batch	86	MTHF	[38]

Notes: [a]The reaction was carried out in two steps, removing water between them.
FBR: Fixed-bed reactor, *MW*: Microwave irradiation, *dioxane*: 1,4-dioxane.

3.5.3. MTHF production from MF

As stated before in this chapter, MTHF can also be produced from platform molecules derived from pentoses. This route, as it is shown in Figure 3.5.3, starts from FUR and can proceed either from the MF route or the THFA route.

Vapor-phase hydrogenation of MF to MTHF was reported with nearly quantitative yield over a Pd/C catalyst at low reaction temperatures (398–473 K) and moderate H_2/MF molar ratios (10–25) [39]. When higher reaction temperatures or other metal phases such as Pt, Ru, Ni or Cu were used the formation of 2-pentanone was largely favored over MTHF production. This trend was also observed over a commercial Pt(5%)/C catalyst which, operating at 353 K and a H_2/MF ratio of 10, enabled 80% MTHF [40]. However, for higher reaction temperatures the MTHF yield drastically decreased according to the increase in the 2-pentanone yield. Aqueous phase MF conversion to MTHF was also reported over a Pt/C catalyst after 1 h reaction time at 433 K under 3 MPa H_2 pressure [41]. In good agreement with the previously discussed results, along with the

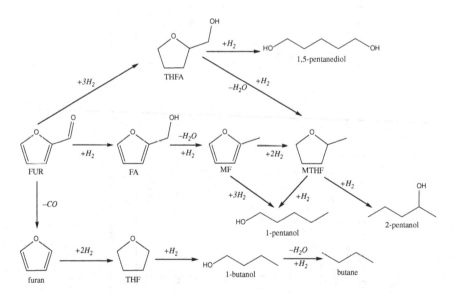

Figure 3.5.3. MTHF production form FUR and some side reactions.

achieved 24% MTHF yield 15% pentanone and 10% pentanol yields were also produced.

In liquid-phase reaction, using 2-propanol as solvent, a Pd(3%)/C catalyst provided 100% MTHF yield from MF operating at 493 K under 3.5 MPa H_2 after 5 h reaction time [42]. Another study corroborated the activity of Pd/C for the hydrogenation of MF to MTHF achieving 97% yield at 368 K and showing its further reaction to produce 2-pentanone with increasing reaction temperatures [43]. In addition, deuterium labeling analysis of the products revealed the progressive substitution of H atoms on the furan ring and on the methyl group for deuterium atoms with increasing reaction temperatures.

Ni based catalysts were also reported to activate this reaction. Operating at 453 K under 1.6 MPa H_2 (MF/H_2 ratio of 10) Ni/SiO_2 catalysts showed up to 92% MTHF yield in a fixed-bed reactor [44]. The activity of the catalysts, however, decreased over 8 h on stream, being this effect more acute for the lowest Ni loaded catalyst.

3.5.4. Direct conversion of FUR to MTHF

The challenging direct synthesis of MTHF from FUR has been also attempted. One-stage hydrogenation of FUR to MTHF was reported in the presence of a structured bed of a copper-chromite-based catalyst and noble metals (mainly, Pd and/or Pt) supported on activated carbon [45]. The process consists of a continuous gas phase hydrogenation of FUR, over a fixed-bed catalyst located in a jacketed tubular reactor. At 448–523 K, 5 mL h^{-1} FUR and 15 L h^{-1} H_2 flows, 91 wt.% MTHF was achieved. Moreover, by using a continuous fixed-bed catalyst (Pd/C) in gas phase, under 1 MPa H_2 pressure at 518 K with a 20 g h^{-1} FUR feed, FUR was fully converted producing 49 wt.% MTHF in the reaction products.

A commercially viable continuous gas-phase process was patented which performed the reaction in two steps: in a first reactor FUR was hydrogenated to MF over a Ba/Mn promoted copper chromite catalyst at 448 K under 0.1 MPa and a H_2/FUR molar ratio of 2. In a second reactor, MF was transformed into MTHF by using a Ni/Al$_2$O$_3$–SiO$_2$ catalyst at 388 K with a H$_2$/MF molar ratio of 2 [46]. This system allowed a 50% MF recovery after the first step, which could be hydrogenated to MTHF afterwards with an up to 87% MTHF recovery in the second reactor. Aside from the desired products MF and MTHF, furfuryl alcohol (FA) and THFA were formed as by-products in the first and second reactors respectively.

Another two-step process was employed for MTHF production from FUR in supercritical CO$_2$, combining a copper chromite catalysts in the first reactor with a Pd(5%)/C catalyst in the second. This process provided incomparable flexibility, allowing a real-time, fast control of the reaction products (FA, THFA, MF, MTHF or furan) by modifying the temperature of each reactor [47]. Thus, up to 82% MTHF yields could be obtained operating with a 1 mL min^{-1} CO$_2$ flow (pumped at 263 K and 5.8 MPa), 0.05 mL min^{-1} FUR flow under 15 MPa pressure. In the first reactor, operating at 513 K, FUR was converted into MF with 90% yield. Then, in the second reactor at 573 K, the produced MF was hydrogenated to MTHF facilitating an overall 82% yield.

A similar double-bed reaction system was studied in a single reactor under atmospheric pressure at 453 K and a H$_2$/FUR ratio of 29. The first catalyst, Cu$_2$Si$_2$O$_5$(OH)$_2$, selectively converted the FUR into MF and the second catalyst, Pd/SiO$_2$, hydrogenated the furan ring of MF to MTHF with an overall 97% yield [48]. Interestingly, the relative weight of the catalysts presented an optimum value requiring double amount of Pd/SiO$_2$ than Cu$_2$Si$_2$O$_5$(OH)$_2$. Besides, the separate catalyst beds provided much better MTHF yields (85%) than a bimetallic Pd–Cu catalyst (1.6%), a physical mixture of Cu$_2$Si$_2$O$_5$(OH)$_2$ and Pd/SiO$_2$ (14.9%) and the inverse catalyst arrangement (1.6%), highlighting the strong dependency of the reaction on the hydrogenation sequence and the required active sites for each step of the reaction.

Liquid-phase reaction has also been recently applied to this reaction. Using a Rh-ReO$_x$/SiO$_2$ catalyst a two-step conversion of aqueous FUR solutions to MTHF with up to 27% yield was reported under 6 MPa H$_2$ [49]. In the first step of the reaction, carried out at 323 K for 2 h, FUR was fully converted to THFA. In the second step, carried out at 393 K for 24 h, THFA was converted to MTHF and other unspecified products (48% yield). In this reaction set-up, the first low temperature step, FUR hydrogenation to THFA, was necessary in order to minimize FUR polymerization reactions that take place at higher reaction temperatures. Then, the higher reaction temperatures required for the hydrogenation of THFA could be applied without the mentioned polymerization issue.

Over a Pt(10%)/C catalyst up to 41% MTHF yields were reported from a 1-propanol solution of FUR operating at 453 K under 3.3 MPa H_2 [50]. The MTHF yield continuously increased with the increase in the contact time, indicating the slow rate of this reaction. Besides, this research showed that FUR decarbonylation to furan was a major reaction pathway only at low H_2 pressures. For high H_2 pressures, the hydrogenation to MF and its further reactions were the main reactions.

Using 2-propanol as solvent and hydrogen donor molecule the direct conversion of FUR to MF and MTHF was carried out under N_2 atmosphere using $Cu–Pd/ZrO_2$ catalysts [51]. The mechanism of these reactions was found to proceed through FA and the selectivity of the process could be tailored by modifying the Cu–Pd ratio on the catalyst: monometallic Cu favored the formation of MF (31%) and Pd the production of MTHF (33%). In addition, a $Cu(10\%)–Pd(5\%)/ZrO_2$ catalyst facilitated a 79% MTHF yield while a physical mixture of $Cu(10\%)/ZrO_2$ and $Pd(3\%)/ZrO_2$ only reached 48% MTHF yield, suggesting that bimetallic promotion effects were important for the activity of these catalysts.

Another example of CTH reaction mechanisms for the conversion of FUR to MTHF can be found in the literature. Pd/Fe_2O_3 catalysts were found to promote the conversion of FA to MTHF with up to 42% yields operating at 453 K under N_2 atmosphere in 7.5 h [52]. The previous step, FUR conversion to FA, could also be catalyzed by Pd/Fe_2O_3 achieving 34–57% yields, rendering an overall 24% MTHF yield.

Table 3.5.2 summarizes some other references where MTHF is obtained in moderate-low yields, normally as a by-product as

Table 3.5.2. Literature overview of MTHF production as side product from FUR and FA using heterogeneous catalysts.

Entry	Catalyst	Feed	T (K)	P (MPa)	Y_{MTHF} (%)	Ref.
1	Pt/MWCNT	4 wt.% FUR in 2-propanol	373–523	2	8	[53]
2	Ni-Cu/SBA-15	5 wt.% FUR in H_2O	433	4	17	[54]
3	Pd/C	5 wt.% FUR in H_2O	448	8	10	[55]
4	Cu/Al_2O_3	0.2 mol/L FUR in ethyl acetate	363	5	10	[56]
5	$Ni-Cu/Al_2O_3$	21 wt.% FUR in 2-propanol	573	4	10–12	[57]
6	Pd/C	5 wt.% FUR in 2-propanol	493	3.5	18	[42]
7	Pd/C	2.5 wt.% FUR in H_2O	448	8	17	[41]
8	Pd/C	5 wt.% FA in H_2O +0.25 wt.% H_3PO_4	433	8	26	[41]

the objective of the research was the selective production of other chemicals such as THFA, FA or MF.

Finally, the most ambitious attempt was reported by Sen and Yang, who studied the direct synthesis from pentose sugars and lignocellulosic biomass by using a soluble rhodium catalyst and HI/HCl + NaI additive in the presence of H_2 [58]. By feeding corn stover (40% glucan and 24% xylan), a maximum 63% MTHF yield was attained. However the economic and environmental viability of this process requires confirmation due to the use of an expensive catalyst and no environmental friendly reaction medium.

References

1. R.V.J. Christian, H.D. Brown, R.M. Hixon, Derivatives of γ-valerolactone, 1,4-pentanediol, and 1,4-di-(b-cyanoethoxy)-pentane, *Journal of American Chemical Society*, 69 (1947) 1961–1963.

2. S.F. Paul, Alternative fuel, WO 9743356 A1, 1997.

3. U.S.D. of Energy, Alternative fuel transportation programe; P-series fuels; Final rule, in: F. Register (Ed.), U.S. Government Printing Office, Washington, NW, 1999, pp. 26822–26829.

4. D.J. Hayes, S.W. Fitzpatrick, M.H.B. Hayes, J.R.H. Ross, The biofine process — Production of levulinic acid, furfural, and formic acid from lignocellulosic feedstocks, in: B. Kamm, P.R. Gruber (Eds.), *Biorefineries-Industrial Processes and Products*, Wiley-VCH Verlag GmbH, 2006, pp. 139–164.

5. D.J. Hayes, An examination of biorefining processes, catalysts and challenges, *Catalysis Today*, 145 (2009) 138–151.

6. J.C. Serrano-Ruiz, J.A. Dumesic, Catalytic routes for the conversion of biomass into liquid hydrocarbon transportation fuels, *Energy & Environmental Science*, 4 (2011) 83–99.

7. J.C. Serrano-Ruiz, A. Pineda, A.M. Balu, R. Luque, J.M. Campelo, A.A. Romero, J.M. Ramos-Fernández, Catalytic transformations of biomass-derived acids into advanced biofuels, *Catalysis for Biorefineries*, 195 (2012) 162–168.

8. G.W. Huber, S. Iborra, A. Corma, Synthesis of transportation fuels from biomass: Chemistry, catalysts, and engineering, *Chemical Reviews*, 106 (2006) 4044–4098.

9. D.M. Alonso, J.Q. Bond, J.A. Dumesic, Catalytic conversion of biomass to biofuels, *Green Chemistry*, 12 (2010) 1493–1513.

10. J. Zhang, S. Wu, B. Li, H. Zhang, Advances in the catalytic production of valuable levulinic acid derivatives, *ChemCatChem*, 4 (2012) 1230–1237.

11. I.T. Horváth, H. Mehdi, V. Fabos, L. Boda, L.T. Mika, γ-Valerolactone-a sustainable liquid for energy and carbon-based chemicals, *Green Chemistry*, 10 (2008) 238–242.

12. R. Aul, B. Comanita, A green alternative to THF, *Manufacturing Chemist*, 78 (2007) 33–34.
13. V. Antonucci, J. Coleman, J.B. Ferry, N. Johnson, M. Mathe, J.P. Scott, J. Xu, Toxicological assessment of 2-methyltetrahydrofuran and cyclopentyl methyl ether in support of their use in pharmaceutical chemical process development, *Organic Process Research & Development*, 15 (2011) 939.
14. D.F. Aycock, Solvent applications of 2-methyltetrahydrofuran in organometallic and biphasic reactions, *Organic Process Research & Development*, 11 (2007) 156–159.
15. J.T. Kuethe, K.G. Childers, G.R. Humphrey, M. Journet, Z. Peng, A. Rapid, Large-scale synthesis of a potent cholecystokinin (CCK) 1R receptor agonist, *Organic Process Research & Development*, 12 (2008) 1201–1208.
16. I.N. Houpis, D. Shilds, U. Nettekoven, A. Schnyder, E. Bappert, K. Weerts, M. Canters, W. Vermuelen, Utilization of sequential palladium-catalyzed cross-coupling reactions in the stereospecific synthesis of trisubstituted olefins, *Organic Process Research & Development*, 13 (2009) 598–606.
17. X.-L. Du, Q.-Y. Bi, Y.-M. Liu, Y. Cao, H.-Y. He, K.-N. Fan, Tunable copper-catalyzed chemoselective hydrogenolysis of biomass-derived γ-valerolactone into 1,4-pentanediol or 2-methyltetrahydrofuran, *Green Chemistry*, 14 (2012) 935.
18. V. Pace, P. Hoyos, L. Castoldi, P. Domínguez De María, A.R. Alcántara, 2-Methyltetrahydrofuran (2-MeTHF): A biomass-derived solvent with broad application in organic chemistry, *ChemSusChem*, 5 (2012) 1369–1379.
19. J.J. Bozell, L. Moens, D.C. Elliott, Y. Wang, G.G. Neuenscwander, S.W. Fitzpatrick, R.J. Bilski, J.L. Jarnefeld, Production of levulinic acid and use as a platform chemical for derived products, *Resources, Conservation and Recycling*, 28 (2000) 227–239.
20. W.R.H. Wright, R. Palkovits, Development of heterogeneous catalysts for the conversion of levulinic acid to y-valerolactone, *ChemSusChem*, 5 (2012) 1657–1667.
21. D.M. Alonso, S.G. Wettstein, J.A. Dumesic, Gamma-valerolactone, a sustainable platform molecule derived from lignocellulosic biomass, *Green Chemistry*, 15 (2013) 584.
22. M.J. Climent, A. Corma, S. Iborra, Conversion of biomass platform molecules into fuel additives and liquid hydrocarbon fuels, *Green Chemistry*, 16 (2014) 516–547.
23. M. Vasiliu, K. Guynn, D.A. Dixon, Prediction of the thermodynamic properties of key products and intermediates from biomass, *The Journal of Physical Chemistry C*, 115 (2011) 15686–15702.
24. P.M. Ayoub, J.-P. Lange, Process for converting levulinic acid into pentanoic acid, WO 2008/142127 A1, 2008.
25. P.P. Upare, M. Lee, S.-K. Lee, J.W. Yoon, J. Bae, D.W. Hwang, U.-H. Lee, J.-S. Chang, Y.K. Hwang, Ru nanoparticles supported graphene oxide catalyst for hydrogenation of bio-based levulinic acid to cyclic ethers, *Catalysis Today*, 265 (2015) 174–183.

26. M.G. Al-Shaal, A. Dzierbinski, R. Palkovits, Solvent-free γ-valerolactone hydrogenation to 2-methyltetrahydrofuran catalysed by Ru/C: A reaction network analysis, *Green Chemistry*, 16 (2014) 1358–1364.

27. M. Li, G. Li, N. Li, A. Wang, W. Dong, X. Wang, Y. Cong, Aqueous phase hydrogenation of levulinic acid to 1,4-pentanediol, *Chemical Communications*, 50 (2014) 1414–1416.

28. T. Mizugaki, Y. Nagatsu, K. Togo, Z. Maeno, T. Mitsudome, K. Jitsukawa, K. Kaneda, Selective hydrogenation of levulinic acid to 1,4-pentanediol in water using a hydroxyapatite-supported Pt-Mo bimetallic catalyst, *Green Chemistry*, 17 (2015) 5136–5139.

29. G.A. Olah, A.P. Fung, R. Malhorta, Synthetic methods and reactions. 99. Preparation of cyclic ethers over superacidic perfluorinated resinsulfonic acid (Nafion-H) catalyst, *Synthesis*, 6 (1981) 474–476.

30. A. Yamaguchi, N. Hiyoshi, O. Sato, K.K. Bando, Y. Masuda, M. Shirai, Thermodynamic equilibria between polyalcohols and cyclic ethers in high-temperature liquid water, *Journal of Chemical & Engineering Data*, 54 (2009) 2666–2668.

31. P.P. Upare, J.-M. Lee, Y.K. Hwang, D.W. Hwang, J.-H. Lee, S.B. Halligudi, J.-S. Hwang, J.-S. Chang, Direct hydrocyclization of biomass-derived levulinic acid to 2-methyltetrahydrofuran over nanocomposite copper/silica catalysts, *ChemSusChem*, 4 (2011) 1749–1752.

32. C.-H. Zhou, X. Xia, C.-X. Lin, D.-S. Tong, J. Beltramini, Catalytic conversion of lignocellulosic biomass to fine chemicals and fuels, *Chemical Society Reviews*, 40 (2011) 5588.

33. R. Palkovits, Pentenoic acid pathways for cellulosic biofuels, *Angewandte Chemie International Edition*, 49 (2010) 4336–4338.

34. I. Obregón, I. Gandarias, N. Miletić, A. Ocio, P.L. Arias, One-pot 2-methyltetrahydrofuran production from levulinic acid in green solvents using Ni-Cu/Al$_2$O$_3$ catalysts, *ChemSusChem*, 8 (2015) 3483–3488.

35. I. Obregón, I. Gandarias, M.G. Al-Shaal, C. Mevissen, P.L. Arias, R. Palkovits, The role of the hydrogen source on the selective production of γ-valerolactone and 2-methyltetrahydrofuran from levulinic acid, *ChemSusChem*, 9 (2016) 2488–2495.

36. J.M. Bermudez, J.A. Menendez, A.A. Romero, E. Serrano, J. Garcia-Martinez, R. Luque, Continuous flow nanocatalysis: Reaction pathways in the conversion of levulinic acid to valuable chemicals, *Green Chemistry*, 15 (2013) 2786–2792.

37. D.C. Elliott, J.G. Frye, Hydrogenated 5C compound and method of making, US 5883266, 1999.

38. T. Mizugaki, K. Togo, Z. Maeno, T. Mitsudome, K. Jitsukawa, K. Kaneda, One-pot transformation of levulinic acid to 2-methyltetrahydrofuran catalyzed by Pt–Mo/H-β in water, *ACS Sustainable Chemistry & Engineering*, 4 (2016) 682–685.

39. P. Biswas, J.-H. Lin, J. Kang, V.V. Guliants, Vapor phase hydrogenation of 2-methylfuran over noble and base metal catalysts, *Applied Catalysis A: General*, 475 (2014) 379–385.

40. J. Kang, X. Liang, V.V. Guliants, Selective hydrogenation of 2-methylfuran and 2,5-dimethylfuran over atomic layer deposited platinum catalysts on multiwalled carbon nanotube and alumina supports, *ChemCatChem*, 9 (2016) 282–286.

41. M. Hronec, K. Fulajtarová, T. Liptaj, Effect of catalyst and solvent on the furan ring rearrangement to cyclopentanone, *Applied Catalysis A: General*, 437–438 (2012) 104–111.

42. N.S. Biradar, A.A. Hengne, S.N. Birajdar, R. Swami, C.V. Rode, Tailoring the product distribution with batch and continuous process options in catalytic hydrogenation of furfural, *Organic Process Research & Development*, 18 (2014) 1434–1442.

43. J. Kang, A. Vonderheide, V.V. Guliants, Deuterium-labeling study of the hydrogenation of 2-methylfuran and 2,5-dimethylfuran over carbon-supported noble metal catalysts, *ChemSusChem*, 8 (2015) 3044–3047.

44. F. Ding, Y. Zhang, G. Yuan, K. Wang, I. Dragutan, V. Dragutan, Y. Cui, J. Wu, Synthesis and catalytic performance of Ni/SiO_2 for hydrogenation of 2-methylfuran to 2-methyltetrahydrofuran, *Journal of Nanomaterials*, 2015 (2015) 1–6.

45. T. Wabnitz, D. Breuninger, J. Heimann, R. Backes, R. Pinkos, Process for one-stage preparation of 2-methyltetrahydrofuran from furfural over a catalyst, US2010/0099895A1, 2010.

46. I. Ahmed, Process for the preparation of 2-methylfuran and 2-methyltetrahydrofuran, US 6.852.868 B2, 2005.

47. J.G. Stevens, R.A. Bourne, M.V. Twigg, M. Poliakoff, Real-time product switching using a twin catalyst system for the hydrogenation of furfural in supercritical CO_2, *Angewandte Chemie*, 49 (2010) 8856–8859.

48. F. Dong, Y. Zhu, G. Ding, J. Cui, X. Li, Y. Li, One-step conversion of furfural into 2-methyltetrahydrofuran under mild conditions, *ChemSusChem*, 8 (2015) 1534–1537.

49. S. Liu, Y. Amada, M. Tamura, Y. Nakagawa, K. Tomishige, One-pot selective conversion of furfural into 1,5-pentanediol over a Pd-added Ir–ReOx/SiO2 bifunctional catalyst, *Green Chemistry*, 16 (2014) 617.

50. J. Luo, M. Monai, H. Yun, L. Arroyo-Ramírez, C. Wang, C.B. Murray, P. Fornasiero, R.J. Gorte, The H_2 pressure dependence of hydrodeoxygenation selectivities for furfural over Pt/C catalysts, *Catalysis Letters*, 146 (2016) 711–717.

51. X. Chang, A.-F. Liu, B. Cai, J.-Y. Luo, H. Pan, Y.-B. Huang, Catalytic transfer hydrogenation of furfural to 2-methylfuran and 2-methyltetrahydrofuran over bimetallic copper-palladium catalysts, *ChemSusChem*, 9 (2016) 3330–3337.

52. D. Scholz, C. Aellig, I. Hermans, Catalytic transfer hydrogenation/ hydrogenolysis for reductive upgrading of furfural and 5-(hydroxymethyl) furfural, *ChemSusChem*, 7 (2014) 268–275.

53. C. Wang, Z. Guo, Y. Yang, J. Chang, A. Borgna, Hydrogenation of furfural as model reaction of bio-oil stabilization under mild conditions using

multiwalled carbon nanotube (MWNT)-supported Pt catalysts, *Industrial & Engineering Chemistry Research*, 53 (2014) 11284–11291.

54. Y. Yang, Z. Du, Y. Huang, F. Lu, F. Wang, J. Gao, J. Xu, E.J. García-Suárez, Conversion of furfural into cyclopentanone over Ni–Cu bimetallic catalysts, *Green Chemistry*, 15 (2013) 1932–1940.

55. M. Hronec, K. Fulajtarová, Selective transformation of furfural to cyclopentanone, *Catalysis Communications*, 24 (2012) 100–104.

56. A.J. Garcia-Olmo, A. Yepez, A.M. Balu, A.A. Romero, Y. Li, R. Luque, Insights into the activity, selectivity and stability of heterogeneous catalysts in the continuous flow hydroconversion of furfural, *Catalysis Science & Technology*, 6 (2016) 4705–4711.

57. S. Srivastava, G.C. Jadeja, J. Parikh, Synergism studies on alumina-supported copper-nickel catalysts towards furfural and 5-hydroxymethyl furfural hydrogenation, *Journal of Molecular Catalysis A: Chemical*, 426 (2017) 244–256.

58. W. Yang, A. Sen, One-step catalytic transformation of carbohydrates and cellulosic biomass to 2,5-dimethyltetrahydrofuran for liquid fuels, *ChemSusChem*, 3 (2010) 597–603.

Cyclopentanone and its Derived Biofuels

Manuel López Granados*

*Sustainable Chemistry and Energy Group (EQS),
Institute of Catalysis and Petrochemistry (CSIC),
C/Marie Curie 2, Cantoblanco 28049 Madrid, Spain*

3.6.1. Introduction

Cyclopentanone (CPONE) is a specialty chemical that can be used in the synthesis of pharmaceuticals, fungicides, rubber chemicals, flavors and fragrances. Due to its chemical similarity with cyclohexanone, it has also been proposed as a feedstock for the synthesis of polyamides via δ-valerolactam [1–5]. The synthesis of adipic acid from CPONE has also been technically demonstrated [6]. Currently, CPONE can be obtained via several petrochemical routes such as vapor-phase cyclization of 1,6-hexanediol or adipic esters (dimethyl adipate), oxidation of cyclopentanol, dehydrogenation of cyclopentanol, isomerization of cyclopentene oxide and liquid-phase oxidation of cyclopentene [1–5].

3.6.2. The renewable route to cyclopentanone

By appropriate selection of the reactions conditions and of the catalyst, CPONE can be produced by hydrogenation of furfural in liquid aqueous phase. As it will be shown, the reaction requires the intermediate formation of FOL but the key is the presence of

---

*Corresponding author: mlgranados@icp.csic.es

water which results in the rearrangement of the furfuryl alcohol, initially formed by hydrogenation of the carbonyl group. If water is not present, the rearrangement of FOL does not take place and FOL is preferentially hydrogenated to THFA, MF and MTHF. Rearrangement of 2-furylcarbinols into 4-hydroxycyclopentenones in the presence of acids was first described by Piancatelli *et al.* in the 70s of the 20th century [7], the latter being present in pharmacologically active natural compounds [8].

A remarkable advantage of this reaction is that it must be conducted in water medium. The first step for the production of furfural from hemicellulose is the formation of a dilute aqueous solution of furfural. This means that this primary furfural-water product is a good candidate for obtaining cyclopentanone. An expensive and energy-demanding distillation step to produce purer furfural is not required for the CPONE reaction, thus making this route very attractive.

This reaction was first described by Hronec *et al.* only few years ago [2] and since then a wide variety of catalysts, based on different noble and non-noble metals, have been revealed as effective for this reaction. Table 3.6.1 compiles this information conducted in batch operation mode.

Comparison between the different investigations reported in Table 3.6.1 is not straightforward because they were conducted under different reaction conditions, especially different catalyst loading and pressure of H_2. But in most of the cases the highest yield presented in the table was the outcome of an optimization of the conditions regarding furfural, catalyst conc., temperature, pressure and time of reaction. Among these catalysts those described in entries 8, 11, 14, 15 and 16 displayed a CPONE yield above 90%. The catalyst based on a Cu–Ni–Al based hydrotalcite (entry 8) reached a 96% yield of CPONE but required 25 wt.% relative to furfural. Reutilization tests showed that the catalyst was not stable. The catalyst based on Ru supported on MIL-101, a Cr-based metal-organic framework, (entry 11) displayed a remarkable stability (stable for six runs) at a high yield of CPONE although at a low furfural concentration and at wt. catalyst/furfural = 1. The Pd–Cu/C catalyst (entry 14)

Table 3.6.1. Summary of the catalytic properties of the different catalysts tested in the hydrogenation of furfural to CPONE.

Entry	Catalyst	Reaction conditions					FUR conv (%)	CPONE yield (%)	Reutilization	Ref.
		Furfural conc. (wt.%)	Wt. cat./FUR	P (MPa)	T (K)	Time (h)				
1	5 wt.% Pd/C	5	0.1	3	433	1	98	68	n.r.	[2]
2	5 wt.% Pt/C	5	0.05	8	433	0.5	100	76	n.r.	[2, 11]
		5	0.1	3	433	1	96.5	51	n.r.	
		5	0.1	8	448	0.5	100	40	2nd cycle stable	
3	5 wt.% Ru/C	5	0.1	8	448	1	100	57	n.r.	[2]
4	Commercial Ni (NiSAT® 320RS)	5	0.1	8	448	0.5	95	67	n.r.	[1]
6	Commercial Ni (G-134-A)	5	0.1	8	448	0.5	100	57	1st run	[11]
							99	60	2nd run	
7	Ni–Co/SBA-15 (10 wt.% Ni at Cu/Ni = 0.5)	5	0.4	4	433	4	100	62	n.r.	[12]

(Continued)

Table 3.6.1. (*Continued*)

Entry	Catalyst	Furfural conc. (wt.%)	Reaction conditions				FUR conv (%)	CPONE yield (%)	Reutilization	Ref.
			Wt. cat./FUR	P (MPa)	T (K)	Time (h)				
8	Cu-Mg-Al (1:14:5) hydrotalcite	6	0.25	4	413	8	100	96	1st run	[13]
							85	74	2nd run	
9	20 wt.% Ni–Y zeolite	5	0.3	4	423	9	96	86	n.r.	[14]
10	Cu–Zn–Al$_2$O$_3$	3	0.42	4	423	6	98	60	1st run	[15]
							95	53.3	2nd run	
11	3 wt.% Ru-MIL-101	1	1	4	433	2.5	100	96	Stable after five runs	[16]
12	Ni-Raney	11	0.13	1	453	4	98	45	Slight deactivation for the 2nd run	[17]
		11[a]	0.13	1	453	4	98	39		
13	5 wt.% Cu-Co	2	0.26	2	443	1	95	67	Stable after five runs	[18]
14	5%Pd-10%Cu/C (wt.%)	5	0.01	3	433	1	98	92	n.r.	[9]
15	CuNi0.5@C	5	0.4	5	423	5	> 99	97	Slight deactivation in four run tests	[19]

(*Continued*)

Table 3.6.1. (*Continued*)

Entry	Catalyst	Furfural conc. (wt.%)	Reaction conditions				FUR conv(%)	CPONE yield (%)	Reutilization	Ref.
			Wt. cat./FUR	P (MPa)	T (K)	Time (h)				
16	0.1 wt.%Au/TiO$_2$	5	0.5	4	433	1.2	>99	>99	Slight deactivation observed in a five run test[b]	[6]
			0.05				>99	>99		
17	0.26 wt.% Ru/CNT	1	0.04	1	433	5	99	90	1st cycle	[20]
							83	79	3rd cycle	

Notes: [a]Methanol was used as source of H$_2$ under N$_2$ pressure, (wt. methanol/FUR = 1.7).
[b]Productivity after 540 h in test conducted under continuous flow was 426 kg of CPONE/kg Au, catalyst was fully regenerated when deactivated by methanol washing.
n.r. not reported.

showed a remarkable yield (92%) using a very low concentration of catalyst (10%) but no information regarding the stability of the catalyst was disclosed. Entry 15 concerns data of a bimetallic Cu–Ni catalyst dispersed in a carbon matrix prepared by decomposition of a Cu metal-organic framework (at. Ni/Cu = 2), yield as high as 97% was accomplished. In a four consecutive run test catalyst exhibited a slight although continuous deactivation. Remarkably, an anatase (TiO_2) supported gold catalyst, prepared with only 0.1 wt.% of gold, (entry 16) exhibited an outstanding CPONE yield (>99.9%) with satisfactory stability. Actually a long-term stability test (540 h) were conducted under continuous flow operation in a fixed catalyst bed and showed that, although the catalyst deactivated slightly with time on stream, it can be fully regenerated by flowing methanol to remove deposits of coke.

There is a general consensus on the main features of the reaction route (Figure 3.6.1). The first step is the reduction of FUR to FOL, which reorganizes to 4-hydroxy-2-cyclopentenone (HCP). The formation of HCP involves the Piancatelli rearrangement of FOL which requires the presence of water and acidity to proceed. HCP is then successively hydrogenated to 2-cyclopentenone (CPENONE) and to CPONE. A different route has been proposed by Hronec *et al.* [1,9] that also goes through FOL rearrangement to CPENONE mediated by water and acidity but that does not involve the formation of HCP. An important remark here is that HCP can also be hydrogenated to 1,3-cyclopentanediol instead than to CPONE,

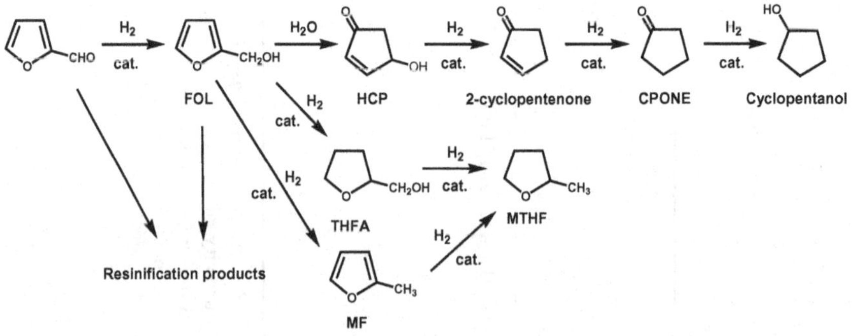

Figure 3.6.1.   Reaction network for the hydrogenation of FUR to CPONE.

the former diol could replace other diols in the synthesis of polyester, polyurethanes and polyether [10]. This latter selective transformation requires the use THF as solvent because when water was the solvent HCP is selectively transformed to CPONE [10].

In order to achieve high CPONE yield its further hydrogenation to cyclopentanol (CPOL) must be inhibited. Yang *et al.* proposed that the presence of CPENONE inhibits the hydrogenation [12] and once all CPENONE is transformed to CPONE, CPOL is formed at the expenses of CPONE. Hronec *et al.* claims that a furfuryl alcohol-derived resin deposited at the surface of the catalyst is responsible of the inhibition [11]. The conventional hydrogenation of FUR to FOL, MF and MTHF must also be prevented but, as already indicated, FOL rearrangement is favored under the presence H_2O. Heavy products can also be formed by resinification of FUR and FOL, which explain the lack of carbon balance frequently observed in this reaction.

The most accepted mechanism of transformation of FOL to CPONE is through Piancatelli rearrangement of FOL to HCP which is described in Figure 3.6.2(a) for the specific case of FOL [8,21]. The reaction is catalyzed by acids and the first step is the protonation of the alcohol group and further dehydration to a carbocation II. The next step involves the nucleophilic attack of a water molecule that generates a cationic intermediate which undergoes the ring opening to 1,4-dihydroxypentadienyl cation (III) which finally experiences a 4π-conrotatory cyclization to render the cation that subsequently is transformed to HCP. The latter is then hydrogenated to CPENONE and finally to CPONE. It is important to remind that this mechanism has been derived from reaction conducted with homogeneous acid catalysts.

Hronec *et al.* have proposed an alternative reductive rearrangement pathway for FOL that directly renders CPENONE but without the formation of HCP (Figure 3.6.2(b)) [9, 22]. They proposed that this new route takes place under high H_2 partial pressure (1–8 MPa), the presence of active and selective metal catalysts and efficient mass transfer of H_2 to the surface of the catalyst (kinetic regime). Under these conditions they proposed that this route is faster than the classical Piancatelli route and predominates. We have to bear in

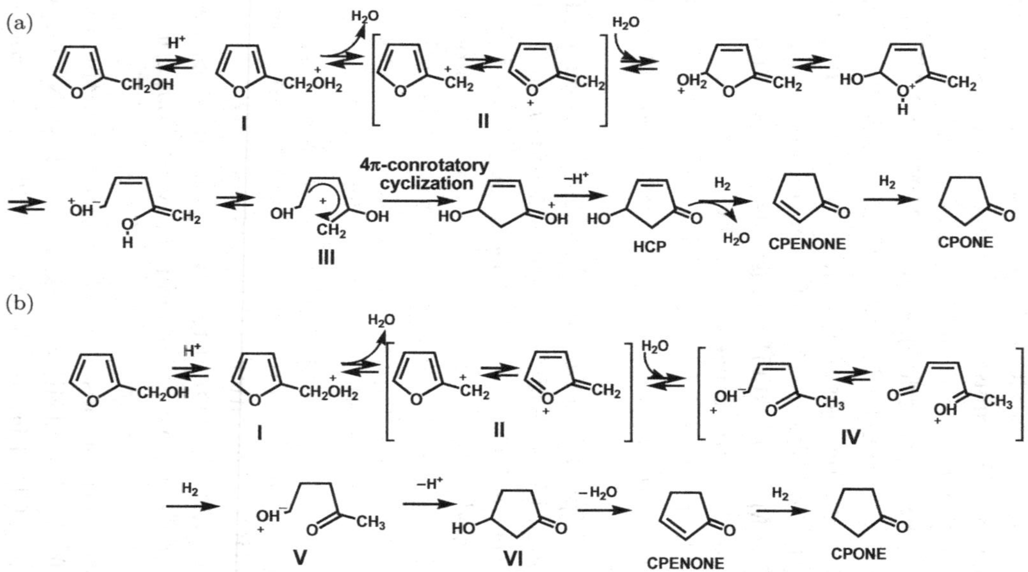

Figure 3.6.2. Different mechanisms proposed for the transformation of FOL to CPONE: (a) via Piancatelli rearrangement and (b) via reductive rearrangement.

mind that the classical Piancatelli rearrangement of furylcarbinols to HCP takes place in the absence of H_2. Both mechanisms shared the intermediates I and II but the latter undertakes an addition of water (intermediate IV) which is hydrogenated to V. The latter, after releasing a proton (intermediate VI) and water, yields CPENONE which is successively hydrogenated to CPONE.

3.6.3. Biofuels derived from cyclopentanone

Recently a new process to produce liquid biofuels from CPONE has been demonstrated. This route implies two steps, first one the self-aldol condensation of CPONE to yield 2-cyclopentylidene-cyclopentanone and the second one the hydrodeoxygenation (HDO) of the latter to bi-(cyclopentane) (BCP) (see Figure 3.6.3) [12, 23]. This liquid fuel has a high density $(0.866 \text{ g mL}^{-1})$ and net heat of combustion $(42.6 \text{ MJ} \cdot \text{kg}^{-1}; 37.0 \text{ MJ} \cdot \text{L}^{-1})$ and can be used as a renewable high-density fuel or as an additive to increase the volumetric heating value of conventional biobased jet fuels and for blending with conventional biodiesel to increase its mileage per liter [23].

For the first step, the self-aldol condensation of CPONE, basic catalysts have been explored. Heterogeneous catalysts have demonstrated higher yield of 2-cyclopentylidene-cyclopentanone than when using homogeneous NaOH [12, 24]. Among the different solid catalysts tested, namely, CaO, CaO–CeO$_2$, MgO, KF/Al$_2$O$_3$, MgO$_2$–ZrO$_2$ mixed oxides and LiAl and MgAl hydrotalcites, the latter, MgAl hydrotalcites, showed the highest yield (86%) after 8 h at

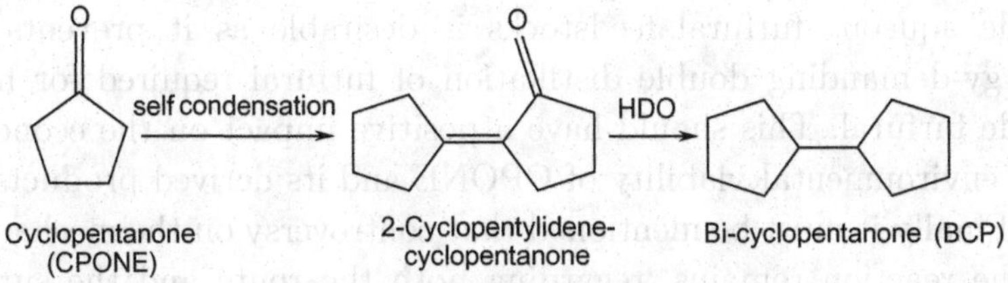

Cyclopentanone   2-Cyclopentylidene-   Bi-cyclopentanone (BCP)
(CPONE)   cyclopentanone

Figure 3.6.3. Simplified reaction scheme of the synthesis of bi-cyclopentane (BCP) from cyclopentanone (CPONE).

423 K and 10 wt.% of catalyst (higher than that obtained with NaOH) [23]. The second step of the process, HDO of the self-condensation product has also been demonstrated using Pd, Ru and Ni based catalysts [12, 23]. A 93% yield was obtained with residual formation of cracking products (C_1–C_5 alkanes) and partially hydrodeoxygenated C_{10} oxygenates (2-cyclopentylcyclopentanone and 2-cyclopentylcyclopentanol) under continuous mode operation in a fixed-bed continuous flow reactor at 503 K and 6 MPa of H_2 and using the Ni/SiO_2 catalyst; besides neither significant deactivation nor Ni leaching after 24 h on stream was observed [23].

A variation of this process has also been proposed but, rather than starting with the self-aldol condensation of CPONE, the process initiates with a furfural–CPONE aldol condensation. More details for the latter process are given in Chapter 3.12.

3.6.4. Summary and future prospects

CPONE can be selectively produced from furfural by hydrogenation of furfural in aqueous solutions. A wide variety of catalysts based on metals (Cu, Ni, Pd, Pt, Ru and Au) have shown activity for this reaction. Stability of the catalyst is the main aspect to be considered because all have shown quite good activity and selectivity. Among those tested, an anatase (TiO_2) supported gold catalyst has resulted in a CPONE yield > 99.9% with satisfactory stability in long-term stability test (540 h).

So far, investigations have been conducted using aqueous solutions of high-grade furfural. To my knowledge no investigation on real low-grade furfural directly obtained from biomass has been accomplished and such investigation is required. Processing this low-grade aqueous furfural feedstocks is desirable as it prevents the energy-demanding double distillation of furfural required for high-grade furfural. This should have a positive impact on the economic and environmental viability of CPONE and its derived products.

Finally it must be mentioned that controversy on the mechanism of the reaction remains, regarding both the route and the surface chemical species involved. A definitive clarification of the mechanism is essential for a correct design of optimum catalytic system.

List of abbreviations

CPENONE	2-Cyclopentenone
CPONE	Cyclopentanone
CPOL	Cyclopentanol
FOL	Furfuryl alcohol
FUR	Furfural
HCP	4-Hydroxy-2-cyclopentenone
MF	2-Methylfuran
MTHF	2-Methyltetrahydrofuran
THFA	Tetrahydrofurfuryl alcohol

References

1. M. Hronec, K. Fulajtarova, T. Liptaj, Effect of catalyst and solvent on the furan ring rearrangement to cyclopentanone, *Applied Catalysis A: General*, 437 (2012) 104–111.
2. M. Hronec, K. Fulajtarová, Selective transformation of furfural to cyclopentanone, *Catalysis Communications*, 24 (2012) 100–104.
3. T. Takahashi, K. Ueno, T. Kai, Vapor phase Beckmann rearrangement of cyclopentanone oxime over high silica HZSM-5 zeolites, *Microporous Materials*, 1 (1993) 323–327.
4. P. Sudarsanam, L. Katta, G. Thrimurthulu, B.M. Reddy, Vapor phase synthesis of cyclopentanone over nanostructured ceria-zirconia solid solution catalysts, *Journal of Industrial and Engineering Chemistry*, 19 (2013) 1517–1524.
5. J. Vojtko, New way of production of cyclopentanone, *Petroleum and Coal*, 47 (2005) 1–4.
6. G.S. Zhang, M.M. Zhu, Q. Zhang, Y.M. Liu, H.Y. He, Y. Cao, Towards quantitative and scalable transformation of furfural to cyclopentanone with supported gold catalysts, *Green Chemistry*, 18 (2016) 2155–2164.
7. G. Piancatelli, A. Scettri, Heterocyclic steroids-III. The synthetic utility of a 2-furyl steroid, *Tetrahedron*, 33 (1977) 69–72.
8. C. Piutti, F. Quartieri, The piancatelli rearrangement: New applications for an intriguing reaction, *Molecules*, 18 (2013) 12290–12312.
9. M. Hronec, K. Fulajtárová, I. Vávra, T. Soták, E. Dobročka, M. Mičušík, Carbon supported Pd-Cu catalysts for highly selective rearrangement of furfural to cyclopentanone, *Applied Catalysis B: Environmental*, 181 (2016) 210–219.
10. G. Li, N. Li, M. Zheng, S. Li, A. Wang, Y. Cong, X. Wang, T. Zhang, Industrially scalable and cost-effective synthesis of 1,3-cyclopentanediol with furfuryl alcohol from lignocellulose, *Green Chemistry*, 18 (2016) 3607–3613.

11. M. Hronec, K. Fulajtárova, M. Mičušik, Influence of furanic polymers on selectivity of furfural rearrangement to cyclopentanone, *Applied Catalysis A: General*, 468 (2013) 426–431.

12. Y. Yang, Z. Du, Y. Huang, F. Lu, F. Wang, J. Gao, J. Xu, Conversion of furfural into cyclopentanone over Ni-Cu bimetallic catalysts, *Green Chemistry*, 15 (2013) 1932–1940.

13. H. Zhu, M. Zhou, Z. Zeng, G. Xiao, R. Xiao, Selective hydrogenation of furfural to cyclopentanone over Cu-Ni-Al hydrotalcite-based catalysts, *Korean Journal of Chemical Engineering*, 31 (2014) 593–597.

14. C.Y. Liu, R.P. Wei, G.L. Geng, M.H. Zhou, L.J. Gao, G.M. Xiao, Aqueous-phase catalytic hydrogenation of furfural over Ni-bearing hierarchical Y zeolite catalysts synthesized by a facile route, *Fuel Processing Technology*, 134 (2015) 168–174.

15. J. Guo, G. Xu, Z. Han, Y. Zhang, Y. Fu, Q. Guo, Selective conversion of furfural to cyclopentanone with Cuznal catalysts, *ACS Sustainable Chemistry and Engineering*, 2 (2014) 2259–2266.

16. R. Fang, H. Liu, R. Luque, Y. Li, Efficient and selective hydrogenation of biomass-derived furfural to cyclopentanone using Ru catalysts, *Green Chemistry*, 17 (2015) 4183–4188.

17. Y. Xu, S. Qiu, J. Long, C. Wang, J. Chang, J. Tan, Q. Liu, L. Ma, T. Wang, Q. Zhang, In situ hydrogenation of furfural with additives over a Raney®-Ni catalyst, *RSC Advances*, 5 (2015) 91190–91195.

18. X.L. Li, J. Deng, J. Shi, T. Pan, C.G. Yu, H.J. Xu, Y. Fu, Selective conversion of furfural to cyclopentanone or cyclopentanol using different preparation methods of Cu-Co catalysts, *Green Chemistry*, 17 (2015) 1038–1046.

19. Y. Wang, S. Sang, W. Zhu, L. Gao, G. Xiao, CuNi@C catalysts with high activity derived from metal-organic frameworks precursor for conversion of furfural to cyclopentanone, *Chemical Engineering Journal*, 299 (2016) 104–111.

20. Y. Liu, Z. Chen, X. Wang, Y. Liang, X. Yang, Z. Wang, Highly selective and efficient rearrangement of biomass-derived furfural to cyclopentanone over interface-active Ru/carbon nanotubes catalyst in water, *ACS Sustainable Chemistry and Engineering*, 5 (2017) 744–751.

21. G. Piancatelli, A. Scettri, S. Barbadoro, A useful preparation of 4-substituted 5-hydroxy-3-oxocyclopentene, *Tetrahedron Letters*, 17 (1976) 3555–3558.

22. M. Hronec, K. Fulajtárova, T. Soták, Highly selective rearrangement of furfuryl alcohol to cyclopentanone, *Applied Catalysis B: Environmental*, 154–155 (2014) 294–300.

23. J. Yang, N. Li, G. Li, W. Wang, A. Wang, X. Wang, Y. Cong, T. Zhang, Synthesis of renewable high-density fuels using cyclopentanone derived from lignocellulose, *Chemical Communications*, 50 (2014) 2572–2574.

24. D. Liang, G. Li, Y. Liu, J. Wu, X. Zhang, Controllable self-aldol condensation of cyclopentanone over MgO-ZrO$_2$ mixed oxides: Origin of activity & selectivity, *Catalysis Communications*, 81 (2016) 33–36.

Chapter 3.7

Levulinic Acid and γ-Valerolactone

Rafael Mariscal* and David Martín Alonso[†,‡]

*Group of Sustainable Energy and Chemistry (EQS),
Institute of Catalysis and Petrochemistry (ICP-CSIC),
C/Marie Curie 2, Cantoblanco, 28049 Madrid, Spain

[†]Department of Chemical and Biological Engineering,
University of Wisconsin-Madison, 1415 Engineering Drive,
Madison, WI, USA

[‡]Glucan Biorenewables LLC, 505 South Rosa Road,
Suite 112, Madison, WI 53719, USA

3.7.1. Introduction

Furfural is a product that can be commercially obtained from biomass. Specifically, it derives from hemicellulose (23–32% of the lignocellulose) highly branched polymer composed mainly of C_5 sugars, mostly xylose. Although LA and GVL can be obtained directly from carbohydrates resulting from the hydrolysis of lignocellulose, in this chapter we will refer only to those processes where furfural is the starting reagent. Neither will it be treated, as has been done elsewhere in this book, the different ways to obtain competitively furfural from biomass.

Therefore, the structure followed in this chapter is to review the production of levulinic acid starting from furfural, taking into account the peculiarities of these processes and paying special attention to its hydrogenation to GVL the most important product from LA. Furfural and LA compounds belong to the 12 molecules platform with more future in biorrefineries [1].

Both, LA and GVL, are key products in any scheme of a biorefinery as they can integrate the conversion of cellulose and hemicellulose into renewable chemicals. These products can be obtained by different routes from the biomass [2–4].

The purpose of this chapter is to provide a broad view of the production of LA and GVL as platform molecules starting directly from furfural, in addition to their use to produce liquid biofuels and chemicals. This chapter will present the different strategies that have been addressed so far with particular attention to the catalytic systems tested and the most important challenges that researchers have to face. Although GVL can be considered as a derivative or intermediate of the process of hydrogenation of LA it deserves a separate section because its enormous importance and potential.

3.7.2. Production of LA and derived chemicals and biofuels

3.7.2.1. *LA synthesis*

LA (4-oxopentanoic acid) is a product with interesting direct applications. A number of routes for LA synthesis involve petro-chemical feedstocks, but LA can also be obtained from cellulose or hemicellulose. The inhibitive production costs have precluded the commercialization of LA as a commodity chemical [2, 5]. The main limitations are the low yields obtained and the high purification costs. The acid dehydration of cellulose has been proposed as one of the most promising solutions to decrease the LA price and enable its commercialization as a commodity chemical. The Biofine process was the first commercial attempt to obtain biomass-derived LA from the glucose present in cellulose. This process employs the aqueous H_2SO_4-catalyzed hydrolysis of cellulose to release the glucose with further dehydration to LA (50–75% yield) [2, 5–7]. The idea of the process was to get the LA down to \$0.5/kg. Segetis has been one of the mayor players on LA production and upgrading in the last years, and it has been recently acquired by GFBiochemicals [8], which is currently the main player in the LA market with a 10,000 MT per year plant located at Caserta and plans to build a larger commercial facility [9].

While the research to improve the LA production is focused on the dehydration of cellulose, currently, LA is produced as a specialty chemical in batch mode by direct conversion of furfuryl alcohol. The production is mainly located in China, where small specialty chemicals, flavors and pharmaceutical companies offer the LA in their products portfolios at a price ranging from \$5–10/kg. The main challenges to reduce the LA price is to reduce the cost of the feedstock (furfural or furfuryl alcohol), improving the yields, reduce or eliminate separation steps and to move from batch to a continuous process.

The route for production of LA from FUR involves several steps typically conducted in different reactors (Figure 3.7.1). FUR is first produced from C_5 carbohydrates and then hydrogenated to FOL, which is then transformed into LA via acid-catalyzed ring opening in H_2O. This process has been thoroughly reviewed [7, 10–14]. The ring opening of FOL can also be conducted in the presence of alcohols rather than H_2O [10, 14], the alcoholysis subsequently affords alkyl levulinates, which can be later hydrolysed to produce LA or used directly as a final product.

The hydration of FOL to LA is an acid-catalyzed process. In addition to sulfuric acid, different strong acid ion-exchange resins and solid acid catalysts have been also used as heterogeneous catalysts for the conversion of FOL to LA [10, 11]. One of the challenges of this reaction, is that FOL is not stable in acid media, leading to

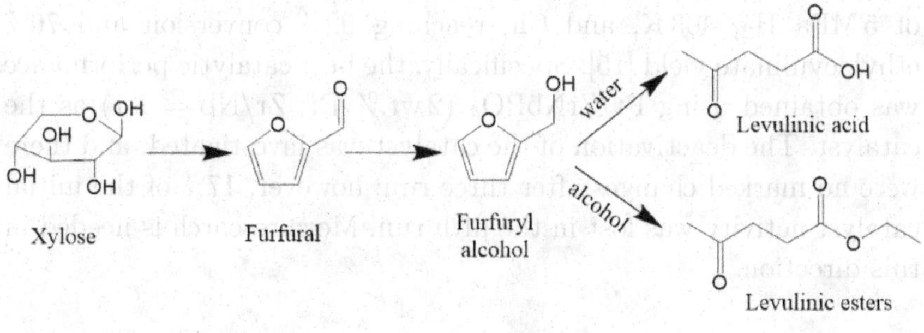

Figure 3.7.1. Production of LA from biomass.

the formation of humins or degradation products that significantly decrease the yields. Working at low FOL concentration has been used to decrease the degradation, but it increases the production and purification cost. Because the final product, LA, is stable in acid media, working in semi-batch reactor has been proposed as a solution [10]. It has also been proposed a biphasic system that consists of an organic layer (FOL in sec-butyl phenol) and an aqueous layer (with the acid catalysts: 1 M H_2SO_4 or Amberlyst-15). The FOL is retained in the organic phase where it is more stable, the small amount of FOL that partitions into the aqueous layer reacts to produce LA which is retained preferable in the aqueous layer. Operating at 398 K, after 1 h of reaction time, a LA yield of 76% is attained, whereas with the single aqueous phase conditions, in the presence of 1 M H_2SO_4, the yield is only 50% [7]. No information was provided regarding the deactivation or reuse of solid catalysts because the research focused on screening different feedstocks and solid acid catalysts for biomass transformation. Another option, is the use of aprotic solvents. Amberlyst-70, Nafion-brand resins and acid zeolites (ZSM-5) are reported as catalysts. By feeding 1 wt.% FOL and 2 wt.% LA in GVL/water solvent (80/20 w/w), a LA yield of 80% can be obtained at 423 K after 1 h using HZSM-5 as a solid catalyst. At lower temperature, 398 K, under similar experimental conditions, the use of Amberlyst-70 achieved a 71% LA yield [11].

Remarkably, the one-pot direct transformation of FUR into alkyl levulinates, not requiring the isolation of FOL, has been reported by Chen *et al.* using a bifunctional catalyst comprising Pt nanoparticles supported on a ZrNb binary phosphate solid acid under conditions of 5 MPa H_2, 403 K, and 6 h, reaching 92% conversion and 76% ethyl levulinate yield [15]. Specifically, the best catalytic performance was obtained using $Pt/ZrNbPO_4$ (2 wt.% Pt; Zr/Nb = 1:1) as the catalyst. The deactivation of the catalyst was investigated, and there were no marked changes after three run; however, 17% of the initial catalyst activity was lost in the fifth run. More research is needed in this direction.

3.7.2.2. *Derived chemicals and biofuels*

Because of its high functionality (keto and carboxyl groups), LA is considered to be an attractive chemical platform for obtaining higher value-added chemicals, liquid fuels, and fuel additives [1, 16, 17]. Figure 3.7.2 shows the main products that can be generated from LA.

A brief discussion of the most relevant chemicals and their usefulness is given [1]. Tetrahydrofuran (THF) is a widely used green solvent that is currently obtained from petrochemical feedstocks, but it may also have a biomass origin. Thus, THF can be obtained from furan, but it can also be obtained through the decarboxylation of levulinic acid and subsequent ring closing. Succinic acid (SAc), a chemical platform for obtaining a number of derivatives [18, 19], is currently produced by the catalytic hydrogenation of maleic anhydride (a petrochemical). However, SAc can also be produced through the selective oxidation of LA. δ-aminolevulinic acid (DALA) is a completely biodegradable broad spectrum herbicide. DALA has also been found to possess insecticide activity and is a component in photodynamic therapy for cancer treatment. The conventional synthesis of DALA has several steps, and the key area for improvement is to scale the process up to commercial levels [5]. Recently, Xiao *et al.* reported the transformation of levulinic acid into pyrrolidones (fine chemicals) via reductive amination with formic acid as the hydrogen source under mild conditions in aqueous solution [20]. Finally, we must mention the production of diphenolic acid (DPA), which has wide applications in the production of polymers and other materials. DPA is easily prepared via the condensation reaction of LA with two molecules of phenol [5].

Several approaches have been proposed for obtaining liquid transportation fuels and additives from LA. These approaches do not specifically start from FUR, and in most cases they require prior isolation of LA; thus, for the sake of simplicity, these transformations will simply be briefly summarized. Nevertheless, they have been included in this review as LA can be obtained from FUR and because, in some cases, the direct synthesis of these species from FUR has been

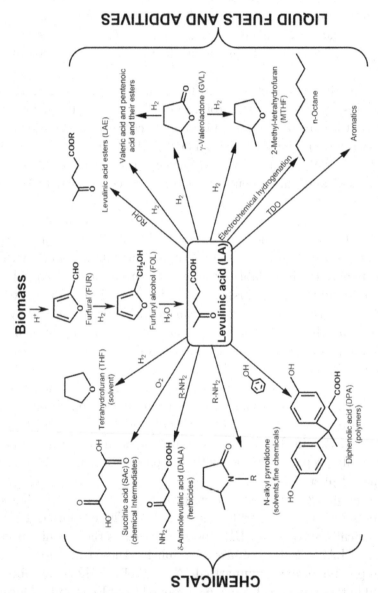

Figure 3.7.2.   Levulinic acid (LA) as a platform for fuels and chemicals.

accomplished. The four approaches are described in the following subsections.

3.7.2.3. *Esterification to levulinate esters*

Levulinic acid esters (LAE), particularly methyl, ethyl and *n*-butyl levulinate (short-chain fatty esters), can be used as a blending component in biodiesel. Although these esters possess a very low cetane number, they are suitable for use as gasoline and diesel transportation fuel additives because of their numerous excellent properties such as low toxicity, high lubricity, flash point stability, and moderate flow properties under low temperature conditions [21, 22]. Most investigations of this reaction have been conducted in alcoholic medium by starting either from isolated LA or directly from C_6 carbohydrates. This approach has been extensively reviewed elsewhere and more details can be found in other publications [23–27].

Another more FUR-oriented route, involves the alcoholysis of FOL. This is prone to polymerization, then ethanolysis/alcoholysis allows high yields of levulinate esters. Most of the relevant research has been conducted with isolated FOL and by using homogeneous acid catalysts such as HCl or H_2SO_4 or a variety of solid acid catalysts such as strongly acidic resins [10,14,28], acidic zeolites like ZSM-5 [29], Al-TUD-1 mesoporous aluminosilicates [30,31], and ionic liquids functionalized with acidic anions [32–34]. Solid catalysts are, in principle, preferred because they can be reutilized and present less corrosion and downstream problems.

Interestingly, direct conversion of FUR was reported to be possible through one-pot transformation of FUR into alkyl levulinates involving initial hydrogenation of FUR to FOL, which proceeded via ring opening in the presence of an alcohol. A bifunctional catalyst composed of a Pt/ZrNb binary phosphate solid acid was reported to be very efficient and moderately stable with use at 5 MPa H_2, 403 K, and 6 h, reaching 92% conversion and 76% ethyl levulinate yield [15]. The possibility of performing this reaction by starting from hemicellulosic biomass still remains a challenge.

3.7.2.4. *Selective hydrogenation*

Figure 3.7.3 clearly illustrates the hydrogenation pathways to the relevant biofuels that can be obtained from hydrogenation of LA, i.e., either valeric acid, GVL or MTHF. The different intermediate products are classified as a function of their H and O contents [35]. The energetic density increases as the oxygen content decreases or the hydrogen content increases. Routes requiring the isolation of GVL, the most stable intermediate, are also possible as mentioned in the ensuing Section 3.7.3.

2-Methyltetrahydrofuran (MTHF) has been approved for use as a component of P-Series type fuels by the US DOE [36] since certain properties of MTHF make it suitable for gasoline blending without engine modifications. These include the octane number of 74, better miscibility with common hydrocarbon-based fuels than alcoholic additives, less susceptibility to polymerization, low volatility, and lower emission of contaminants [37]. MTHF is currently produced via hydrogenation of 2-Methylfuran. However, an alternative method for producing MTHF is via the hydrogenation of LA through a multistep sequence of reactions in which GVL is first produced; a deeper reduction gives rise to 1,4-pentanediol, and finally, ring closure by dehydration affords MTHF.

The liquid-phase hydrogenation of LA to MTHF in one-pot was first demonstrated using a bimetallic Re–Pd/C catalyst. Yields close to 70% were achieved in a continuous process (6 h, 523 K, 100 MPa H_2) [38]. Leitner *et al.* proposed an efficient and flexible process capable of tuning the selectivity for GVL, 1,4-PDO, and MTHF [39] by selecting the composition of the catalytic system (i.e., by using a Ru-containing precursor complex, a set of mono-, bi- and tri-dentate phosphine ligands, and ionic and/or acidic additives). Thus, by adding an acidic ionic liquid, the selectivity to MTHF was greatly improved to furnish a yield of 92% upon complete conversion of LA (18 h, 433 K, 10 MPa H_2). The catalyst was reutilized in four consecutive batches without any loss of the catalytic activity. Catalytic transfer hydrogenation (CTH) has also been explored. This approach is particularly interesting because it

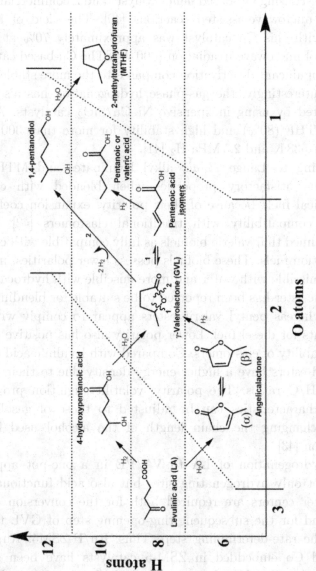

Figure 3.7.3. LA hydrogenation pathways to different 5-carbon-atom biofuels (adapted from Ref. [35]).

involves the use of formic acid, which can be obtained during the production of LA from C_6 sugars. Bermdez *et al.* recently reported selective CTH using Cu-based nanocatalysts and a commercial Pd/C system in microwave-assisted reactions [40]. The yield of MTHF achieved with this Cu catalyst was approximately 70% at 323 K after 0.5 h of microwave irradiation (300 W). The Cu-based catalysts exhibited significant deactivation compared to the more stable Pd/C catalyst. Interestingly, the gas-phase hydrogenation has also been demonstrated by using inexpensive Ni–Cu/SiO$_2$ catalysts. A high yield of MTHF (89%) and high stability for more than 300 h was reported at 538 K and 2.5 MPa H$_2$ [41].

According to Lange *et al.*, alkyl levulinates and MTHF do not exhibit satisfactory properties when blended with current petrochemical fuels because of their polarity, expansion coefficient and poor compatibility with traditional elastomers [42]. These authors claimed that valeric biofuels as fully compatible with current transportation fuels. These biofuels possess lower polarities, making them less miscible with water and more miscible with hydrocarbons: ethyl valeric ester has been reported to be suitable for blending with gasoline, whereas pentyl valeric esters appear to comply with the requirements for diesel fuel. Lower polarity also has positive effects on the durability of elastomers. Compared with levulinic acid esters, valeric acid esters have a higher energy density due to their higher C/O and H/C ratios. The polarity, volatility, ignition properties and flow characteristics can be adjusted to those of gasoline or diesel by changing the chain length of the alcohol used in the esterification [43].

The hydrogenation of LA to VA/VE in a one-pot approach requires not only hydrogenating sites but also acid functionalities, as the latter centers are required both for the conversion of LA to GVL and for the subsequent ring-opening step of GVL to VA, which is the rate-determining step. Thus, Ru/H-ZSM5, Ru/SBA-SO$_3$H, and Co embedded in ZSM-5 catalysts have been shown to generate good yields of the targeted products (46%, 94%, and 97%, respectively) at 3–5 MPa H$_2$ and 473–523 K [35, 44, 45]. The presence of alcohols is required, otherwise the yield of VA and its esters is lower. Deactivation is a major reported problem for Ru catalysts, where coke deposition, dealumination of zeolites [35], and leaching of the sulfonic sites of the SBA-SO$_3$H catalyst [44] were identified as the main sources of deactivation. In contrast, embedding Co nanoparticles within the HZSM-5 cages prevents Co leaching, resulting in a very stable catalyst with consequent maintenance of the yield of VA and its esters above 90% for eight cycles [45].

3.7.2.5. *Electrochemical hydrogenation to n-octane*

The electrochemical transformation of LA into *n*-octane at room temperature has been demonstrated [46]. Overall, the process is a net reduction, but in practice, the process requires two electrochemical reactions, namely, electrochemical reduction of keto compounds and electrochemical oxidative decarboxylation-dimerization of a carboxylic acid (known as the Kolbe reaction). Figure 3.7.4 describes the two possible pathways proposed. One route involved the electroreduction of LA into valeric acid and the subsequent Kolbe electrolysis of VA in *n*-octane (A–B route). The other route is the Kolbe electrolysis of LA into 2,7-octadione and the subsequent electroreduction into *n*-octane (C–D pathway). It was

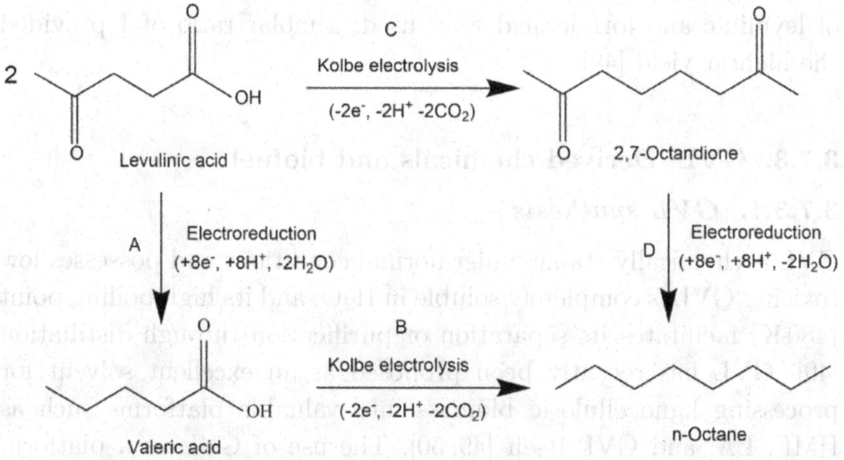

Figure 3.7.4. Reaction network of the electrochemical transformation of LA into *n*-octane (adapted from Ref. [46]).

demonstrated that the first route exhibited better selectivity and coulombic efficiency. Interestingly, this process can be conducted in aqueous solutions, and therefore, the n-octane spontaneously separates from the medium without any separation and purification steps. Admittedly, there is still considerable room for improvement and optimization of the electrochemical process in terms of the energy efficiency of the electrochemical process in comparison to the existing chemical/catalytic routes.

3.7.2.6. *Thermal deoxygenation to liquid aromatic fuels*

Thermal deoxygenation (TDO) of LA into a mixture of aromatic liquids has been proposed as an alternative pathway for producing biofuels. Specifically, the TDO procedure involves neutralizing LA with $Ca(OH)_2$ to obtain the corresponding salt. This salt is subjected to high temperatures (623–723 K) to form cyclic and aromatic products with a low oxygen–carbon ratio, thereby improving the energy density. These TDO products can be enhanced by hydrogenation and dehydration using Ru and Pt catalysts and can then be used as hydrocarbon fuels [47]. Later, these authors also reported achieving high yields of deoxygenated hydrocarbons when mixtures of levulinic and formic acid were used; a molar ratio of 1 provided the highest yield [48].

3.7.3. GVL: Derived chemicals and biofuel

3.7.3.1. *GVL synthesis*

GVL is chemically stable under normal conditions and possesses low toxicity. GVL is completely soluble in H_2O, and its high boiling point (480 K) facilitates its separation or purification through distillation [49]. GVL has recently been proposed as an excellent solvent for processing lignocellulosic biomass into valuable platforms such as HMF, LA, and GVL itself [49, 50]. The use of GVL as a platform for the production of biofuels and chemicals has been exhaustively reviewed elsewhere [49, 51]. However, the commercial use of GVL is still limited, primarily because of its high production costs which is directly linked to the production of LA discussed earlier.

Extensive research has been conducted on the transformation of LA or esters into GVL involving the use of Ru, Pt, and inexpensive Cu-based catalysts and relatively moderate H_2 pressure (1–5.5 MPa) and temperature (403–473 K). This route has been the subject of excellent reviews and interested readers are directed to these reviews and references therein for further details as this is not a direct synthesis from FUR [43, 49, 52, 53]. The main challenge in the H_2 reduction of LA is replacing commercially pure levulinic acid by streams obtained directly from lignocellulose. In this case, formic acid and mineral acids (required as catalysts) such as sulphuric acid are present, and their presence significantly affects the reaction rate. The incorporation of Re into the Ru/C catalyst has been shown to stabilize the catalyst, although the GVL yield was not sufficiently high [54]. Other strategies to minimize the impact of these acid contaminant problems arising from the presence of sulphuric acid include the extraction of levulinic acid from aqueous solutions with alkyl phenol solvents, reactive extraction to produce levulinate esters and even the replacement of mineral acids by solid acid catalysts in the synthesis of LA from cellulosic sugars.

GVL can also be synthesized from FOL, but this involves an additional step for production and subsequent isolation of FOL from FUR [33]. These authors used a combination of two catalysts: an ionic liquid with a sulphonic acid functionalization (1-butyl sulphonic acid 3-methyl imidazolium) and a Ru/C catalyst. High efficiency (>99% conversion) and high selectivity (approximately 86% of GVL) were achieved at 403 K, with ca. 5 MPa H_2 and methanol as the solvent.

Remarkably, the direct conversion of FUR into GVL has been reported by Román-Leshkov *et al.* [55] via one-pot conversion of FUR into GVL through a cascade scheme with three steps (Figure 3.7.5) as follows: (i) a Meerwein–Ponndorf–Verley (MPV) hydrogenation of FUR to FOL, in which butanol is used as the solvent and hydrogen as the donor; butyl furfuryl ether can also be formed; (ii) ring-opening hydration of FOL (or butyl furfuryl ether, BFE) in LA (or butyl levulinate, BL); and finally, (iii) the MPV hydrogenation of LA

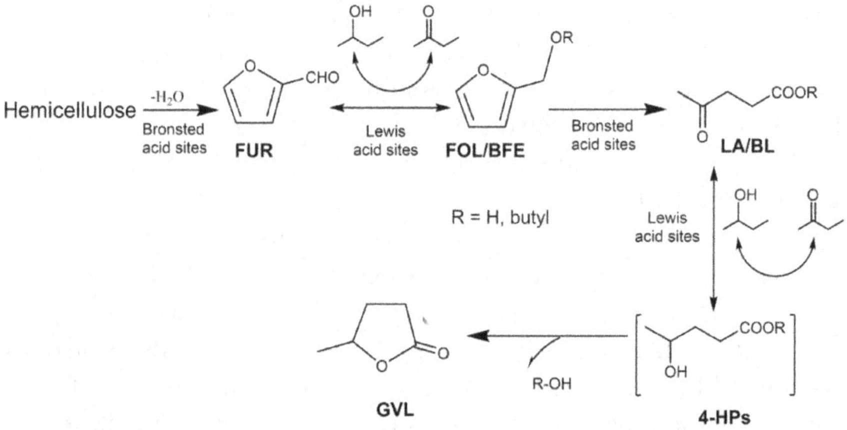

Figure 3.7.5. Domino reaction for the synthesis of GVL from cellulose with a combination of Lewis and Brönsted acid sites (adapted from Ref. [55]).

(or butyl levulinate, BL) into GVL (through 4-hydroxypentanoic acid or its butyl ester intermediate). A GVL yield as high as 68% was obtained with 5 wt.% FUR in butanol containing H_2O, at 393 K over 24 h using Zr-Beta zeolite (with Lewis acid sites) as the hydrogen transfer catalyst and Al-MFI zeolite (with nanosheet morphology an Brønsted acid sites) as the ring-opening hydration catalyst. This domino-like reaction involving the use of a hydrogen donor alcohol, which precludes the need for H_2 at high pressure, does not use a non-precious metal catalyst but rather utilizes considerably less expensive zeolites. Unfortunately, the GVL yield declined progressively over successive cycles (three cycles), and more investigation is required to make this direct synthesis possible. The initial yield was substantially (but not completely) recovered when the catalysts were calcined at 823 K for 10 h in air. This result suggests that fouling or poisoning by coke deposits has a great impact on the catalyst deactivation.

3.7.3.2. *Chemicals and fuels derived from GVL*

GVL has been proposed as a feedstock for producing chemicals, particularly monomers for polymers that are either currently derived from oil or that possess specific chemical properties and excellent acceptance by consumers. Figure 3.7.6 summarizes the major routes

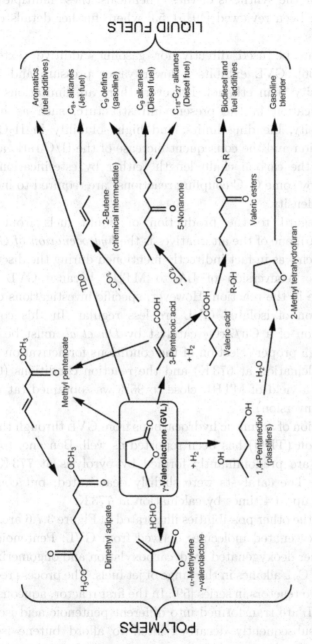

Figure 3.7.6. GVL as a chemical platform for the production of fuels and chemicals (adapted from Ref. [49]).

described for the synthesis of these chemicals; these multiple possibilities have been reviewed [49, 51, 53] where further details can be found.

GVL can be used directly for gasoline additive (extender) admixing [56]. GVL exhibits a lower vapor pressure and higher energy density than ethanol. However, there are limitations to its direct application in the present infrastructure, such as its low energy density, blending limits, and high solubility in H_2O [16]. Deoxygenation and the consequent increase of the H/C ratio and/or increase of the carbon chain length (either by esterification with alcohols or by some C–C coupling reactions) are required to increase the energy density.

With regard to the production of liquid fuels from GVL (Figure 3.7.6), one of the alternatives is the *hydrogenation of GVL to MTHF*, which was in fact indirectly mentioned during the discussion of the direct conversion of LA to MTHF because GVL is an intermediate in the reaction. However, specific investigations on the hydrogenation of isolated GVL are less regular. In this context, the evaluation of a Cu/ZrO_2 catalyst by *Du et al.* must be noted [57]. Through proper selection of the conditions for activation of the catalyst (calcination at 673 K) and the reaction conditions (6 MPa H_2, 513 K), a yield of MTHF close to 95% was obtained (at nearly full GVL conversion).

Production of aromatic hydrocarbons from GVL through thermal deoxygenation (TDO) has been reported as well. Benzene, toluene, and xylene are predominantly formed by pyrolysis at 773 K with ZSM-5 [58]. The catalysts were slightly deactivated, but could be regenerated up to 4 times by calcination at 773 K.

Finally, the other possibilities illustrated in Figure 3.7.6 are based on two deoxygenated molecules derived from GVL. Pentenoic acid can be further deoxygenated by decarboxylation and oligomerized to a mixture of C_{8+} alkanes in the range of jet fuels. The process requires the use of two reactors in series [59]. In the first reactor, aqueous solutions of GVL are transformed into different pentenoic acid isomers, which are subsequently decarboxylated to afford butene isomers and CO_2 in the presence of silica–alumina catalyst; the optimum conditions were found to be 3.6 MPa and 648 K for achieving butene yields above 99%. The catalyst was deactivated by coke formation; coke can be removed by calcination in air and in the presence of water vapor, which favors catalyst stability. In the second reactor, after water is removed by condensation, butene is oligomerized in the presence of CO_2 (the co-product of the decarboxylation of pentenoic acid) over another Amberlyst 70 at 423 K.

Valeric acid also presents two simultaneous possibilities: esterification to alkyl esters (biodiesel) or deoxygenation and oligomerization via 5-nonanone to C_9 alkanes/olefins (diesel/gasoline fuels) or to a mixture of C_{18}–C_{27} alkanes (diesel fuel). Serrano-Ruiz *et al.* also reported that solid Pd/Nb_2O_5 (also a bifunctional catalyst with both acid and metal sites) catalyzes the transformation of GVL to VA at moderate temperatures and pressures (598 K, $1.2\,h^{-1}$ WHSV and 3.5 MPa H_2) [60]. The conversion was accomplished with concentrated solutions of VA (50%) in water, and therefore, esterification must be later performed, but a 92% yield of VA at complete conversion can be achieved by controlling the H_2 pressure and the Pd loading. The incorporation of SiO_2 into the Nb oxide support is required to improve the thermal stability of the catalytic system and to consequently prevent the deactivation by sintering of the Pd and support particles [61]. In a later work, these authors reported improved performance of a carbon-supported Pd–NbCe catalyst for this reaction [62]. The importance of dispersing the active phases (acidic and metallic) on the carbon support is evidenced by an increase in GVL conversion while maintaining similarly high selectivity values to VA from 60% aqueous GVL, reaching a yield of approximately 90% at 598 K and 3.5 MPa. This improvement is due to an increase in the surface acidity of the catalyst, and the interaction between noble metal and cerium oxide is favored, facilitating the hydrogenation of pentenoic acid. A 30% deactivation was observed during the first hours, reaching a steady state until 200 h.

5-nonanone is obtained from valeric acid by a ketonization process, which was first described by Dumesic *et al.* [60]. They demonstrated the process starting from GVL using a dual-bed reactor: the first reactor produces pentanoic acid from GVL using

0.1 wt.% Pd/Nb_2O_5, and then, in the second ceria–zirconia catalytic bed, VA is transformed to 5-nonanone by ketonization at 698 K and 2 MPa. The authors report a final 5-nonanone yield of 84%, with a 6% yield of other lower ketones. Separations during the process were simplified because 5-nonanone is hydrophobic and spontaneously separates from the aqueous solvent.

3.7.4. Main challenges and conclusions

Both, LA and GVL can be used as platform molecules to produce several high value chemicals and biofuels. Thus, they constitute one of the main exponents of the high possibilities that the furfural has as a key product in future biorefineries. However, several challenges remain to commercialize these products. LA can be produced from the cellulosic fraction of the biomass in a single step and so this route attracts most of the attention; However, because the C6 sugars are easier to ferment and do have more possibilities, converting the C5 fraction in the lignocellulosic biomass into valuable products is necessary for a biorefinery to success. To increase the interest in the C5 route to produce LA and GVL, further research is necessary to better understand the reaction mechanism and kinetics involved in the one-pot reaction systems, in a way that LA and GVL can be produced from furfural in a single step increasing the yield and reducing the separation and purification costs. Increasing the stability and activity of the catalysts involve in these one-pot reaction is also critical.

Finally one of the main challenges to unlock the development of furfural based biorefineries is to reduce the cost of producing furfural from the actual biomass. One way to reduce the cost of the feedstock, is to utilize low purity furfural. For example, solution of furfural/water could be used as feedstock eliminating expensive distillation steps to purify furfural. This option implies that the separation of the final products is less costly than the separation of the furfural itself. Another option, when the furfural is produced in organic solvents, is to use the same solvents to perform further transformations, in a way that only a final purification step is necessary. Developing new catalyst that is tolerant to biomass derived impurities, such as lignin or carboxylic acid will be critical not only to produce LA and GVL, but other chemicals.

References

1. J.J. Bozell, G.R. Petersen, Technology development for the production of biobased products from biorefinery carbohydrates — The US Department of Energy's "top 10" revisited, *Green Chemistry*, 12 (2010) 539–554.
2. D.W. Rackemann, W.O. Doherty, The conversion of lignocellulosics to levulinic acid, *Biofuels, Bioproducts and Biorefining*, 5 (2011) 198–214.
3. F.D. Pileidis, M.M. Titirici, Levulinic acid biorefineries: New challenges for efficient utilization of biomass, *ChemSusChem*, 9 (2016) 562–582.
4. Z. Zhang, Synthesis of γ-valerolactone from carbohydrates and its applications, *ChemSusChem*, 9 (2016) 156–171.
5. J.J. Bozell, L. Moens, D.C. Elliott, Y. Wang, G.G. Neuenscwander, S.W. Fitzpatrick, R.J. Bilski, J.L. Jarnefeld, Production of levulinic acid and use as a platform chemical for derived products, *Resources, Conservation and Recycling*, 28 (2000) 227–239.
6. B. Kamm, P.R. Gruber, M. Kamm, *Biorefineries — Industrial processes and products: Status Quo and future directions*, Wiley-VCH Verlag GmbH & Co., Weinheim, 2006.
7. J.A. Dumesic, D.M. Alonso, E.I. Gürbüz, S.G. Wettstein, Production of levulinic acid, furfural, and gamma valerolactone from C5 and C6 carbohydrates in mono- and biphasic systems using gamma-valerolactone as a solvent, US Patent No. US8399688 (2013).
8. http://www.biofuelsdigest.com/bdigest/2016/19/gfbiochemicals-acquires-se ogetis-enters-the-us-market/.
9. http://www.gfbiochemicals.com/company/.
10. J.P. Lange, W.D. van de Graaf, R.J. Haan, Conversion of furfuryl alcohol into ethyl levulinate using solid acid catalysts, *ChemSusChem*, 2 (2009) 437–441.
11. G.M. González Maldonado, R.S. Assary, J. Dumesic, L.A. Curtiss, Experimental and theoretical studies of the acid-catalyzed conversion of furfuryl alcohol to levulinic acid in aqueous solution, *Energy & Environmental Science*, 5 (2012) 6981–6989.
12. M. Otsuka, Y. Hirose, T. Kinoshita, T. Masawa, Manufacture of levulinic acid, US Patent No. US3752849 (1973).
13. B. Capai, G. Lartigau, Preparation of levulinic acid, US Patent No. US5175358 (1992).
14. W.D. Van De Graaf, J.P. Lange, Process for the conversion of furfuryl alcohol into levulinic acid or alkyl levulinate, US Patent No. US7265239 (2007).
15. B. Chen, F. Li, Z. Huang, T. Lu, Y. Yuan, G. Yuan, Integrated catalytic process to directly convert furfural to levulinate ester with high selectivity, *ChemSusChem*, 7 (2014) 202–209.
16. D.M. Alonso, J.Q. Bond, J.A. Dumesic, Catalytic conversion of biomass to biofuels, *Green Chemistry*, 12 (2010) 1493–1513.

17. B.V. Timokhin, Levulinic acid in organic synthesis, *Russian Chemical Reviews*, 68 (1999) 73–84.
18. A. Cukalovic, C.V. Stevens, Feasibility of production methods for succinic acid derivatives: A marriage of renewable resources and chemical technology, *Biofuels, Bioproducts and Biorefining*, 2 (2008) 505–529.
19. I. Bechthold, K. Bretz, S. Kabasci, R. Kopitzky, A. Springer, Succinic acid: A new platform chemical for biobased polymers from renewable resources, *Chemical Engineering and Technology*, 31 (2008) 647–654.
20. Y. Wei, C. Wang, X. Jiang, D. Xue, J. Li, J. Xiao, Highly efficient transformation of levulinic acid into pyrrolidinones by iridium catalysed transfer hydrogenation, *Chemical Communications*, 49 (2013) 5408–5410.
21. J. Zhang, S.B. Wu, B. Li, H.D. Zhang, Advances in the catalytic production of valuable levulinic acid derivatives, *ChemCatChem*, 4 (2012) 1230–1237.
22. E. Christensen, A. Williams, S. Paul, S. Burton, R.L. McCormick, Properties and performance of levulinate esters as diesel blend components, *Energy and Fuels*, 25 (2011) 5422–5428.
23. J.A. Melero, G. Morales, J. Iglesias, M. Paniagua, B. Hernández, S. Penedo, Efficient conversion of levulinic acid into alkyl levulinates catalyzed by sulfonic mesostructured silicas, *Applied Catalysis A: General*, 466 (2013) 116–122.
24. M. Mascal, E.B. Nikitin, High-yield conversion of plant biomass into the key value-added feedstocks 5-(hydroxymethyl)furfural, levulinic acid, and levulinic esters via 5-(chloromethyl)furfural, *Green Chemistry*, 12 (2010) 370–373.
25. C. Chang, G. Xu, X. Jiang, Production of ethyl levulinate by direct conversion of wheat straw in ethanol media, *Bioresource Technology*, 121 (2012) 93–99.
26. S. Saravanamurugan, A. Riisager, Zeolite catalyzed transformation of carbohydrates to alkyl levulinates, *ChemCatChem*, 5 (2013) 1754–1757.
27. A. Démolis, N. Essayem, F. Rataboul, Synthesis and applications of alkyl levulinates, *ACS Sustainable Chemistry & Engineering*, 2 (2014) 1338–1352.
28. J. Dumesic, D. Alonso, J. Bond, T. Root, M. Chia, Method to produce, recover and convert furan derivatives from aqueous solutions using alkylphenol extraction, US Patent No. US8389749 (2013).
29. J.P. Lange, E. van der Heide, J. van Buijtenen, R. Price, Furfural: A promising platform for lignocellulosic biofuels, *ChemSusChem*, 5 (2012) 150–166.
30. P. Neves, M.M. Antunes, P.A. Russo, J.P. Abrantes, S. Lima, A. Fernandes, M. Pillinger, S.M. Rocha, M.F. Ribeiro, A.A. Valente, Production of biomass-derived furanic ethers and levulinate esters using heterogeneous acid catalysts, *Green Chemistry*, 15 (2013) 3367–3376.
31. P. Neves, S. Lima, M. Pillinger, S.M. Rocha, J. Rocha, A.A. Valente, Conversion of furfuryl alcohol to ethyl levulinate using porous aluminosilicate acid catalysts, *Catalysis Today*, 218–219 (2013) 76–84.
32. P. Demma Carà, R. Ciriminna, N.R. Shiju, G. Rothenberg, M. Pagliaro, Enhanced heterogeneous catalytic conversion of furfuryl alcohol into butyl levulinate, *ChemSusChem*, 7 (2014) 835–840.

33. A.M. Hengne, S.B. Kamble, C.V. Rode, Single pot conversion of furfuryl alcohol to levulinic esters and γ-valerolactone in the presence of sulfonic acid functionalized ILs and metal catalysts, *Green Chemistry*, 15 (2013) 2540–2547.

34. Z. Zhang, K. Dong, Z. Zhao, Efficient conversion of furfuryl alcohol into alkyl levulinates catalyzed by an organic–inorganic hybrid solid acid catalyst, *ChemSusChem*, 4 (2011) 112–118.

35. W. Luo, U. Deka, A.M. Beale, E.R.H. Van Eck, P.C.A. Bruijnincx, B.M. Weckhuysen, Ruthenium-catalyzed hydrogenation of levulinic acid: Influence of the support and solvent on catalyst selectivity and stability, *Journal of Catalysis*, 301 (2013) 175–186.

36. Y. Kar, H. Deveci, Importance of P-series fuels for flexible-fuel vehicles (FFVs) and alternative fuels, *Energy Sources, Part A*, 28 (2006) 909–921.

37. S. Bayan, E. Beati, Furfural and its derivatives as motor fuels, *Chimica e Industria*, 23 (1941) 432–434.

38. D.C. Elliott, J.G. Frye, Hydrogenated 5-carbon compound and method of making, US Patent No. US5883266 (1999).

39. F.M.A. Geilen, B.H. Engendahl, A. Harwardt, W. Marquardt, J. Klankermayer, W. Leitner, Selective and flexible transformation of biomass-derived platform chemicals by a multifunctional catalytic system, *Angewandte Chemie, International Edition*, 49 (2010) 5510–5514.

40. J.M. Bermudez, J.A. Menéndez, A.A. Romero, E. Serrano, J. Garcia-Martinez, R. Luque, Continuous flow nanocatalysis: Reaction pathways in the conversion of levulinic acid to valuable chemicals, *Green Chemistry*, 15 (2013) 2786–2792.

41. P.P. Upare, J.-M. Lee, Y.K. Hwang, D.W. Hwang, J.-H. Lee, S.B. Halligudi, J.-S. Hwang, J.-S. Chang, Direct hydrocyclization of biomass-derived levulinic acid to 2-methyltetrahydrofuran over nanocomposite copper/silica catalysts, *ChemSusChem*, 4 (2011) 1749–1752.

42. J.P. Lange, R. Price, P.M. Ayoub, J. Louis, L. Petrus, L. Clarke, H. Gosselink, Valeric biofuels: A platform of cellulosic transportation fuels, *Angewandte Chemie International Edition*, 49 (2010) 4479–4483.

43. G. Dautzenberg, M. Gerhardt, B. Kamm, Bio based fuels and fuel additives from lignocellulose feedstock via the production of levulinic acid and furfural, *Holzforschung*, 65 (2011) 439–451.

44. T. Pan, J. Deng, Q. Xu, Y. Xu, Q.X. Guo, Y. Fu, Catalytic conversion of biomass-derived levulinic acid to valerate esters as oxygenated fuels using supported ruthenium catalysts, *Green Chemistry*, 15 (2013) 2967–2974.

45. P. Sun, G. Gao, Z. Zhao, C. Xia, F. Li, Stabilization of cobalt catalysts by embedment for efficient production of valeric biofuel, *ACS Catalysis*, 4 (2014) 4136–4142.

46. P. Nilges, T.R. Dos Santos, F. Harnisch, U. Schröder, Electrochemistry for biofuel generation: Electrochemical conversion of levulinic acid to octane, *Energy & Environmental Science*, 5 (2012) 5231–5235.

47. T.J. Schwartz, A.R.P. Van Heiningen, M.C. Wheeler, Energy densification of levulinic acid by thermal deoxygenation, *Green Chemistry*, 12 (2010) 1353–1356.

48. P.A. Case, A.R.P. Van Heiningen, M.C. Wheeler, Liquid hydrocarbon fuels from cellulosic feedstocks via thermal deoxygenation of levulinic acid and formic acid salt mixtures, *Green Chemistry*, 14 (2012) 85–89.

49. D.M. Alonso, S.G. Wettstein, J.A. Dumesic, Gamma-valerolactone, a sustainable platform molecule derived from lignocellulosic biomass, *Green Chemistry*, 15 (2013) 584–595.

50. D. Fegyverneki, L. Orha, G. Láng, I.T. Horváth, Gamma-valerolactone-based solvents, *Tetrahedron*, 66 (2010) 1078–1081.

51. K. Yan, Y. Yang, J. Chai, Y. Lu, Catalytic reactions of gamma-valerolactone: A platform to fuels and value-added chemicals, *Applied Catalysis, B: Environmental*, 179 (2015) 292–304.

52. W.R.H. Wright, R. Palkovits, Development of heterogeneous catalysts for the conversion of levulinic acid to γ-valerolactone, *ChemSusChem*, 5 (2012) 1657–1667.

53. M.J. Climent, A. Corma, S. Iborra, Conversion of biomass platform molecules into fuel additives and liquid hydrocarbon fuels, *Green Chemistry*, 16 (2014) 516–547.

54. D.J. Braden, C.A. Henao, J. Heltzel, C.C. Maravelias, J.A. Dumesic, Production of liquid hydrocarbon fuels by catalytic conversion of biomass-derived levulinic acid, *Green Chemistry*, 13 (2011) 1755–1765.

55. L. Bui, H. Luo, W.R. Gunther, Y. Román-Leshkov, Domino reaction catalyzed by zeolites with Brønsted and Lewis acid sites for the production of γ-valerolactone from furfural, *Angewandte Chemie International Edition*, 52 (2013) 8022–8025.

56. I.T. Horváth, H. Mehdi, V. Fábos, L. Boda, L.T. Mika, γ-valerolactone-a sustainable liquid for energy and carbon-based chemicals, *Green Chemistry*, 10 (2008) 238–242.

57. X.-L. Du, Q.-Y. Bi, Y.-M. Liu, Y. Cao, H.-Y. He, K.-N. Fan, Tunable copper-catalyzed chemoselective hydrogenolysis of biomass-derived γ-valerolactone into 1,4-pentanediol or 2-methyltetrahydrofuran, *Green Chemistry*, 14 (2012) 935–939.

58. Y. Zhao, Y. Fu, Q.X. Guo, Production of aromatic hydrocarbons through catalytic pyrolysis of γ-valerolactone from biomass, *Bioresource Technology*, 114 (2012) 740–744.

59. J.Q. Bond, D.M. Alonso, D. Wang, R.M. West, J.A. Dumesic, Integrated catalytic conversion of γ-valerolactone to liquid alkenes for transportation fuels, *Science*, 327 (2010) 1110–1114.

60. J.C. Serrano-Ruiz, D. Wang, J.A. Dumesic, Catalytic upgrading of levulinic acid to 5-nonanone, *Green Chemistry*, 12 (2010) 574–577.

61. H.N. Pham, Y.J. Pagan-Torres, J.C. Serrano-Ruiz, D. Wang, J.A. Dumesic, A.K. Datye, Improved hydrothermal stability of niobia-supported Pd catalysts, *Applied Catalysis A: General*, 397 (2011) 153–162.

62. R. Buitrago-Sierra, J.C. Serrano-Ruiz, F. Rodríguez-Reinoso, A. Sepúlveda-Escribano, J.A. Dumesic, Ce promoted Pd-Nb catalysts for γ-valerolactone ring-opening and hydrogenation, *Green Chemistry*, 14 (2012) 3318–3324.

Chapter 3.8

Amination of Furfural

Pedro Maireles-Torres[*,‡] and Pedro L. Arias[†]

*Universidad de Málaga,
Departamento de Química Inorgánica,
Cristalografía y Mineralogía (Unidad Asociada al ICP-CSIC),
Facultad de Ciencias, Campus de Teatinos,
29071 Málaga, Spain

†Department of Chemical and Environmental Engineering,
Engineering School of the University of the Basque Country
(UPV/EHU), Alameda Urquijo s/n, 48013 Bilbao, Spain

3.8.1. Introduction

The reductive amination of furfural gives rise to furfurylamine (FAM) and tetrahydrofurfurylamine (THFAM), which are colorless to pale yellow liquids, with high boiling points, with an ammonia odour. Their main physical properties are shown in Table 3.8.1.

3.8.2. Production of FAM and THFAM

The hydrogenation of furfural phenylhydrazone, furfural oxime, furfuryl azide or furonitrile can yield FAM, but expensive reagents are required for the preparation of such derivatives from FUR, and complicated reaction steps are involved in their preparation in most of cases [1]. For this reason, the reductive amination of furfural (FUR) over metal catalysts is a more economical process, which allows the conversion of the carbonyl function to an amine. The pathway

‡Corresponding author: maireles@uma.es

Table 3.8.1. Physical properties of furfurylamine (FAM) and tetrahydrofurfurylamine (THFAM).

	B. p. (K)	M. p. (K)	Density (293 K) (g/mL)	Solubility in water (298 K) (wt.%)	Vapor pressure (mm Hg)	Flash point (K)
FAM: C_5H_7ON	428	198	1.057	∞	10 (423 K)	310
THFAM: $C_5H_{11}ON$	426	—	0.98	∞	—	316

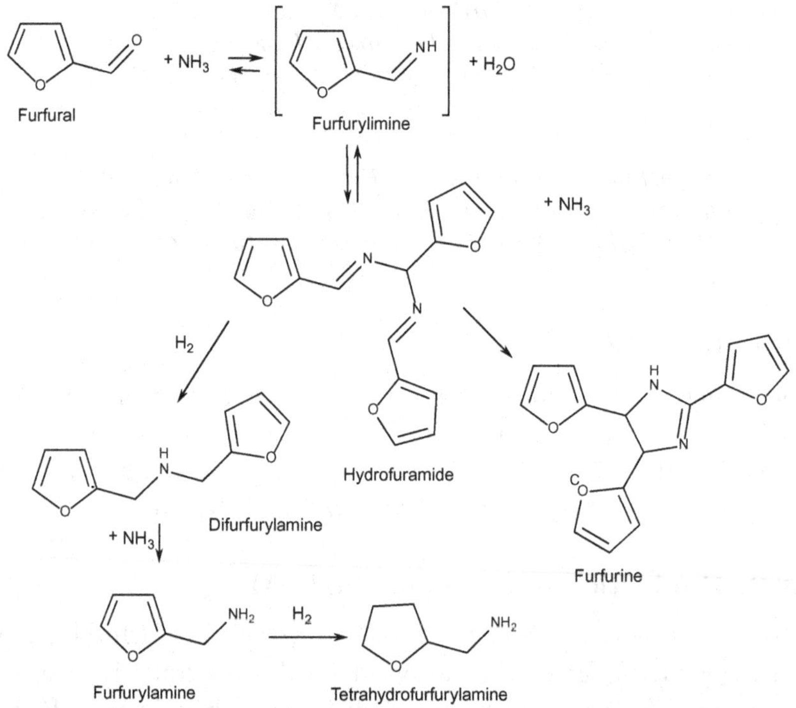

Figure 3.8.1. Conversion of furfural to FAM and THFAM (adapted from Ref. [2]).

proposed is not conventional, the contact between FUR and ammonia leads to the formation of furfurylimine, which is unstable, and readily condensates to give rise to the trimeric hydrofuramide (Figure 3.8.1). The hydrogenation of this trimeric compound yields difurfurylamine which is subsequently aminated to FAM. The formation of furfurine by decomposition of the hydrofuramide, which occurs at temperatures higher than 388 K, must be avoided [1, 2]. The main drawback in the synthesis of FAM is associated with the undesired formation of difurfurylamine and the stable furfurine, and to circumvent it the use of primary amines as co-solvent, an excess of ammonia or gradual admission of FUR during reaction have been proposed.

Thus, FAM was prepared by direct hydrogenation of FUR in the presence of ammonia-saturated cold ethanol, over a Raney nickel catalyst [3]. Nevertheless, FAM yield was not higher than 80%, and generates difurfurylamine as by-product. Nickel supported on a diatomite has been also tested as catalyst, and by using ethanol, ammonia and butyl amine as auxiliary primary amine at 388–393 K and 2.5 MPa H_2, FAM yields higher than 90% can be attained [4]. However, without butyl amine, the yield was only 70%. The use of high H_2 pressure and similar temperatures, over hydrogenating metal catalysts (Co and Ni), has also been reported in the patent literature, although, in some cases, hydrofuramide is used instead of FUR [5, 6].

Other catalytic systems not based on nickel have also been investigated and this is the case of a Co:Re:Mo catalyst with an atomic ratio of 1:0.03:0.015, prepared by impregnation [1]. The reaction is performed in liquid phase, using dioxane as solvent and moderated temperatures, in a batch reactor, at 348 K and 9.0 MPa, a dioxane/FUR weight ratio of 9 and NH_3/FUR molar ratio of 9. A FAM yield of 97% was achieved after 3 h of reaction time. By increasing the temperature until 423 K, with a pressure of 8.0 MPa, a dioxane/FUR weight ratio of 2 and ammonia/FUR molar ratio of 1.3, FAM is hydrogenated to THFAM with a yield of 89% after 8 h of reaction time. In both cases, FUR conversion was 100%.

A poly(N-vinyl-2-pyrrolindone)-capped ruthenium supported hydroxyapatite catalyst has also exhibited a suitable catalytic activity in the reductive amination of furfural to furfurylamine [7]. A 60% yield of FAM in an ammonia aqueous solution (25%) at 0.4 MPa H_2 pressure was achieved at 373 K after 2 h. Compared to the use of other supports, the ruthenium particle sizes were smaller on hydroxyapatite, which explains its better performance. Recently, Chatterjee *et al.* have evaluated the catalytic behavior of Rh- and Ru-based catalysts on different supports, for the reductive amination of furfural to furfurilamina, using an aqueous solution of ammonia and molecular hydrogen [8]. A Rh/Al_2O_3 catalyst allowed to attain a FAM yield of 91.3% at 353 K, 2 MPa H_2 pressure, after 2 h, by using a FUR:catalyst weight ratio of 100 and a FUR/NH_3 molar ratio of 0.03. It was found that the selectivity to FAM mainly depends on the metal nature rather than the support used. The optimization of different experimental variables shed light on the reaction mechanism, in such a way that the hydrogen pressure seems to play a key role in determining the formation of a Schiff base type intermediate, from which furfurylamine is generated via hydrogenolysis. However, the hydrogenation of furfurylimine to FAM emerges as a major pathway, favored at high H_2 pressure, and the conversion of the Schiff base type intermediate, formed between furfurylimine and furfural, to FAM and secondary amines as the minor one. By increasing the H_2 pressure beyond 2 MPa, the formation of THFAM is detected, together with furfuryl alcohol and 2-methylfuran. On the other hand, the reusability of catalysts must be improved, since ammonia tolerance of the metal catalyst remains a challenging issue. Nevertheless, the excellent performance and versatility of this catalyst is proven in the reductive amination of a large variety of aldehydes besides furfural.

Concerning kinetic data and the mechanism of reductive amination of FUR on the active sites of catalysts employed in this catalytic process, there is no information in the literature. Similarly, deactivation and reutilization studies are also absent.

3.8.3. Applications of FAM and THFAM

Furfurylamine and tetrahydrofurfurylamine are important chemicals with a broad spectrum of applications, such as the synthesis of herbicides, pesticides, fibers, piperidine derivatives and pharmaceuticals [2]. In this sense, FAM is a key intermediate for the production of the diuretic furosemide, which can be accomplished by treatment of 2,4-dichloro-5-sulfamoylbenzoic acid with excess furfurylamine in the absence of solvent at 393–403 K [9]. Moreover, FAM has been found to have unique properties in engine cleaning formulations [10, 11]. The formation of THFAM has also been proposed as protection and recovery method of plant tissue from damage upon exposure to chilling temperatures [12,13]. These hydrogenated derivatives of FUR enter in the composition employed for removing polymer residue of a photosensitive etching-resistant layer, being both environmentally friendly [14]. Among the derivatives of FAM, piperidine, used as solvent and in the pharmaceutical industry, can be obtained by catalytic hydrogenation of FAM or THFAM in the presence of cobalt-based catalysts in an organic solvent [1,15]. Furfurylamine has been also proposed for blending with gasoline, due to its blending research octane number of 194 [16], giving rise to a superior antiknock activity.

On the other hand, the oxidation of FAM with hydrogen peroxide is the last synthesis step for the conversion of pentoses into 3-pyridinol, which is an important intermediate for the synthesis of herbicides, insecticides and the cholinergic drugs of the pyridostigmine type [9].

Finally, optically active FAM derivatives have attracted the interest of organic chemists as building blocks for the preparation of a large spectrum of nitrogen containing natural products [17], and substituted THFAM was found to possess potent antidepressant activity in animals [18].

References

1. T. Ayusawa, S. Mori, T. Aoki, R. Hamana, Process for producing furfurylamine and/or tetrahydrofurfurylamine, US4598159, 1986.
2. H.E. Hoydonckx, W.M. Van Rhijn, W.M. Van Rhijn, D.E. De Vos, P.A. Jacobs, *Furfural and Derivatives*, Wiley-VCH Verlag GmbH & Co. KGaA, Weinheim, 2012.
3. C.F. Winans, Process for preparing furfurylamines, US2109159, 1938.
4. A. Bouniot, Process for preparing primary furfurylamines, US3812161, 1974.
5. C. Sly, Hydrogenation of hydrofuramide to furfurylamine, US2112715, 1938.
6. H. Adkins, Preparation of furfurylamines, US2175585, 1939.
7. S. Nishimura, K. Mizuhori, K. Ebitani, Reductive amination of furfural toward furfurylamine with aqueous ammonia under hydrogen over Ru-supported catalyst, *Research on Chemical Intermediates*, 42 (2016) 19–30.
8. M. Chatterjee, T. Ishizaka, H. Kawanami, Reductive amination of furfural to furfurylamine using aqueous ammonia solution and molecular hydrogen: An environmentally friendly approach, *Green Chemistry* 18 (2016) 487–496.

9.   A.L. Harreus, *Ullmann's Encyclopedia of Industrial Chemistry*, Wiley-VCH, Weinheim, 2002.
10.   L.J. Adams, T.R. Fruda, P.D. Hughett, Composition for cleaning an internal combustion engine, US5340488, 1994.
11.   L.J. Adams, P.D. Hughett, Engine cleaner composition, method and apparatus with acetonitrile, US5858942, 1999.
12.   C.C. Shin, Methods for protection and treatment of plants exposed to chilling temperatures, US5244864, 1993.
13.   C.C. Shin, N.A. Favstritsky, A. Dadgar, Cryoprotectant composition, US5276006, 1994.
14.   H.S. Choi, D. Kim, Composition for removing polymer residue of photosensitive etching-resistant layer, US7858572 B2, 2010.
15.   S. Mori, T. Aoki, R. Hamana, Y. Nomura, Process for the production of piperidine, US4605742, 1986.
16.   V.E. Tarabanko, M.Y. Chernyak, I.L. Simakova, K.L. Kaigorodov, Y.N. Bezborodov, N.F. Orlovskaya, Antiknock properties of furfural derivatives, *Russian Journal of Applied Chemistry*, 88 (2015) 1778–1782.
17.   W.-S. Zhou, Z.-H. Lu, Y.-M. Xu, L.-X. Liao, Z.-M. Wang, Synthesis of optically active α-furfuryl amine derivatives and application to the asymmetric syntheses, *Tetrahedron*, 55 (1999) 11959–11983.
18.   I. Monkovic, Y.G. Perron, R. Martel, W.J. Simpson, J.A. Gylys, Substituted tetrahydrofurfurylamines as potential antidepressants, *Journal of Medicinal Chemistry*, 16 (1973) 403–407.

Chapter 3.9

On the Oxidation of Furfural to Furoic Acid

Michela Signoretto* and Federica Menegazzo
*CATMAT Lab, Department of Molecular Sciences and Nanosystems,
Ca' Foscari University Venice and INSTM Consortium RU Ve,
Via Torino 155, 30172 Venezia Mestre, Italy*

3.9.1. Introduction

Exploring the sub-products from furfural, which can be obtained from biomass processing, as the replacements of the fossil resources is greatly attractive. In fact furan derivatives such as furfural, obtained by the acid-catalyzed dehydration of carbohydrates, has been described as key substances that bridge petroleum based chemistry to carbohydrate industrial chemistry because of the wide range of chemical intermediates and end products that can be produced from it. The potential variety of industrial applications is of enormous interest for the development of future biorefineries. In contrast to biofuels, transition from petroleum to biomass feedstocks for the production of chemicals seems to be more feasible, although this will require to overcome some significant technological challenges.

Among many other promising options, an important valorization of the biomass-derived furanics is their selective oxidation to a number of highly functionalized carboxylic acids, dialdehydes, hydroxyl acids, which may well serve as monomers for innovative biobased products, such as polymers or resins.

*Corresponding author: miky@unive.it

Overall, the oxidation of furfural is challenging due to the high reactivity of both substrate and formed products and intermediates. Oxidative conditions must be strong enough to allow oxidations to occur, but mild enough to secure an acceptable selectivity and a negligible by-products formation. Catalytic oxidation has been used to convert furfural to many chemical intermediates and end products. Most of the current studies focus on the selective oxidation of furfural to diacids or acid anhydrides, especially to: furoic acid, fumaric acid [1, 2], succinic acid [2–8], maleic acid [9–13], maleic anhydride [6, 10, 14–35]. Another interesting example of furfural oxidation is its transformation to 1,4-butanediol [36].

The present review is focused on the transformation from furfural to 2-furoic acid, which is the first down-line oxidation derivative of furfural.

FURFURAL   2-FUROIC ACID

3.9.2. 2-Furoic acid

2-Furoic acid (furan-2-carboxylic acid, pyromucic acid; 2-furan-carboxylic acid; α-furancarboxylic acid; α-furoic acid; 2-carboxy-furan) is a heterocyclic acid, consisting of a five-membered aromatic ring and a carboxylic acid group. It was first described by Scheele in 1780 as the first derivative of the compound furan. Its name is derived from the Latin word *furfur*, meaning bran [37]. 2-Furoic acid is a white to light brown solid at room temperature. Its molecular formula is $C_5H_4O_3$, its average mass 112.084 Da, its melting point is between 128–132°C, its boiling point between 230°C and 232°C and its solubility in water is 4 g/100 g (at 25°C). It has a distinct odor described as sweet, oily, herbaceous and earthy [38].

It has market applications in the pharmaceutical, agrochemical, fragrance, flavor industries [39], and mainly in food manufacturing sector. In particular, 2-furoic acid is most widely found in food products as a preservative, acting as a bactericide and fungicide. It is also considered an acceptable flavoring ingredient and therefore it is often used as a starting material for furoate esters productions. It is an intermediate in pharmaceutical industry and it is generally converted to furoyl chloride to be used in the production of drugs and insecticides. Other uses for 2-furoic acid include nylon preparation, biomedical research, optic technologies. In fact, 2-furoic acid crystals can be used in optical devices due to its favorable properties such as non-linear optical material.

In terms of demand, North America was the leading region for 2-furoic acid market. The demand is huge due to growing consumption from food industry. North America was followed by Europe owing to demand of 2-furoic acid in medical applications and optic technology. Asia Pacific is expected to exhibit higher demand in the near future due to various manufacturing activities.

3.9.3. Production of 2-furoic acid

Furoic acid is industrially produced by the Cannizzaro disproportionation reaction of furfural in aqueous sodium hydroxide solution [40, 41]. In particular, the process starts with a reaction between furfural and the base to yield furfuryl alcohol and sodium 2-furancarboxylate (reaction 1). After removal of furfuryl alcohol, sodium 2-furancarboxylate is acidified with sulfuric acid to yield furoic acid and sodium hydrogen sulfate (reaction 2). The addition of sulfuric acid is required to neutralize the solution and to yield precipitated furoic acid crystals.

$$2C_5H_4O_2 + NaOH \longrightarrow C_5H_6O_2 + C_5H_3O_3Na \quad \text{(reaction 1)}$$

$$C_5H_3O_3Na + H_2SO_4 \longrightarrow C_5H_4O_3 + NaHSO_4 \quad \text{(reaction 2)}$$

The Cannizzaro reaction is highly exothermic and therefore the temperature of the process must be carefully controlled. The furfural is cooled to 2°C before the sodium hydroxide solution is added. Such addition must be carried out slowly under vigorous stirring in order to keep the temperature lower than 20°C. Since the Cannizzaro reaction is slow, after completion of sodium hydroxide addition the stirring must be continued for at least one hour. In the course of the reaction, considerable quantities of sodium 2-furancarboxylate precipitate in the form of scale-like crystals. These are separated by filtration and then washed with acetone. The dry furoate salt is dissolved in a minimum amount of water and then treated with sulfuric acid to form raw furoic acid and sodium hydrogen sulfate [39]. The triple point pressure of furoic acid is high (10.3 torr) and as the impurities (polymers of furfuryl alcohol) are essentially nonvolatile. Therefore the purification of the raw furoic acid is best carried out by sublimation. Thus, passing a hot carrier gas over the raw furoic acid selectively vaporizes the desired compound while leaving the nonvolatile impurities behind. According to the reactions (1) and (2), if there were no losses due to unwanted byproducts, the theoretical yield of furoic acid amounts to 58% of the furfural input [39]. Alternative processes are not actually viable on an industrial scale.

3.9.4. Furfural chemo-oxidation

As regard as literature investigations, 2-furoic acid is mainly obtained by oxidation of furfural. However, a straight oxidation of furfural with oxygen over a catalyst is very difficult as the furfural undergoes not only oxidation to furoic acid but also competitive nucleus oxidation resulting often in ring cleavage and unwanted by-products [39].

A series of chemical oxidants have been developed for this reaction [42, 43], such as chromium species [41, 44–47], chlorite [48–50], $KMnO_4$ [51, 52], MnO_2 [53], NaOCl [54], hydrogen peroxide [55–57] and molecular oxygen [58–63].

Furfural can be oxidized [46] to the corresponding carboxylic acid with good yield (85%) with some "non-aqueous" chromium (V) complexes such as $(Phen)H_2CrOCl_5$, $(Phen)CrOCl_3$ (Phen = 1,10-phenonthrolin), $(Bipy)H_2CrOCl_5$ and $(Bipy)CrOCl_3$ (Bipy = α, α'-bipyridyl) The oxidation takes place very readily (0.5 h) at room temperature in CH_2Cl_2 under anhydrous conditions.

The same reaction was investigated [45] using a chromium(VI) reagent (quinolinium dichromate, QDC, $(C_9H_7N^+H)_2Cr_2O_7^{2-}$) in sulfuric acid and a kinetic study was performed. The reactions were carried out under a nitrogen atmosphere, in acidic medium [50% (v/v) acetic acid–water] to study the effect of dielectric constant on the rates of the reactions. The reaction mixtures remained homogeneous in the solvent systems used. The oxidation of 2-furaldehyde by QDC resulted in the formation of the corresponding acid, without further oxidation.

The dichromate ion was suggested to be the predominant species in these oxidation reactions and the stoichiometry of the reaction conforms to the following overall equation:

$$3C_5H_4O_2 + 2Cr^{VI} + 3H_2O \longrightarrow 3C_5H_4O_3 + 2Cr^{III} + 6H^+$$

The mechanistic pathway involves the formation of the ester of the aldehyde hydrate, followed by its slow oxidative decomposition.

Some authors [49] addressed their attention to the use of the inexpensive sodium chlorite, which reacts with aldehydes under very mild conditions to give carboxylic acids:

$$RCHO + HClO_2 \longrightarrow RCO\,OH + HOCl$$

However, hypochlorite ion must be removed in order to avoid side reactions, since the redox pair $HOCl/Cl^-$ is a more powerful oxidant than $ClO_2^-/HOCl$. Another drawback is the oxidation of ClO_2^- to ClO_2 according to

$$HOCl + 2ClO_2^- \longrightarrow 2ClO_2 + Cl^- + OH^-$$

As HOCl scavenger, they used 35% H_2O_2, which reduces HOCl without formation of organic side products according to

$$HOCl + H_2O_2 \longrightarrow HCl + O_2 + H_2O$$

Best reaction conditions were achieved by working in a weakly acidic medium, where oxidation was rapid with no competitive reduction of $HClO_2$ to HOCl. Under these conditions, any chlorine dioxide is reduced to chlorous acid. Under optimized conditions furfural gives mainly 2-furoic acid (82% isolated yields) with small quantities of maleic acid as a side product [49].

Other methods reported to obtain 2-furoic acid involve oxidation of furfural using strong oxidative reactants such as $KMnO_4$ [51, 52], MnO_2 [53], or sodium hypochlorite [54].

Unfortunately, the given oxidizing agents represent a safety risk and none of the former syntheses can be considered green syntheses.

Investigations have been aimed at achieving selective oxidation to furoic acid with cheaper and less polluting and toxic oxidative reactants such as hydrogen peroxide and molecular oxygen, that is by an eco-friendly method.

Some patents reported for example the selective oxidation of furfural to furoic acid by using hydrogen peroxide as oxidant. Reactions occurs in the presence of bases such as secondary and tertiary amines or pyridine, giving yields between 40% and 80% [55–57].

A number of catalysts based on noble metals supported on different metal oxides have been explored using molecular oxygen as oxidant. In 1946, Dunlop claimed in a patent [58] a process for producing furoic acid from furfural which consists of suspending furfural and the catalyst in an aqueous medium maintained in an alkaline reacting condition while supplying an excess of oxygen. Furfural is oxidized to a salt of furoic acid and then acidified to obtain furoic acid. Alternatively, the process is featured by incrementally introduction of furfural into an aqueous solution of an alkaline material containing the catalyst while continuously blowing an oxygen containing gas through the suspension and maintaining the alkalinity of the solution by incremental addition of an alkaline reacting material. Furfural is oxidized to furoic acid and the latter combined with the alkaline material to form the furoate. The process operated at temperatures in the range 35–100°C, preferably at 50–55°C. The catalyst is made in major proportion of a base metal oxide (such as copper, iron, nickel, cobalt, titanium, cerium, thorium, bismuth and antimony) and a minor proportion of a noble metal oxide (such as silver, gold, platinum and palladium). The life of such a catalyst is claimed unlimited due to its continuous regeneration by intimate admixture with oxygen. The amount of oxidized furfural is therefore dependent upon the amount of oxygen containing gas used and not upon the quantity of catalyst used.

More recently, 2-furoic acid was obtained from furfural by oxidation of aqueous alkaline medium with molecular oxygen under mild conditions over a platinum catalyst supported on active carbon [59]. Furfural was purified by distillation before use. During a typical catalytic test, an aqueous solution of furfural and the catalyst were introduced in the reactor. At the beginning of the reaction, oxygen was supplied to the reactor and the pH was adjusted by addition of 2 M sodium hydroxide. When the reaction is carried out without adjusting the pH value, it stops when the pH becomes too acid. Increasing the pH, by progressively adding aqueous NaOH, leads to an increase of catalytic activity. However, above pH $= 10$ the selectivity decreases due to the formation of furfuryl alcohol, indicating that furfural is subjected to the Cannizzaro reaction. Therefore the best pH was found to be in the range between 8 and 9. The authors investigated the effects of the main reaction parameters. Proper catalytic ratio (2%), temperature (65°C) initial concentration of furfural (0.25 M) were determined. The authors verified the effect of promoters and found that $Pb(OAc)_2$ addition to the 5%Pt/C increased furfural conversion from 44 to 88% and increased selectivity to furoic acid from 46 to 92%. Best results were obtained with a co-catalytic ratio of 15%. Also $(BiO)_2CO_3$ addition was investigated but with lower results. The lead/platinum on carbon catalytic system was not subjected to deactivation and it can be reused 10 times in a discontinuous system. In order to maintain a high catalytic activity is better to keep the concentration of furfural constant in the reaction mixture. Therefore the authors concluded that developing the process into a continuous one with constant addition of furfural and progressive extraction of the reaction mixture lead to an improvement of the performance.

Taking into account the obtained results a reaction scheme was proposed too. The authors proposed that the aldehyde is first hydrated and the Pb^{2+} ions act as adsorption sites for the hydrated aldehyde species. Being in an ionic form, Pb^{2+} is softer than Pt which is essentially present in a reduced form. The π-electrons of the furanic ring are involved in the chemisorption: reorganization of the adsorbed hydrate results in the formation of furoic acid chemisorbed on the lead sites; the platinum atoms act as a sink for hydride ions liberated during rearrangement. Furoic acid desorption is facilitated at basic pH through the formation of the sodium furoate. Chemisorbed hydride reacts with oxygen, giving rise to OH^-, which yields water with the H^+ released during the rearrangement of the hydrated aldehyde. To prevent catalyst deactivation by furoate chemisorption, a strong base is required to maintain a high pH that results in the formation of soluble furoate species.

The same authors subsequently optimized the reaction conditions [60]. They studied the influence of five factors using a Doehlert matrix and determined the optimal operating conditions:

Platinum mass/furfural mass × 100: 1.7%
Lead diacetate mass/platinum on carbon mass: 20%
Initial furfural concentration: 0.8 M
Temperature: 55°C

After 1 h of reaction they obtained that the whole of the furfural is selectively transformed into furoic acid. However unfortunately the addition of base is a serious drawback for the environmental sustainability of the process.

Silver and copper oxides mixtures were also investigated as catalysts for furfural oxidation to 2-furoic acid [61–63].

The highest selectivity (96%) was achieved with the use of a $Ag/CuO–CeO_2$ catalyst (50°C). This catalyst could be reused for an unlimited number of runs, but accidental deactivation of the catalyst was possible, requiring rejuvenation by transfer of the catalyst to an alkaline medium and passing oxygen through the medium.

More recently, some authors studied catalytic oxidation of furfural to furoic acid on CuO and Ag_2O/CuO samples [63]. The latter always gave better results, but CuO has many special advantages, such as low cost, easy preparation and regeneration and therefore it represents a better choice for an industrial production. The best temperature for the reaction was found to be 70°C, while higher temperatures would decrease the adsorption of oxygen and give faster side reactions. The acid can react with the catalyst to form corresponding silver and copper carboxylates which can inhibit the catalytic reaction. Therefore, it is essential to neutralize the formed acid. Meanwhile, alkali will promote the undesired Cannizzaro reaction of aldehydes. The optimal pH value was found to be 13. Surprisingly, when deionized water was replaced with running water, no furoic acid could be obtained. The authors verified that Ca^{2+} inhibited the reaction and Mg^{2+} just made the product darker while Na^+, K^+, Cu^{2+} and Fe^{3+} did little damage. A proper size of catalyst must also be used. The desired oxidation occurs on a 150 nm CuO and Ag_2O/CuO. On the contrary, a furfural Diels–Alder adduct was obtained on a 30 nm CuO and Ag_2O/CuO, since they catalyze the Diels–Alder reaction more efficiently than the oxidation of aldehydes.

Under optimized conditions, the obtained yields of furoic acid were 87% and 92%, respectively for CuO and Ag_2O/CuO. The main issue of the process was however the aging of the catalyst. In fact, the recycled catalyst cannot be used directly because some organic matter was adsorbed on the surface and part of the CuO had been reduced to Cu_2O and Cu. The recycled catalyst must be first washed with hot deionized water to get rid of most organic adsorbates and then calcined again to oxidize the reduced Cu_2O and Cu to CuO.

An alternative to direct furfural oxidation is the oxidative esterification to methyl furoate in the presence of oxygen and methanol. Methyl furoate can provide furoic acid through further hydrolysis of the ester. The same alkyl furoates (methyl or ethyl) find applications as flavor and fragrance component in the fine chemical industry. Traditionally, the furoate ester is prepared by oxidizing furfural with potassium permanganate, preferably using acetone as solvent, and reacting the furoic acid so formed with methyl or ethyl alcohol, in the presence of sulfuric acid. It has been shown that furfural can be converted into methyl furoate by oxidative esterification in the presence of a base ($NaCH_3O$) in CH_3OH under mild conditions on a Au/TiO$_2$ reference catalyst purchased by the World Gold Council (WGC) [64]. Gold has always been able to fascinate humanity and it plays a central role in the modern society. However, unlike other noble metals, gold had never been considered for catalysis due to its chemical inertness. Only since 1987, gold has been shown to be highly active if deposited as nanoparticles over oxidic supports [65]. In particular, the first studies have demonstrated that surface adsorption and reactivity of gold can be enhanced by creating defective surface structures through downsizing of gold nanoparticles. When gold nanoparticles, with sizes of less than about 5 nm, are supported on oxides, very active catalysts are produced. Nanoparticulated gold catalysts are active under mild conditions, even at ambient temperature or below, and this feature makes them unique. In particular, nanodispersed Au has been recognized as a very good catalyst for selective oxidations with molecular O_2 [66].

Kegnæs *et al.* [67] found that furfural can be oxidized to the corresponding methyl ester in high yields when supported nanoparticles (Au/TiO$_2$ with gold particles having 4–8 nm size) and a base (KOMe) were used as catalysts. Deng *et al.* [68] studied pyrolysed complexes with a noble-metal-free approach and explored different metals (such as Co, Mn) and ligands (such as 1,10 phenantroline and 2,2-dipyridine) to investigate the optimum catalytic efficiency in the esterification of furfural in methanol. High yield and selectivity were obtained under optimized conditions in the presence of the CoxOy-N@C catalyst and by adding a base. A one-pot process involving the oxidative condensation and the hydrogenation of furfural with aliphatic alcohols catalyzed by metallic platinum has also been reported [69]. The authors examined several typical supported Pt catalysts and bases as catalysts for the reaction and found best results when working in the presence of 5% Pt/HT (HT = hydrotalcite Mg/Al ratio = 3) and of potassium carbonate either in the furfural–ethanol–O$_2$ system, or in the furfural–n-propanol–O$_2$ one.

However, furfural esterification can be efficiently carried out over Au-based catalyst even without using a base as co-catalyst. Such systems appear to be the most promising catalyst for the production of alkyl furoates.

A base-free synthesis of methyl furoate on gold catalysts was firstly reported by Corma and co-workers [70]. The authors investigated a Au/CeO$_2$ catalytic system and reported an excellent oxidation activity even if increasing both temperature and pressure conditions with respect the activity test with the base.

Very good catalytic performances were observed by Signoretto *et al.* [71] over a gold-supported sulfated zirconia catalyst, especially if compared with the Au/TiO_2 reference catalyst. It was proposed that the enhanced activity was due to the presence of Au clusters able to dissociate O_2 producing atomic O with basic properties that could activate CH_3OH. Afterwards a series of Au/ZrO_2 catalysts calcined at different temperature (from 150 up to 650°C) were investigated in order to modulate the size of the gold nanoparticles, demonstrating that it is required for good catalytic performances the presence of highly dispersed gold clusters able to activate atomic oxygen [72]. In particular, oxygen is activated on the Au clusters and the reaction proceeds through direct oxidation of the substrate into the desired product. Moreover, the stability of these new catalysts was studied. It is possible to completely recover the catalytic performances only if the organic residue of the exhausted sample is removed from both gold and zirconia sites. Such results suggested that also the support plays a role in the furfural esterification reaction. Therefore, different oxidic supports commonly used in catalysis were examined: titania, ceria and zirconia [73]. It has been found [74] that both chemical and morphological properties of the samples, such as (i) high dispersion of Au, (ii) specific surface area of the support, and (iii) proper surface sites on the support itself, influence the esterification processes. While the surface area of the support influenced the conversion, surface sites affected the selectivity in the process. In particular, zirconium proved to be the ideal support for Au based oxidation of furfural as it is active, selective, recyclable, and applicable in a biomass based renewables producing industry.

The preparation of Au/CeO_2 was optimized too [75]. Very efficient Au/CeO_2 catalysts for furfural oxidative esterification were prepared by deposition of gold colloids using polyvinyl alcohol as protective agent [76]. Sol-immobilized catalysts do not require any preliminary calcination to be activated and can be recovered by simple filtration: no oxidation of the exhausted catalyst is required for at least six catalytic runs. Their activity is due to the presence of gold nanoparticles and to the ceria capability to supply activated oxygen.

As regard as reactions, many variables of the processes, such as reaction time, temperature, pressure, nature of the oxidant, have been optimized [77]. A considerable effect of the reaction temperature has been evidenced in the range here investigated (60–140°C). Oxygen pressure can be lowered without significant changes in the catalytic performances and molecular oxygen can be replaced by the more economic air, still at very low relative pressure.

3.9.5. Other techniques

An alternative to chemo-oxidation is the furfural photo-oxidation to furoic acid. It has been tested on a photogenerated Fe catalyst derived from iron organometallic complexes at room temperature. The process has been explored in methanol in the presence of hydrogen peroxide. The catalyst can be recovered from the reaction mixture by a precipitation and recrystallization protocol, and can be reused three times without deactivation [78].

Another alternative to chemo-oxidation is the electrochemical oxidation to furoic acid. However such electrochemical process is found to result in a competitive decomposition of furoic acid as the oxidation potentials of furoic acid and furfural are extremely close, so that it is not possible to selectively produce furoic acid by this route. Belgsir and co-workers demonstrated the simultaneous electrosynthesis of furoic acid and furfuryl alcohol from aqueous furfural by performing cyclic voltammetry and long-term preparative electrolysis experiments on noble (Au and Pt) and non-noble (Pb, Cu and Ni) metal electrodes [79, 80]. According to the authors, an 80% yield of furoic acid was achieved through the electrooxidation of furfural, and a 55% yield of furfuryl alcohol was obtained by electroreduction.

Some strategies using biotransformation and enzymes have also been reported: resting cells of *Nocardia corallina* displayed the oxidation of furfural to furoic acid [81]. Similarly, some enzymatic methods involving *oxidases* and *chloroperoxidases* have been evaluated [82,83]. A particular peracid-based route for the oxidative valorization of furfural was also proposed. Some authors [84] successfully explored the use of lipases as biocatalysts for the *in situ* production of organic peracids, which may subsequently perform the selective oxidation of biomass-derived furanics and finally afford a range of useful building blocks. In such investigation, peracids were generated *in situ* in catalytic amounts using lipases as biocatalysts and alkyl esters as acyl donors upon addition of aqueous hydrogen peroxide under very mild reaction conditions. The biocatalytic promiscuity of lipases enables them to accept hydrogen peroxide as a nucleophile (instead of water or alcohols) in nonaqueous solutions, affording organic peracids. Subsequently, these peracids are able to mildly (40°C) oxidize furfural to obtain furoic acid in moderate to excellent yields. For example, reactions conducted in neat ethyl acetate, acting both as acyl donor and solvent, led to lower yields in furoic acid than processes performed with a mixture of tert-butanol and ethyl acetate. Moreover, further addition of higher amounts of oxidant led to even higher yields in furoic acid ($5\,\mathrm{gL}^{-1}$, 91%) with excellent selectivity (100%).

3.9.6. Conclusions and future challenges

The highly functionalized molecular structure of furfural makes it a desired raw material for the sustainable production of value-added chemicals containing oxygen atoms. Catalytic oxidation has been used to convert furfural to many chemical intermediates and end products. The first down-line oxidation derivative of furfural is 2-furoic acid. The latter has market applications in food, pharmaceutical, agrochemical, fragrance and flavor industries. Other uses for 2-furoic acid include nylon preparation, biomedical research, optic technologies. 2-Furoic acid is industrially produced by the Cannizaro disproportionation reaction of furfural in aqueous sodium hydroxide solution. According to what is reported in literature investigations, 2-furoic acid is mainly obtained by chemo-oxidation of furfural. Other techniques, such as photo-oxidation, electrochemical oxidation and enzymatic methods have been reported too. A series of reagents have been developed for this reaction, such as chromium species, chlorites, $KMnO_4$, MnO_2, $NaOCl$, H_2O_2 and O_2. A number of catalysts based on noble metals supported on different metal oxides have been investigated. However, the main issue of the process is the stability and recyclability of the catalysts, mainly due to organic matter adsorbed on the surface. In order to find application in a large-scale production and to make it a really sustainable one, the process needs further optimization, starting from the composition and the microstructure of the catalyst.

References

1. C.F. Cross, E.J. Bevan, F. Briggs, Einwirkung des caro'schen reagens auf furfurol, *Chemische Berichte*, 33 (1900) 3132–3138.
2. Y. Tachibana, T. Masuda, M. Funabashi, M. Kunioka, Chemical synthesis of fully biomass-based poly(butylene succinate) from inedible-biomass-based furfural and evaluation of its biomass carbon ratio, *Biomacromolecules*, 11 (2010) 2760–2765.
3. M. Taniyama, *Toho Reiyon Kenkyu Hokoku*, 1 (1954) 40–46.
4. V.I. Krupenskii, *Nauchn. Tr. -Leningr. Lesotekh. Akad. Im. S. M. Kirova*, 158 (1973) 68–71.
5. V.I. Krupenskii, *Russian Journal of General Chemistry*, 66 (1996) 1874–1875.
6. E.P. Grunskaya, L.A. Badovskaya, V.V. Poskonin, Y.F. Yakuba, Catalytic oxidation of furan and hydrofuran compounds. Oxidation of furfural by hydrogen peroxide in the presence of sodium molybdate, *Chemistry of Heterocyclic Compounds*, 34 (1998) 775–780.
7. H. Choudhary, S. Nishimura, K. Ebitani, Highly efficient aqueous oxidation of furfural to succinic acid using reusable heterogeneous acid catalyst with hydrogen peroxide, *Chemistry Letters*, 41 (2012) 409–411.
8. H. Choudhary, S. Nishimura, K. Ebitani, Metal-free oxidative synthesis of succinic acid from biomass-derived furan compounds using a solid acid catalyst with hydrogen peroxide, *Applied Catalysis A*, 458 (2013) 55–62.
9. E.R. Nielsen, Vapor phase oxidation of furfural, *Industrial Engineering Chemistry*, 41 (1949) 365–368.

10. S. Shi, H. Guo, G. Yin, Synthesis of maleic acid from renewable resources: Catalytic oxidation of furfural in liquid media with dioxygen, *Catalysis Communications*, 12 (2011) 731–733.

11. H. Guo, G. Yin, Catalytic aerobic oxidation of renewable furfural with phosphomolybdic acid catalyst: An alternative route to maleic acid, *Journal of Physical Chemistry C*, 115 (2011) 17516–17522.

12. N.A. Milas, W.L. Walsh, Catalytic oxidations. I. oxidations in the furan series, *Journal of the American Chemical Society*, 57 (1935) 1389–1393.

13. S. Shi, H. Guo, G. Yin, Synthesis of maleic acid from renewable resources: Catalytic oxidation of furfural in liquid media with dioxygen, *Catalysis Communications*, 12 (2011) 731–733.

14. V.V. Guliants, M.A. Carreon, Vanadium-phosphorus-oxides: From fundamentals of *n*-Butane oxidation to synthesis of new phases, *Catalysis*, 18 (2005) 1–45.

15. G. Centi, F. Trifiro, J.R. Ebner, V.M. Franchetti, Mechanistic aspects of maleic anhydride synthesis from C4 hydrocarbons over phosphorus vanadium oxide, *Chemical Reviews*, 88 (1988) 55–80.

16. W.V. Sessions, Catalytic oxidation of furfural in the vapour phase, *Journal of the American Chemical Society*, 50 (1928) 1696–1698.

17. E. Nielsen, Catalytic oxidation of furfural, US 2421428 A (1947).

18. D.R. Kreile, V.A. Slavinskaya, M.V. Shimanskaya, E.Y. Lukevits, The reactivity of furan compounds in vapor-phase catalytic oxidation, *Chemistry of Heterocyclic Compounds*, 5 (1972) 429–430.

19. M.S. Murthy, K. Rajamani, Kinetics of vapour phase oxidation of furfural on vanadium catalyst, *Chemical Engineering Science*, 29 (1974) 601–609.

20. K. Rajamani, P. Subramanian, M.S. Murthy, Kinetics and mechanism of vapor phase oxidation of furfural over tin vanadate catalyst, *Industrial and Engineering Chemistry Process Design and Development*, 15 (1976) 232–234.

21. V.A. Slavinskaya, D.R. Kreile, E. Dzilyuma, D. Sile, Incomplete catalytic oxidation of furan compounds, *Chemistry of Heterocyclic Compounds*, 13 (1977) 710–721.

22. H. Guo, G. Yin, Catalytic aerobic oxidation of renewable furfural with phosphomolybdic acid catalyst: An alternative route to maleic acid, *Journal of Physical Chemistry C*, 115 (2011) 17516–17522.

23. N. Alonso-Fagfflndez, M. Lopez Granados, R. Mariscal, M. Ojeda, Selective conversion of furfural to maleic anhydride and furan with VO_x/Al_2O_3 catalysts, *ChemSusChem*, 5 (2012) 1984–1990.

24. J. Lan, Z. Chen, J. Lin, G. Yin, Catalytic aerobic oxidation of renewable furfural to maleic anhydride and furanone derivatives with their mechanistic studies, *Green Chemistry*, 16 (2014) 4351–4358.

25. A. Hashem, E. Kleinpeter, The chemistry of 2(5H)-furanones, *Advances in Heterocyclic Chemistry*, 81 (2001) 107–165.

26. X. Li, X. Lan, T. Wang, Selective oxidation of furfural in a bi-phasic system with homogeneous acid catalyst, *Catalysis Today*, 276 (2016) 97–104.

27. J.W.E. Glattfeld, G. Leavell, G.E. Spieth, D. Hutton, The C4-Saccharinic acids. The preparation of 2,3 dihydroxybutyric acid lactone.

3-hydroxyisocrotonic acid lactone. An attempt to prepare 2,2′-dihydroxyiso-butyric acid, *Journal of the American Chemical Society*, 53 (1931) 3164–3171.

28. C.C. Price, J.M. Judge, γ-crotonolactone, *Organic Syntheses*, 5 (1973) 255.

29. J.E. Garst, G.L. Schmir, Hydrolysis of 2-methoxyfuran, *The Journal of Organic Chemistry*, 39 (1974) 2920–2923.

30. W.Y. Yu, H. Alper, Palladium-catalyzed cyclocarbonylation of terminal and internal alkynols to 2(5H)-furanones, *The Journal of Organic Chemistry*, 62 (1997) 5684–5687.

31. H.D. Mansilla, J. Baeza, S. Urzúa, M. Maturana, J. Villaseñor, N. Durfin, Acid-catalysed hydrolysis of rice hull: Evaluation of furfural production, *Bioresource Technology*, 66 (1998) 189–193.

32. R. Cao, C. Liu, L. Liu, A Convenient synthesis of 2(5H)-Furanone, *Organic Preparations and Procedures International*, 28 (1996) 215–216.

33. V.V. Poskonin, Catalytic oxidation reactions of furan and hydrofuran compounds. Characteristics and synthetic possibilities of the reaction of furan with aqueous hydrogen peroxide in the presence of compounds of niobium (ii) and (v), *Chemistry of Heterocyclic Compounds*, 45 (2009) 1177–1183.

34. L.A. Badovskaya, V.M. Latashko, V.V. Poskonin, E.P. Grunskaya, Z.I. Tyukhteneva, S.G. Rudakova, S.A. Pestunova, A.V. Sarkisyan, Catalytic oxidation of furan and hydrofuran compounds. 7. production of 2(5H)-furanone by oxidation of furfural with hydrogen peroxide and some of its transformations in aqueous solutions, *Chemistry of Heterocyclic Compounds*, 38 (2002) 1040–1048.

35. A. Gassama, C. Ernenwein, N. Hoffmann, Synthesis of surfactants from furfural derived 2[5H]-furanone and fatty amines, *Green Chemistry*, 12 (2010) 859–865.

36. F. Li, T. Lu, B.Chen, Z. Huang, G. Yuan, Pt nanoparticles over TiO_2–ZrO_2 mixed oxide as multifunctional catalysts for an integrated conversion of furfural to 1,4-butanediol, *Applied Catalysis A*, 478 (2014) 252–258.

37. A. Senning, *Elsevier's Dictionary of Chemoetymology. The Whys and Whences of Chemical Nomenclature and Terminology*, Elsevier Science, Amsterdam, 2006.

38. G. Burdock, *Encyclopedia of Food and Color Additives*, CRC Press, Boca Raton, FL, 1996.

39. K.J. Zeitsch, *The Chemistry and Technology of Furfural and its Many By-Products, Sugar Series, Vol. 13*, Elsevier Science, Amsterdam, 2000.

40. W.C. Wilson, 2-Furancarboxylic acid and 2-Furylcarbinol [(2-Furoic acid) (Furfuryl alcohol)], *Organic Syntheses*, 6 (1926) 44–47.

41. C.D. Hurd, J.W. Garrett, E.N. Osborne, Furan reactions. IV. Furoic acid from furfural, *Journal of the American Chemical Society*, 55 (1933) 1082–1084.

42. R. Mariscal, P. Maireles-Torres, M. Ojeda, I. Sádaba, M. Lopez Granados, Furfural: A renewable and versatile platform molecule for the synthesis of chemicals and fuels, *Energy & Environmental Science*, 9 (2016) 1144–1189.

43. A. Corma, S. Iborra, A. Velty, Chemical routes for the transformation of biomass into chemicals, *Chemical Reviews*, 107 (2007) 2411–2502.

44. G.S. Chaubey, S. Das, M.K. Mahanti, Kinetics of the oxidation of hetero-cyclic aldehydes by quinolinium dichromate, *Bulletin of the Chemical Society of Japan*, 75 (2002) 2215–2220.

45. G.S. Chaubey, Kharsyntiew, B., M.K. Mahanti, Oxidation of substituted 2-furaldehydes byquinolinium dichromate: A kinetic study, *Journal of Physical Organic Chemistry*, 17 (2004) 83–87.

46. T.K. Chakraborty, S. Chandrasekaran, Facile oxidation of aldehydes to carboxylic acids with chromium(V) reagents, *Synthetic Communications*, 10 (1980) 951–956.

47. G.G. Kharnaior, G.S. Chaubey, M.K. Mahanti, Kinetics of oxidation of 2-furfural by quinolinium dichromate, *Oxidation Communications*, 24 (2001) 377–381.

48. B.R. Babu, K.K. Balasubramaniam, Simple and facile oxidation of aldehydes to carboxylic acids, *Organic Preparations and Procedures International*, 26 (1994) 123–125.

49. E. Dalcanale, F. Montanari, Selective oxidation of aldehydes to carboxylic acids with sodium chlorite-hydrogen peroxide, *The Journal of Organic Chemistry*, 51 (1986) 567–569.

50. J.A. Moore, E.M. Partain III, Oxidation of furfuraldehydes with sodium chlorite, *Organic Preparations and Procedures International*, 17 (1985) 203–205.

51. H. Bassett, I. Sanderson, Observation upon mechanism of permanganate reduction and the induced oxidation of chlorion, *Journal of the American Chemical Society*, 13 (1936) 207–211.

52. E.V. Aidanova, K.E. Kva, I.F. Ratovsky, F.K. Shmidt, *Khimiya i Tekhnoliya Vody*, 16 (1994) 256260.

53. H. Baba, *Kagaku Kenkyusto Hokoku*, 33 (1957) 168.

54. Y.M. Shapiro, O.A. Pustovarova, E. Baum, V.G. Kulnevich, Reactions of aldehydes of the furan series. Oxidation by sodium hypohalogenides. *Khim. Geterotsikl Soedin*, 11 (1982) 14–63.

55. K.T. Rowbottom, D.A. Cummerson, Oxidation of furfural to furoic acid, GB Patent 2188927 (1986).

56. S. Yokota, JP Patent 03123776 (1989).

57. S. Yokota, K. Matsuoka, JP Patent 04046167 (1990).

58. A.P. Dunlop, Process for manufacturing furoic acid and furoic acid salts, US Patent 2, 407 (1946) 066.

59. P. Verdeguer, N. Merat, L. Rigal, A. Gaset, Optimization of experimental conditions for the catalytic oxidation of furfural to furoic acid, *Journal of Chemical Technology and Biotechnology*, 61 (1994) 97–102.

60. P. Verdeger, N. Merat, A. Gaset, Lead/platinum on charcoal as catalyst for oxidation of furfural. Effect of main parameters, *Applied Catalysis A*, 112 (1994) 1–11.

61. L. Isenhour, Method for the production of furoic acid, US Patent 2, 041 (1936) 184.

62. R.J. Harrison, M. Moyle, 2-Furoic acid, *Organic Syntheses*, 36 (1956) 36.

63. Q.Y.Tian, D.X. Shi, Y.W. Sha, CuO and Ag_2O/CuO catalyzed oxidation of aldehydes to the corresponding carboxylic acids by molecular oxygen, *Molecules*, 13 (2008) 948–957.

64. E. Taaring, I.S. Nielsen, K. Egeblad, R. Madsen, C.H. Christensen, Chemicals from renewables: Aerobic oxidation of furfural and hydroxymethylfurfural over gold catalysts, *ChemSusChem*, 1 (2008) 75–78.

65. M. Haruta, T. Kobayashi, H. Sano, N. Yamada, Novel gold catalysts for the oxidation of carbon monoxide at a temperature far below 0. DEG., *Chemistry Letters*, 16 (1987) 405–408.

66. G.J. Hutchings, M. Brust, H. Schmidbaur, Gold — An introductory perspective, *Chemical Society Review*, 37 (2008) 1759–1765.

67. S. Kegnæs, D. Mielby, U.V. Mentzel, T. Jensen, P. Fristrup, A. Riisager, One-pot synthesis of amides by aerobic oxidative coupling of alcohols or aldehydes with amines using supported gold and a base as catalyst, *Chemistry Communications*, 48 (2012) 2427–2429.

68. J. Deng, H.-J. Song, M.S. Cui, Y.-P. Du, Y. Fu, Aerobic oxidation of hydroxymethylfurfural and furfural by using heterogeneous CoxOy-N@C catalysts, *ChemSusChem*, 7 (2014) 3334–3340.

69. Z. Liu, X. Tong, J. Liu, S. Xue, A smart catalyst system for the valorization of renewable furfural in aliphatic alcohols, *Catalysis Science & Technology*, 6 (2016) 1214–1221.

70. O. Casanova, S. Iborra, A. Corma, Biomass into chemicals: One pot-base free oxidative esterification of 5-hydroxymethyl-2-furfural into 2,5-dimethylfuroate with gold on nanoparticulated ceria, *Journal of Catalysis*, 265 (2009) 109–116.

71. F. Pinna, A. Olivo, V. Trevisan, F. Menegazzo, M. Signoretto, M. Manzoli, F. Boccuzzi, The effects of gold nanosize for the exploitation of furfural by selective oxidation, *Catalysis Today*, 203 (2013) 196–201.

72. M. Signoretto, F. Menegazzo, L. Contessotto, F. Pinna, M. Manzoli, F. Boccuzzi, Au/ZrO_2: An efficient and reusable catalyst for the oxidative esterification of renewable furfural, *Applied Catalysis B*, 129 (2013) 287–293.

73. M. Signoretto, F. Menegazzo, F. Pinna, M. Manzoli, V. Aina, G. Cerrato, F. Boccuzzi, Oxidative esterification of renewable furfural on gold based catalysts: Which is the best support? *Journal of Catalysis*, 309 (2014) 241–247.

74. M. Manzoli, F. Menegazzo, M. Signoretto, D. Marchese, Biomass derived chemicals: Furfural oxidative esterification to methyl-2-furoate over gold catalysts, *Catalysts*, 6 (2016) 107–134.

75. M. Manzoli, F. Menegazzo, M. Signoretto, G. Cruciani, F. Pinna, Effects of synthetic parameters on the catalytic performance of Au/CeO_2 for furfural oxidative esterification, *Journal of Catalysis*, 330 (2015) 465–473.

76. F. Menegazzo, M. Signoretto, T. Fantinel, M. Manzoli, Sol-immobilized vs deposited-precipitated Au nanoparticles supported on CeO_2 for furfural oxidative esterification, *Journal of Chemical Technology and Biotechnology*, 92 (2017) 2196–2205.

77. F. Menegazzo, M. Signoretto, D. Marchese, F. Pinna, M. Manzoli, Structure–activity relationships of Au/ZrO$_2$ catalysts for 5-hydroxymethylfurfural oxidative esterification: Effects of zirconia sulphation on gold dispersion, position and shape, *Journal of Catalysis*, 326 (2015) 1–8.

78. F. Moulines, J. Ruiz, D. Astruc, Oxidation of furfural with H$_2$O$_2$ in the presence of a photogenerated iron catalyst, *Journal of Organometallic Chemistry*, 340 (1988) C13–C14.

79. G. Chamoulaud, D. Floner, C. Moinet, C. Lamy, E.M. Belgsir, Biomass conversion II: Simultaneous electrosyntheses of furoic acid and furfuryl alcohol on modified graphite felt electrodes, *Electrochimica Acta*, 46 (2001) 2757–2760.

80. P. Parpot, A.P. Bettencourt, G. Chamoulaud, K.B. Kokoh, E.M. Belgsir, Electrochemical investigations of the oxidation–reduction of furfural in aqueous medium: Application to electrosynthesis, *Electrochimica Acta*, 49 (2004) 397–403.

81. H.I. Prez, N. Manjarrez, A. Sols, H. Luna, M.A. Ramirez, J. Cassani, Microbial biocatalytic preparation of 2-furoic acid by oxidation of 2-furfuryl alcohol and 2-furanaldehyde with Nocardia coralline, *African Journal of Biotechnology*, 8 (2009) 2279–2282.

82. P.D. Hanke, Enzymatic oxidation of hydroxymethylfurfural, US8183020 (2012).

83. M.P. van Deurzen, F. van Rantwijk, R.A. Sheldon, Chloroperoxidase-catalyzed oxidation of 5-hydroxymethylfurfural, *Journal of Carbohydrate Chemistry*, 16 (1997) 299–309.

84. M. Krystof, M. Perez-Sanchez, P. Dominguez de Maria, Lipase-mediated selective oxidation of furfural and 5-hydroxymethylfurfural, *ChemSusChem*, 6 (2013) 826–830.

Chapter 3.10

Furan, Tetrahydrofuran and Other Furan-derived Chemicals

Francisco Vila*,‡, Manuel Ojeda† and Manuel López Granados†

*IBERCAT SL, C/Faraday 7,
PCM-Edificio CLAID 28049 Madrid, Spain

†Institute of Catalysis and Petrochemistry (CSIC),
Sustainable Chemistry and Energy Group,
C/Marie Curie 2, Campus de Cantoblanco, 28049 Madrid, Spain

3.10.1. Furan

The main industrial interest in the production of furan lies in its usefulness for obtaining tetrahydrofuran (THF) and as a solvent (Figure 3.10.1) [1]. Its unique electronic configuration confers enormous chemical versatility what allows furan to be a reagent in several processes of fine chemical industry (pharmaceuticals and agrochemicals) involving C–C bonds formation (Diels–Alder cycloaddition reactions, alkylation reactions, Grignard reagent, among others) [2]. The synthesis of thiophene and pyrrole has also been proposed [3, 4].

Catalytic decarbonylation of furfural is the main renewable route for furan production. Decarbonylation can be carried out both in the liquid phase and in the gaseous phase by the use of catalysts mainly based on Group VIII metals such as Pd, Pt and Rh. Industrially, supported Pd catalysts have been reported to show great activity and selectivity to furan [1, 5–7]. Unfortunately such behavior is only

‡Corresponding author: franvila@ibercatsl.com

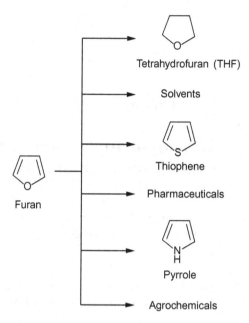

Figure 3.10.1. Main industrial applications of furan.

maintained during few hours on stream leading to fast deactivation, both in gas and liquid phase processes, which currently represents the main technical drawback of this process.

Deactivation phenomena are derived from the high temperature required for the decarbonylation reaction (533–623 K optimal temperature on vapor phase). Deposition of coke over the surface of the catalyst is considered the main source of deactivation. In addition sintering of Pd particles provoked by such temperature also contributes to the activity decay. Regarding the coke deposits, kinetic studies have showed that the formation of coke deposits initiates by the adsorption of two furfural molecules on two adjacent active sites that react with each other resulting in heavy weight molecular species [8]. Actually furan dimers have been detected in the liquid phase after reaction that points these species as responsible for blocking active centers [9].

Three general strategies have been attempted to cope with the deactivation and to lengthening the life of the catalysts. First strategy is the thermal regeneration of the catalyst by gasification of carbonaceous residues. This has been proved to be effective because in this reaction the fouling of the active centers by coke deposits is one of deactivation phenomenon with the largest impact. The main drawback of the gasification of coke is derived from the fact that it is a very exothermic reaction leading to thermal sintering of Pd particles and consequently deactivating the catalyst. The use of steam together with air during the regeneration stage minimizes the thermal effects derived from the gasification by dissipating the heat generated during the gasification of the coke deposits. The second strategic line is to inhibit coke accumulation by *in situ* gasification. Feeding hydrogen as gasifier together with furfural (H_2:furfural molar ratios between 0.5 and 10) has resulted in great improvements in the stability of the catalyst. However it should be noted that higher ratios lead to a hydrogenation of furfural to FOL, THFA, MF and MTHF and of furan to THF. The third approach is the modification of the catalytic formulation to increase stability. Because the use of metal phase has been practically confined to the use of Pd, the incorporation of promoters seems to be more interesting option. The combination of these three strategies rather than a single approach could be applied to improve the stability of the catalyst and accordingly profitability for the industrial process.

3.10.1.1. *Liquid-phase decarbonylation*

Remarkable results have been achieved with Pd-based catalysts (Pd/C and Pd/Al_2O_3) using an atmospheric pressure reactive distillation system working at few degrees below the boiling point of furfural (543 K) [9–12]. The carbon-based catalyst displayed significantly better yields. Thus, Pd/C catalyst doped with K_2CO_3 maintaining a 1:1 mass ratio allows to reach a productivity of 13 kg Furan/g Pd for semi-continuous after ca. 300 h of operation (36 kg Furan/g Pd in continuous tests). Since the liquid phase system operates at temperatures below the boiling temperature of furfural the formation of coke deposits is minimized. In contrast, this thermal limitation detracts from industrial attractiveness by making the system economically less competitive. Probably due to this intrinsic limitation the majority of works reported later focus their interest in studies in the gas phase. Utilization of some other oxidizing catalysts based on iron, manganese, chromium or mixture of them has shown negligible behavior to furan [11]. Influence of pressure above atmospheric has not reported for such reactive distillation system but in principle, as it will be stated below in the vapor phase decarbonylation section, increasing of pressure (hydrogen pressure) has shown no benefit in the yield. On the contrary the yield of THF and of other furfural hydrogenation products like FOL, THFA, MF and MTHF may increase.

3.10.1.2. *Vapor-phase decarbonylation*

Catalysts based on Group VIII metals (Pd, Pt and Rh) have displayed the better results. As in the liquid phase, the incorporation of potassium to the catalyst promotes its activity and stability. Such promotion effect was also observed for metal aluminates, both alkaline and alkaline earth metals and other basic metal oxides such as Ga, La and Y [13–15]. Apparently, increasing the basic properties in the catalyst via either addition of alkaline promoters or by removal of acid centers in the case of the aluminates results in a lower coke formation and consequently in a higher stability of the catalysts. Thus a Pd/Al_2O_3 system promoted by potassium can work under conditions of higher hydrogen concentration (H_2/furfural ratio 20) with no sign of apparent deactivation for more than 100 h of operation and without visible decrease in furan selectivity (99.5% yield to furan) [16]. Similarly, an alumina-supported Pt catalyst (0.7 wt.% Pt) promoted with cesium salts showed high activity and high resistance to deactivation even at a H_2/furfural ratio as low as 0.74. This catalyst achieved a productivity of 131 kg furan/g Pt [17]. The use of catalysts with a noble metal free formulation based on the oxide mixture (manganese, zinc, strontium and cadmium) has also been reported achieving yields of furan similar to those obtained by the supported catalysts of Pd [18, 19]. In this case a remarkable deactivation was observed at 5 h of reaction and unfortunately no further study concerning its regeneration was reported.

3.10.1.3. *Studies of the mechanism of the reaction*

Some studies have been directed revealing the mechanism that governs the reaction and the reasons which cause catalytic deactivation in gas decarbonylation of furfural. Kinetic studies have demonstrated that the reaction requires two contiguous surface active centers [8]. Density functional theory (DFT) studies on Pd surfaces have indicated that furfural approach to the catalytic surface occurs so that the furan ring is held parallel to the surface [20–22]. The chemisorption on the metal centers of Pd occurs through a planar configuration that can give rise to di-coordinated species and acyl species (Figure 3.10.2). In this chemisorption the π-orbitals of the heterocyclic ring with an orientation perpendicular to the surface interact with Pd d-orbitals providing additional stability. Coordination mode of furfural shows dependence with the temperature. Thus at low temperatures the relative population of di-coordinated species is greater, whereas at high temperatures the acyl species become predominant. Higher temperatures will lead to the decomposition of the acyl species giving rise to CO and furan. Based on the type of coordination it can be concluded that the acyl species, predominant at higher temperatures, are the intermediate in furfural decarbonylation. Very similar reaction mechanism and coordination mode for furfural was proposed over Pt (111) based also on DFT studies [23]. On the other hand, the di-coordinated species in the presence of hydrogen can also be hydrogenated leading to hydroxyalkyl intermediate species which in turn could be hydrogenated to give rise to FOL, THFA, MF and MTHF [20–22]. DFT calculations also revealed that although the decarbonylation path is thermodynamically favored, the kinetic

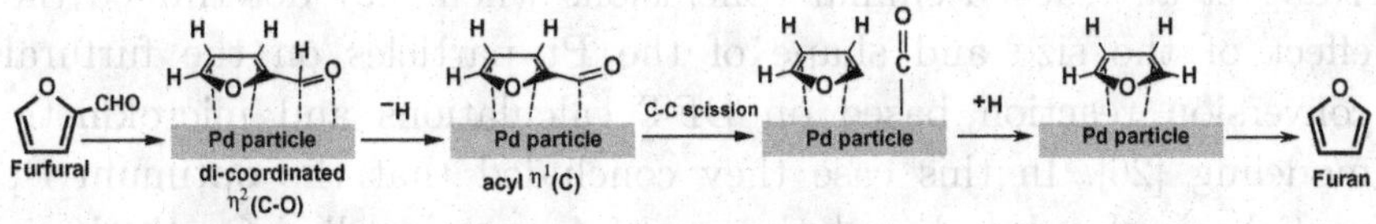

Figure 3.10.2. Mechanism of furfural decarbonylation on Pd from DFT calculations.

barrier for the hydrogenation pathway of furfural is lower than for the decarbonylation. TPD experiments have showed that the adsorbed flat furfural on Pd evolves by the loss of the aldehyde hydrogen to form the intermediate acyl species and the consequent formation of furan and CO through the rupture of the C–C bond [20, 21].

Assuming that furfural coordination mode on the catalyst could selectively determine the pathway, decarbonylation or hydrogenation routes, all parameters able to control this coordination should be examined. Three fundamental parameters that could modify coordination have been reported: particle size and shape, hydrogen concentration and alkali promoters. The nature of the active phase, Pd or Pt, has not been independently considered because identical reaction mechanism and coordination modes have been identified over Pd (111) and Pt (111) for decarbonylation of furfural.

Regarding the effect of particle size Somorjai *et al.* have demonstrated that this reaction is structure-sensitive [24, 25]. In their studies they synthesized Pt nanoparticles supported on silica catalysts with particle size in the range of 1.5–7.1 nm and also with different forms that were tested under atmospheric pressure with H_2: Furfural ratio equal to 10, and temperatures in the range 443–513 K. The results showed that the turnover rate (TOR) as well as selectivity to furan or furfuryl alcohol has a strong dependence on the size and shape of the Pt nanoparticles. Highest selectivity to furan, ca. 96%, was achieved with Pt particles of 1.5 nm. The selectivity to furan decreases to ca. 30% as the particle size increases to Pt particles larger than 3.6 nm. In these cases, the balance of selectivity was almost completed by the formation of furfuryl alcohol (ca. 70%). They also suggested that Pt nanoparticles may have two types of active centers and that the relative population of each would be strongly linked to the size and shape of the particles. Wang *et al.* reached similar conclusions when they determined the effect of the size and shape of the Pt particles on the furfural conversion reaction based on DFT calculations and microkinetic modeling [26]. In this case they concluded that the optimum Pt particle size for decarbonylation was 1.4 nm or smaller. On the basis of the activation energy they indicated that Pt atoms in corners gave rise to decarbonylation whereas those in terraces and steps are selective to hydrogenation products.

On the other hand a dependence between the coordination of the furfural adsorbed on Pd (111) and the surface hydrogen coverage has been stablished [27]. At low concentrations of hydrogen, furfural interacts through the carbonyl group and the aromatic ring whereas larger hydrogen coverage causes an inclination of the furfural adsorbed on the surface and hence the interaction through the ring is partially lost. Based on these observations, they proposed that surface hydrogen coverage played a key role in the selectivity of the furfural conversion reaction by affecting the preferential pathway leading to decarbonylation or hydrogenation products. This assumption could explain many of the disparate results in terms of selectivity.

Finally, an increase in the electron density of Pd particles in Pd-based catalysts promoted by alkali metals have been described by IR studies [16]. From these observations, it has been proposed that the incorporation of alkali oxides determines the mode of adsorption of furfural on Pd active sites and inhibiting the formation of di-coordinated species, intermediate associated with the hydrogenation pathway, favoring the formation of acyl species precursor of the decarbonylation pathway.

3.10.2. Tetrahydrofuran (THF)

THF, with an annual production rate above 430 MT, finds industrial application basically as precursor of polytetramethylene ether glycol (PTMEG) (an elastomer fibber) and as solvent (Figure 3.10.3) [1,28, 29]. Industrially THF is obtained by dehydration and cyclization of 1,4-Butanediol (1,4-BDO) what makes THF a petrochemical because BDO is obtained by different technologies from oil derived chemicals (either acetylene, propylene, butadiene or maleic anhydride) [28].

However, THF can indeed be produced from biomass through a number of different routes. One of them, already in commercial application by BioAmber and Myriant, is the hydrogenation of biobased succinic to THF. The latter process uses the Johnson

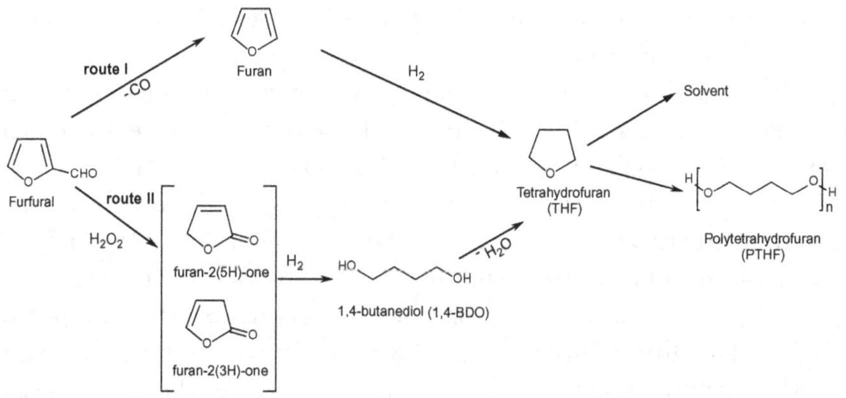

Figure 3.10.3.   Routes for the formation of THF from furfural and THF main applications.

Matthey David technology previously used to transform petrochemical maleic anhydride to BDO/THF via succinic acid [28, 30]. Alternatively, THF could be produced from furfural trough two different approaches (Figure 3.10.3). The first one involves furan as intermediate species whereas the second pathway involves the synthesis of 1,4-BDO from furfural by the formation of furanones as intermediate species. We will concentrate on the latter two processes as they make use of furfural.

3.10.2.1.   *Furan hydrogenation route*

A number of non-noble and noble metal catalysts have demonstrated the capacity of conducting this reaction. Thus 99% furan conversion and 97% THF selectivity were achieved with commercial Ni-Raney (Ni–Cr–Fe) catalysts at 393 K, 7 MPa H_2 and using propanol as the solvent [31]. The reuse for 5 cycles was possible without significant decay on its behavior. Ni–Cu–Cr and Ni/Al_2O_3 catalysts also showed a suitable behavior but yielding both THF and 1,4-BDO, which can be separated by distillation at 373 K [32–34]. 1,4-BDO has applications in the polymer industry itself but since 1,4-BDO is the starting material for the synthesis of THF via petrochemicals, the integration into the system increases the performance of THF. It should be noted that the use of dicarboxylic acids allows increasing the selectivity to 1,4-BDO in detriment of THF. Most favorable operational conditions reported to 1,4-BDO production were 428–458 K, 5.5 MPa H_2 and water/furfural molar ratio between 1 and 3.5 to obtain 48% and 52% of selectivity to THF and 1,4-BDO, respectively. Noble metal-based catalysts such as Re, Ru and Rh (and alloys of them) have shown good behavior. The best performance was reported by using the carbon-supported Rh–Re catalyst (1 wt.% Rh–5 wt.% Re) at 433 K and 3 MPa yielding 75% of THF; remarkably 18% of 1,4-BDO was also obtained [35]. Carbon-supported Pd catalysts (5 wt.% Pd loading) have also been shown to be active [36]. Unlike furfural decarbonylation, this reaction requires lower temperatures but higher pressures of H_2 (373 K, 10 MPa). The type of support plays an important role, thus alumina-supported Pd catalysts have shown reaction rates four times lower than those carbon supported. Nevertheless Pd/C catalyst has shown a progressive loss of activity (a decrease of the surface Pd concentration during 10 cycles of reuse has been observed by XPS).

Studies aiming ate revealing the surface mechanism of the reaction has been conducted by DFT. The energetic barriers of hydrogenation and furan ring opening to different intermediates have been calculated on Pd (111) surfaces [27]. It was concluded the complete hydrogenation of furan to THF occurs via four sequential hydrogenation stages through three intermediates: Hydrofuran (HF), Dihydrofuran (DHF) and Trihydrofuran (TriHF). The DHF and TriHF intermediates are thermodynamically stable and the formation of ring opening byproducts is kinetically very unfavorable route. On the contrary, they determined that the intermediate species HF is an intermediate that can evolve through both hydrogenation route to DHF and ring opening route to 1-butanol. They concluded that partial pressure of hydrogen was a critical factor in controlling selectivity.

Since Pd catalysts show good behavior in both furfural decarbonylation to furan and in furan hydrogenation to THF, the possible direct route of obtaining THF from furfural easily comes to mind. Early studies the hydrogenation of furan with Pd based catalyst in the presence of furfural and CO resulted in a quick deactivation of the catalyst [36]. It was concluded that the chemisorption of furfural and CO over Pd particles is stronger than the chemisorption of furan and then the further hydrogenation of furan would be inhibited in the presence of furfural and/or CO due to blocking of the metallic sites. On the other hand, more recently Medlin *et al.* determined the adsorption energies of furfural and furan on Pd (111) by DFT calculations and concluded that furfural chemisorption on Pd (111) surface was slightly weaker than furan [20]. Other investigation confirms the possibility of carrying out the direct synthesis of THF from furfural hydrogenation by using Pd nanoparticles supported on carbon catalysts. It must be stressed that the reaction was carried out in liquid phase and microwave-assisted to control the temperature at 373 K and that the reaction was not conducted with gas H_2 but by using formic acid as a source of hydrogen (*in situ* decomposition resulted in the release of H_2 and CO_2. Under such operational conditions ca. 90% yield of THF was achieved but no reutilization experiments were reported [37].

Few studies have been devoted to the role of coke/carbon deposits except that reported by Jackson *et al.* in the catalytic deactivation with zirconium oxide supported palladium catalyst [38, 39]. They detected the presence of furan and THF species strongly adsorbed on Pd particles in the initial stages of exposure but under stream the former species evolves to non-volatile carbon resistant to hydrogenation. Only at higher temperatures the partial methanization of these deposits could be observed in a reducing atmosphere. Oxidizing atmospheres lead to greater gasification of deposits but not complete.

3.10.2.2. *Furanone hydrogenation route*

The second route requires three steps rather than the two steps needed for previous one. It goes through the formation of 1,4-BDO from furfural that must be dehydrated later to THF (Figure 3.10.3). The formation of 1,4 BDO requires first the formation of furanones from furfural that must be subsequently reduced to 1,4-BDO [40]. By using a Pt/TiO_2-ZrO_2 catalyst in aqueous acid solution a yield of 85.2% to 1,4-BDO was achieved in a two-step process: (i) oxidation of furfural to furanones (furan-2(5H)-one and furan-3(5H)-one) by H_2O_2 at room temperature; and (ii) further hydrogenation to 1,4-BDO at 393 K and 3.5 MPa. Deactivation of the catalyst by Pt leaching was observed. Since 1,4-BDO is the precursor for THF synthesis the process developed from furfural can be extended as a promising alternative route to furan for the also renewable production of THF.

3.10.3. Thiophene

Thiophene is an important sulfur compound precursor of different chemicals such as drugs, herbicides, polyorganosiloxanes, polymers, stabilizers, vulcanization accelerators, dyes, oil additives, among others (Figure 3.10.4) [3,41]. Thiophene is currently a petrochemical and consequently a non-renewable product but a renewable route involving furan has been demonstrated: catalytic sulphurisation of furans. It should be noted that although the synthetic route has been known for more than 80 years, the number of articles reported remains fairly poor.

In the early investigations, quite poor yields were obtained by using activated carbon, copper–carbon and alumina (promoted and unpromoted) as catalysts, the formation of by-products was inevitable [42]. The best thiophene yield (close to of 40%) was obtained with unpromoted alumina at 673 K. Later, thiophene yields

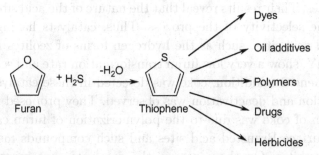

Figure 3.10.4. Synthesis of thiophene from furfural and H_2S and its main industrial applications.

up to 75% were reached at similar temperatures than previous studies (693 K) by using alkali-zeolites, such as NaX, NaY, NaLiX, NaLiY as catalyst [43]. Although the deactivation was observed upon the course of the reaction the studies further reported that regeneration in this case may be possible. More recently alumina-supported Co–Mo catalysts (5% Co and 16% Mo) have been shown to be active in the thiophene synthesis reaction reaching a thiophene yield of 80% under milder conditions (473–523 K, 2.8 MPa and a H_2S/furan molar ratio of 12) [44]. Under similar conditions, the presence of alumina catalyst doped with potassium phosphotungstate ($H_3PW_{12}O_{40}$) the behavior was improved to yield and selectivity greater than 90% [45].

Two mechanisms for the sulphurisation of furan have been proposed. Both proposals focus on the role of Lewis acid sites. The mechanism proposed by Li *et al.* (mechanism (a) in Figure 3.10.5) is based on FTIR techniques and involves five steps [41]: (a) dissociative adsorption of hydrogen sulfide in which SH group is adsorbed on unsaturated cations Al^{3+} and proton is transferred to an oxygen atom of Al_2O_3; (b) intramolecular dehydration yield an unsaturated cation Al^{3+} and sulfide group; (c) furan adsorption over Al^{3+}; (d) ring opening of furan by cleavage of C–O bond and formation of C–S bond; (e) cyclization to thiophene adsorbed over Al^{3+} through C–O bond rupture and C–S bond formation; (f) desorption of thiophene and Al_2O_3 recovery. By DFT studies similar conclusions were achieved for this mechanism [46].

An alternative mechanism for the reaction was proposed by Maskhina *et al.* [42] based on their investigation of the role of surface acidic sites. Their results reveal that the nature of the acid sites determines the selectivity of the process. Thus, catalysts having strong Brønsted acid sites, such as the hydrogen forms of zeolites HZSM-5 and HNaY, show a very low furan transformation rate and selectivity to thiophene. In addition, coke was detected in these catalysts after the reaction and deactivation was observed. They proposed that the formation of coke was due to the polymerization of furan catalyzed by the surface Brønsted acid sites and such compounds may block the active sites. On the contrary, the catalysts containing mainly Lewis acid sites showed a better behavior in the conversion of furan

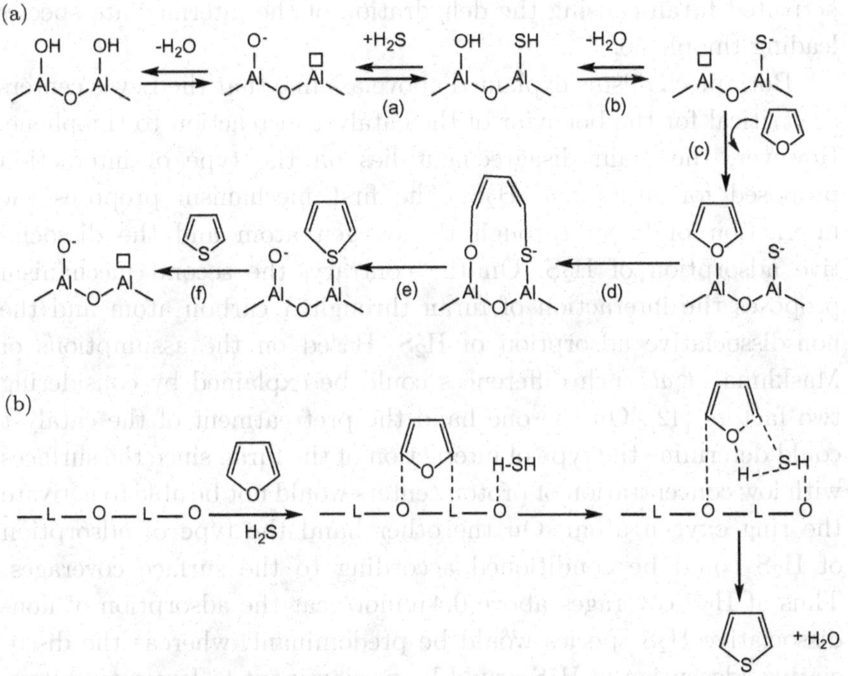

Figure 3.10.5. Different mechanisms proposed for the synthesis of thiophene from furan and H_2S: (a) mechanism involving H_2S dissociative; (b) mechanism involving non-dissociative adsorption of H_2S.

and in the selectivity to thiophene. Specifically catalysts containing Al^{3+} species, such as alumina-supported catalysts, showed better yields to thiophene. Extended studies showed that the B_2O_3/Al_2O_3 catalyst presented the best performance achieving a thiophene yield of more than 90% at 573–673 K. A good correlation between Lewis acid sites concentration and thiophene yield was reported by using Al_2O_3 as catalyst. Consequently the alternative proposed mechanism is as follows (Figure 3.10.5(b)): (a) simultaneous non-dissociative adsorption of furan and hydrogen sulphide on the surface of the catalyst; furan is adsorbed through the interaction between acidic sites (Lewis acid sites) and α carbon atom of furan (with higher electron density to facilitate interaction with electrophilic agents) and H_2S on basic (O^{2-} unsaturated sites); (b) addition of H_2S to the activated furan causing the dehydration of the intermediate species leading thiophene.

Both mechanisms explained above assume that the Lewis centers are critical for the behavior of the catalyst in reaction to thiophene. However, the main disagreement lies on the type of interaction proposed for furan and H_2S. The first mechanism proposes the interaction of furan through the oxygen atom and the dissociative adsorption of H_2S. On the contrary, the second mechanism proposes the interaction of furan through α carbon atom and the non-dissociative adsorption of H_2S. Based on the assumptions of Maskhina *et al.* such differences could be explained by considering two factors [42]. On the one hand the pretreatment of the catalyst could determine the type of interaction of the furan since the surfaces with low concentration of proton centers would not be able to activate the ring oxygen atom. On the other hand the type of adsorption of H_2S could be conditioned according to the surface coverages. Thus at H_2S coverages above 0.4 mmol/gcat the adsorption of non-dissociative H_2S species would be predominant, whereas the dissociative adsorption of H_2S would be predominant to lower coverages. However such assumptions should be further corroborated in order to shed light on the key factors governing the mechanism of this reaction.

3.10.4. Pyrrole

Pyrrole and its derivatives are widely used as solvents and intermediates in the synthesis of pharmaceuticals, medicines, agrochemicals, dyes, photographic chemicals, perfumes and other organic compounds (Figure 3.10.6) [47, 48]. They are also used in metallurgical processes, as catalysts for polymerization process, as corrosion inhibitors, as preservatives, and as solvents for resins and terpenes. Furthermore, they find applications in the intensive study of transition-metal complex catalyst chemistry for uniform polymerization, luminescence chemistry and spectrophotometric analysis. An important derivative of pyrrole is *N*-methylpyrrole, which is a precursor for *N*-methylpyrrole carboxylic acid, a key building-block used

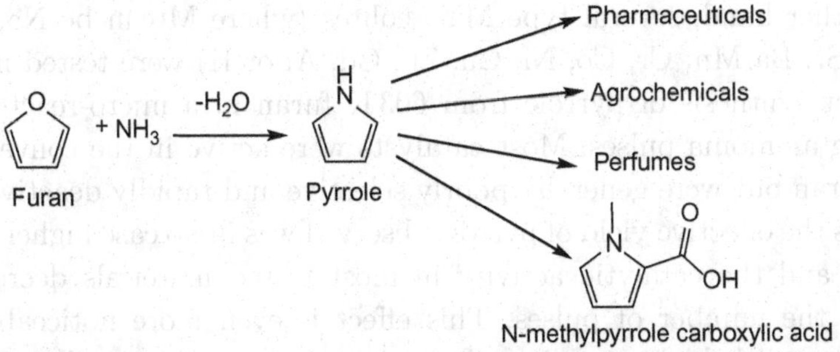

Figure 3.10.6. Synthesis of pyrrole from furan and NH_3 and its main industrial applications.

in pharmaceutical chemistry [48]. Industrially, pyrrole is prepared by treating furan with ammonia in the presence of solid acid catalysts, typically γ-Al_2O_3- or SiO_2–Al_2O_3-based materials (Figure 3.10.4) [49]. Another important route is the so-called Paal–Knorr pyrrole synthesis, which is the condensation of a 1,4-dicarbonyl compound with an excess of a primary amine or ammonia to form a pyrrole ring [49].

Industrially the pyrrole synthesis is carried out either by dehydrogenation of pyrrolidine or from furan treated with ammonia or in the presence of acidic catalysts based on alumina or aluminosilicate [49]. The latter route (Figure 3.10.6) makes pyrrole a renewable chemical as long as furan is obtained from furfural. Although there are several alternative chemical routes of pyrrole synthesis, perhaps the most relevant is the Paal–Knorr synthesis with which pyrrole and numerous derivatives can be obtained from 1,4-dicarbonyl compounds in the presence of ammonia or primary amines. However, despite the versatility offered by the later proposal, the furan route is of greater industrial interest because of the ease of access to furan of renewable origin.

In the first reports of this reaction Al_2O_3 was used as catalyst reaching yields of pyrrole of 30% [4, 50, 51]. Subsequently, the use of Mo and V oxide supported on alumina catalysts (highly acidic material) has been reported in this reaction at atmospheric pressure and temperatures above 643 K, catalytic behavior was also poor [52].

In other hand, several type MY zeolites (where M can be Na, Mg, Ca, Sr, Ba Mn, Cr, Co, Ni, Cu, Zn, Cd, Al or H) were tested in the direct synthesis of pyrrole from 603 K furan in a micro-reactor by using ammonia pulses. Most catalysts were active in the conversion of furan but were generally poorly selective and rapidly deactivated. Thus the effective yield of pyrrole observed was in no case higher than 20% and the catalytic activity in most tested materials decreased with the number of pulses. This effect is even more noticeable in zeolites with higher acidity where the presence of byproducts with higher molecular weight was deduced. Only the catalysts BaY, MgY and ZnY showed close to 100% selectivity and no signs of catalytic deactivation were observed in these cases.

A substantial improvement in the pyrrole yield was achieved by co-feeding stream along with the furan–ammonia mixture [53]. A yield of 63% of pyrrole was obtained at 758–773 K, at a feed ratios of furan:NH_3:H_2O of 1.0:1.8:9.3 and at LHSV for furan of $0.3\,h^{-1}$. If water was not fed, then the yield dramatically decreases to 17%, demonstrating that the water vapor substantially inhibits the polymerization or condensation of furan products that are deposited over the surface, fouling and deactivating the catalyst. It must also be remarked that although the yield remains quite constant with time on stream, the selectivity increases evidencing that the deposition of products by polymerization and condensation of furan decreases with time.

The largest yield of pyrrole (89% at 90.2% conversion) has been reported by Lou *et al.* using a CdO promoted aluminum silicate as catalyst, at 693 K, at a feed furan:NH_3:H_2O mol ratio of 1:8:37 and at a LHSV for furan of $0.2\,h^{-1}$ [54].

Mechanism for the direct synthesis of pyrrole from furan and ammonia has been proposed (Figure 3.10.7). Activation of the furan ring through the interaction between the oxygen atom of the ring and a proton acid site of the catalyst is necessary to drive the reaction. The protonated furan molecule would facilitate the addition of ammonia on α carbon atom giving rise to ring opening by cleavage the C–O bond. Further dehydration and subsequent desorption of pyrrole from the acid center would complete the catalytic cycle [48].

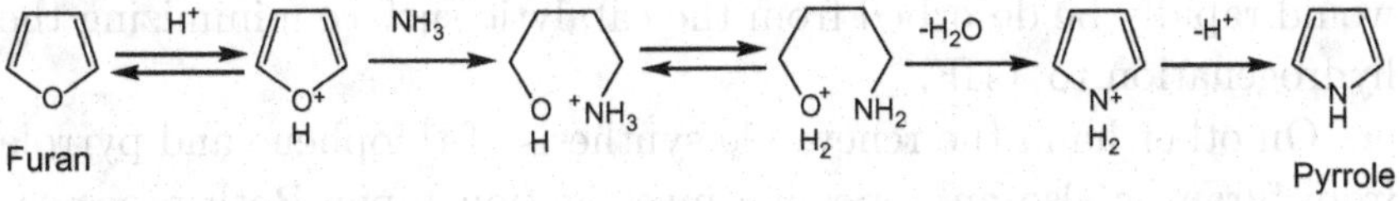

Figure 3.10.7. Proposed reaction mechanism for the acid-catalyzed formation of pyrrole from furan and NH_3.

Summarizing, the direct synthesis of pyrrole from furan requires the use of acid catalysts capable of activating the heterocycle ring. A compromise between the concentration of acid sites and the acid strength of them must be achieved since strong acid sites can give rise to heavy weight derivatives by oligomerization of the starting furan (instead by oligomerization of synthetized pyrrole).

3.10.5. Conclusions and future prospects

THF, a relevant building block of the current chemical industry, can be obtained by renewable chemical routes. The more conventional route involves first the decarbonylation of furfural to furan and then a subsequent hydrogenation process to render THF.

Furfural decarbonylation to furan is readily produced by using Pd-based catalysts and at temperatures of ca. 573 K. Feeding H_2 at low pressures provides important benefits over catalytic durability. Moderate to high hydrogen pressures favor the undesirable hydrogenation of furfural to FOL, MTFA, MF and MTHF. The use of fluidized-type reactors and also the use of doped nanoporous carbons as support are some of other promising developments to improve the durability. Future prospects involve the development of noble metal-free catalysts. It can be envisaged that such catalysts may be modulated to provide the thermal stability required in this process and also to modify the interaction between furan and active phase to make feasible the direct synthesis of THF under low pressure conditions.

Direct synthesis of THF from furfural still remains a challenge, being furan and/or furfuryl alcohol the main products due to the energy differences between the strong furfural–Pd interaction and much weaker furan–Pd interaction. Based on that observation furan would rapidly be desorbed from the catalytic surface minimizing the hydrogenation to THF.

On other hand the renewable synthesis of thiophene and pyrrole from furan, is also an appealing investigation topic. Both processes are far to be mature and extensive research is still needed to achieve reasonably high yields and to develop stable catalysts. The difference in the acid-base properties of main precursor for each route, H_2S and NH_3, is the key. In both cases it is required to clarify and to understand the role of acidic sites of the catalyst. Future prospects for the thiophene route should focus on the catalyst durability (optimization of the acid sites). More difficulties are envisaged for pyrrole synthesis since ammonia, a strong base, may interact more strongly with the acidic sites.

List of abbreviations

1,4-BDO	1,4-Butanediol
FUR	Furfural
FOL	Furfuryl alcohol
MF	2-Methylfuran
MTHF	2-Tetrahydromethylfuran
THFA	Tetrahydrofurfuryl alcohol
THF	Tetrahydrofuran
GBL	γ-Butyrolactone

References

1. K.J. Zeitsch, *The Chemistry and Technology of Furfural and its Many By-Products, Sugar Series, Vol. 13*, Elsevier, The Netherlands, 2000.
2. R.H. Kottke, *Furan Derivatives, Kirk-Othmer Encyclopedia of Chemical Technology, Vol. 12*, John Wiley and Sons, New York, 1998.
3. A.V. Mashkina, A catalytic process for preparation of thiophene from furan and hydrogen sulfide, *Russian Journal of Applied Chemistry*, 84 (2011) 1223–1228.
4. R.B. Bishop, W.I. Denton, Production of pyrrole, in, Google Patents, 1949.
5. D.G. Manly, J.P. ÓHalloran, Process of producing furan, US Patent 3,223,714, 1965.
6. R. Ozer, Vapor-phase decarbonylation of furfural, US Patent 2011/0196126, 2011.

7. H. Singh, M. Prasad, R.D. Srivastava, Metal support interactions in the palladium-catalysed decomposition of furfural to furan, *Journal of Chemical Technology and Biotechnology*, 30 (1980) 293–296.

8. R.D. Srivastava, A.K. Guha, Kinetics and mechanism of deactivation of PdAl$_2$O$_3$ catalyst in the gaseous phase decarbonylation of furfural, *Journal of Catalysis*, 91 (1985) 254–262.

9. K.J. Jung, A. Gaset, J. Molinier, Furfural decarbonylation catalyzed by charcoal supported palladium: Part I — Kinetics, *Biomass*, 16 (1988) 63–76.

10. A.G.P. Lejemble, P. Kalck, G. Merle, B. Molinier, Procédé amélioré de décarbonylation du furfural en vue d'obtenir du furanne., EP 0096913, (1986).

11. P. Lejemble, A. Gaset, P. Kalck, From biomass to furan through decarbonylation of furfural under mild conditions, *Biomass*, 4 (1984) 263–274.

12. K.J. Jung, A. Gaset, J. Molinier, Furfural decarbonylation catalyzed by charcoal supported palladium: Part II — A continuous process, *Biomass*, 16 (1988) 89–96.

13. R. Ozer, K. Li, Vapor-phase decarbonylation process, US Patent 2011/0165561, 2012.

14. R. Ozer, Vapor phase decarbonylation process, US Patent 2012/0165560, 2012.

15. R. Ozer, K. Li, Decarbonylation process, US Patent 2012/0157698, 2012.

16. W. Zhang, Y.L. Zhu, S. Niu, Y.W. Li, A study of furfural decarbonylation on K-doped Pd/Al$_2$O$_3$ catalysts, *Journal of Molecular Catalysis A: Chemical*, 335 (2011) 71–81.

17. M.I.L. Wambach, M. Fischer, Preparation of furan by decarbonylation of furfural, US Patent 4,780,552, 1988.

18. J. Coca, E.S. Morrondo, H. Sastre, Catalytic decarbonylation of furfural in a fixed-bed reactor, *Journal of Chemical Technology and Biotechnology*, 32 (1982) 904–908.

19. J. Coca, E.S. Morrondo, J.B. Parra, H. Sastre, Properties of some catalysts used for the decarbonylation of furfural, *Reaction Kinetics and Catalysis Letters*, 20 (1982) 415–423.

20. S.H. Pang, J.W. Medlin, Adsorption and reaction of furfural and furfuryl alcohol on Pd(111): Unique reaction pathways for multifunctional reagents, *ACS Catalysis*, 1 (2011) 1272–1283.

21. J.W. Medlin, Understanding and controlling reactivity of unsaturated oxygenates and polyols on metal catalysts, *ACS Catalysis*, 1 (2011) 1284–1297.

22. V. Vorotnikov, G. Mpourmpakis, D.G. Vlachos, DFT study of furfural conversion to furan, furfuryl alcohol, and 2-methylfuran on Pd(111), *ACS Catalysis*, 2 (2012) 2496–2504.

23. Ni Zhe-Ming, Xia Ming-Yu, Shi Wei, Qian Ping-Ping, Adsorption and decarbonylation reaction of furfural on Pt(111) surface, *Acta Physica Sinica*, 29 (2013) 1916–1922.

24. K. An, N. Musselwhite, G. Kennedy, V.V. Pushkarev, L. Robert Baker, G.A. Somorjai, Preparation of mesoporous oxides and their support effects

on Pt nanoparticle catalysts in catalytic hydrogenation of furfural, *Journal of Colloid and Interface Science*, 392 (2013) 122–128.

25. V.V. Pushkarev, N. Musselwhite, K.J. An, S. Alayoglu, G.A. Somorjai, High structure sensitivity of vapor-phase furfural decarbonylation/hydrogenation reaction network as a function of size and shape of Pt nanoparticles, *Nano Letters*, 12 (2012) 5196–5201.

26. Q.-X. Cai, J.-G. Wang, Y.-G. Wang, D. Mei, Mechanistic insights into the structure-dependent selectivity of catalytic furfural conversion on platinum catalysts, *Aiche Journal*, 61 (2015) 3812–3824.

27. S. Wang, V. Vorotnikov, D.G. Vlachos, Coverage-induced conformational effects on activity and selectivity: Hydrogenation and decarbonylation of furfural on Pd(111), *ACS Catalysis*, 5 (2015) 104–112.

28. A. Cukalovic, C.V. Stevens, Feasibility of production methods for succinic acid derivatives: A marriage of renewable resources and chemical technology, *Biofuels, Bioproducts and Biorefining*, 2 (2008) 505–529.

29. R.C.A.V. Bridgwater, P.W. Smith, Report "Identification and market analysis of most promising added-value products to be co-produced with the fuels", Deliverable of European project "BIOREF-INTEG", http://www.bioref-integ.eu/fileadmin/bioref-integ/user/documents/D2total_including_D2.1_D2.2_D2.3_.pdf (2010).

30. M. Besson, P. Gallezot, C. Pinel, Conversion of biomass into chemicals over metal catalysts, *Chemical Reviews*, 114 (2014) 1827–1870.

31. S.K.S.K.W. Hutchenson, Hydrogenation process for the preparation of tetrahydrofuran and alkylated derivatives thereof, EP 2417116, 2012.

32. C. Nalepa, Preparation of alkanediols, US Patent 4,476,332, 1984.

33. C.J. Nalepa, Preparation of alkanediols, US Patent 4,475,004, 1984.

34. W.W. Prichard, Conversion of furan to 1,4 butanediol and tetrahydrofuran, US patent 4,146,741, 1979.

35. R.P.R. Fischer, Method of producing 1,4-Butanediol and tetrahydrofuran from furan, US Patent 5,905,159, 1999.

36. C. Godawa, A. Gaset, P. Kalck, Y. Maire, Mise en oeuvre d'un catalyseur actif pour l'hydrogenation selective du furanne en tetrahydrofuranne, *Journal of Molecular Catalysis*, 34 (1986) 199–212.

37. E.J. Garcia-Suarez, A.M. Balu, M. Tristany, A.B. Garcia, K. Philippot, R. Luque, Versatile dual hydrogenation-oxidation nanocatalysts for the aqueous transformation of biomass-derived platform molecules, *Green Chemistry*, 14 (2012) 1434–1439.

38. S. David Jackson, Processes occurring during deactivation and regeneration of metal and metal oxide catalysts, *Chemical Engineering Journal*, 120 (2006) 119–125.

39. S.D. Jackson, A.S. Canning, E.M. Vass, S.R. Watson, Carbon laydown associated with furan hydrogenation over palladium/zirconia, *Industrial and Engineering Chemistry Research*, 42 (2003) 5489–5494.

40. F. Li, T. Lu, B. Chen, Z. Huang, G. Yuan, Pt nanoparticles over TiO_2-ZrO_2 mixed oxide as multifunctional catalysts for an integrated conversion

of furfural to 1,4-butanediol, *Applied Catalysis A: General*, 478 (2014) 252–258.

41. Q. Li, Y. Xu, C. Liu, J. Kim, Catalytic synthesis of thiophene from the reaction of furan and hydrogen sulfide, *Catalysis Letters*, 122 (2008) 354–358.
42. A.V. Mashkina, L.N. Khairulina, Activity of catalysts in thiophene synthesis from furan and hydrogen sulfide, *Kinetics and Catalysis*, 49 (2008) 245–252.
43. P.B. Venuto, P.S. Landis, *Advances in Catalysis*, 18 (1968) 259–371.
44. J. Shaw, W. Sattich, Thiophene compound synthesis method, BE 1008868 (A3), 1996.
45. R.H.G.A.B.B. T.E. Deger, Belgian patent 623801, 1963.
46. H. Song-Qing, Y. Jian-Ye, S. Xin, G. Ai-Ling, H. Jian-Chun, Experimental and theoretical research on catalytic synthesis of thiophene from furan and H_2S, *Brazilian Journal of Chemical Engineering*, 28 (2011) 95–99.
47. V. Estevez, M. Villacampa, J.C. Menendez, Multicomponent reactions for the synthesis of pyrroles, *Chemical Society Reviews*, 39 (2010) 4402–4421.
48. A.L. Harreus, *Ullmann's Encyclopedia of Industrial Chemistry*, Wiley-VCH, Weinheim, 2002.
49. Y.S. Higasio, T. Shoji, Heterocyclic compounds such as pyrroles, pyridines, pyrollidins, piperdines, indoles, imidazol and pyrazins, *Applied Catalysis A: General*, 221 (2001) 197–207.
50. C.L. Wilson, Reactions of furan compounds. Part VI. Formation of indole, pyrrocoline, and carbazole during the catalytic conversion of furan into pyrrol, *Journal of Chemical Society*, 63 (1945) 63–64.
51. J.K. Jurjew, Katalytische Umwandlungen von heterocyclischen Verbindungen, I. Mitteil.: Umwandlungen von Furan in Pyrrol und Thiophen, *Chemische Berichte*, 69 (1936) 440.
52. K. Hatada, M. Shimada, K. Fujita, Y. Ono, T. Keii, Ring transformations of oxygen containing heterocycles into nitrogen containing heterocycles over synthetic zeolites, *Chemistry Letters*, 3 (1974) 439–442.
53. C. Bordner, Production of pyrroles, US 2600289 (A), 1952.
54. Z. Lou, M. Jia, C. Xue, Preparation of pyrrole from furan and ammonia with presence of water vapor, *CIESC Journal*, 54 (2003) 407–413.

Catalytic Oxidation of Furfural to C_4 Diacids-anhydrides and Furanones

Manuel López Granados*

Sustainable Chemistry and Energy Group (EQS),
Institute of Catalysis and Petrochemistry (CSIC),
C/Marie Curie 2, Cantoblanco 28049 Madrid, Spain

3.11.1. Introduction

A variety of C_4 oxygenated products with industrial interest, namely, maleic anhydride, C_4 dicarboxylic acids (like succinic, malic, and fumaric acids) and furanones, can be obtained by oxidation of furfural. Maleic anhydride (MA) is a petrochemical commodity with a market volume of greater than 1,600 kton/year [5–8] and with multiple applications (Figure 3.11.1), including the production of pharmaceuticals, agrochemicals, unsaturated polyester resins, vinyl copolymers, tetra- and hexa-hydrophthalic anhydrides (used as curing agents for epoxy resins), and polyalkenyl succinic anhydrides (like polyisobutylene succinic anhydride, used as a lubricant additive) [6, 7]. Other important derivatives of MA are succinic acid, 1,4 butanediol (BDO) and tetrahydrofuran (THF) which are obtained by hydrogenation processes. Actually around 30% of the global production of MA ends up as BDO [11, 12]. More recently the technical and economic viability of the production of renewable phthalic

*Corresponding author: mlgranados@icp.csic.es

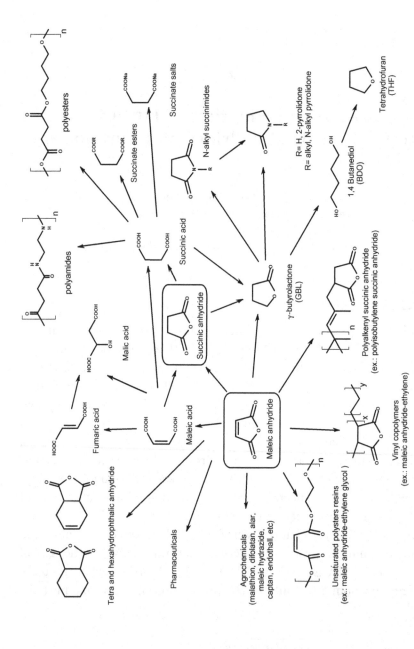

Figure 3.11.1. Summary of the most important uses of maleic anhydride/acid and succinic anhydride/acid.

anhydride has also been demonstrated, the process involved Diels–Alder cycloaddition of MA and furan and further cyclodehydration of the Diels–Alder adduct [14–18].

Maleic acid (MAc), with circa of 25 kton/year of market volume, is produced by hydration of maleic anhydride. It is used as an acidulant in certain beverages and in the production of succinic, fumaric and malic acids (the latter, a food and beverage additive, is also used in the pharmaceutical industry). Fumaric acid has a market volume of approximately $12\,\text{kton}\cdot\text{year}^{-1}$ and is used as an acidulant in baking powders and beverages, as an additive in animal food, and in the synthesis of aspartic acid. It presents fewer environmental restrictions and is a non-toxic reactant. It can substitute maleic anhydride as a monomer for manufacturing unsaturated polyester resins and copolymers because for certain resin applications, substitution of MAc with fumaric acid results in improved hardness [5, 7, 8].

Succinic acid (SAc) with a market volume of around 80 kTon/year in 2011 [11] is nowadays obtained both through the hydrogenation of MA or MAc and the fermentation of biomass sugars (a renewable route). Fermentation route initially accounted for a small share of the global production but it is gaining competitiveness and reaching similar prices to that derived from oil [11, 19, 20]. Figure 3.11.1 highlights the main industrial applications of SAc; more detailed reviews are available elsewhere [12, 21, 22]. Succinic acid-derived salts, esters, and imides have applications in the pharmaceutical, food, and agrochemical industries and as solvents and cooling and de-icing compounds. Polymerization of SAc with diamines or with diols has also been demonstrated to afford polyamides or polyesters with interesting properties, such as commercially available Bionolle® (a polybutylene succinate-based polymer). Hydrogenation and dehydration reactions are in fact involved in one of the current petrochemical routes utilized to produce γ-butyrolactone (GBL), 1,4-butanediol (BDO), and tetrahydrofuran from MA. Consequently the latter products can also be derived from SAc. These petrochemical compounds find numerous applications as solvents and as intermediates in the synthesis of pharmaceuticals, agrochemicals, and polymers. For example, GBL is an intermediate in the synthesis of

2-pyrrolidone, N-alkyl or N-vinyl pyrrolidones and the corresponding succinimides. BDO is mainly used for the production of THF and polybutylene terephthalate [23]. It must also be mentioned that, although still far from commercial application, the dehydrogenation of SA to MA has also been demonstrated [12] what implies that, in practice, the synthesis of any of these two C_4 diacids entails the synthesis of the rest compounds depicted in Figure 3.11.1.

The obtaining of the aforementioned C_4 anhydrides and dicarboxylic acids through the partial oxidation of FUR has been technically demonstrated on the laboratory scale both in gas phase and liquid phase. O_2 from air has been proposed as an oxidant for the gas-phase oxidation while both O_2 and H_2O_2 have been reported in the liquid phase.

3.11.2. Gas-phase oxidation of furfural with O_2

Apparently the main features of this transformation should be already well understood because the earliest investigations on fixed-bed catalytic gas-phase oxidation of FUR date back to the first decades of the last century [1, 24–28]. Thus, Vanadium oxide-based catalysts (V oxide, V–Mo, and V–Bi mixed oxides) are among the best tested catalysts. Short contact times (a few seconds) and temperatures between 473 and 573 K are required to reach the highest yield of MA. Yield as high as 90% has been reported. However, these pioneer works lack of technical details and systematic analysis. Since then, only few technical reports have been published. As a consequence, many key technical details like the catalytic active species or the deactivation routes still remain clusive.

In the temperature range of 493–753 K the FUR oxidation proceeds in a parallel mode to either MA or to CO_x [29–31]. Higher temperatures are required to consecutively oxidize MA to carbon oxides. The reaction rates have been fitted by assuming a redox Mars–van Krevelen mechanism in which the active site is reduced by FUR and reoxidized by gaseous O_2 to reoxidize the catalyst. The reoxidation of the reduced catalyst was proposed to be the rate-determining step. The reaction order was found to be unity for the consumption of FUR and O_2.

Furfural resins are inevitably formed during the gas-phase oxidation of FUR decreasing the selectivity to MA. These heavy products either remain in the catalytic bed or migrates downstream the reactor if the melting point allows it. They can be formed via the homogeneous resinification of FUR itself but also via the surface-catalyzed or homogeneous reactions between FUR and intermediates (or by-products) [2, 32].

Regarding the effect of the variables of operation, the selective route to MA is favored by increasing the reaction temperature because the rate-determining step (the re-oxidation of the reduced vanadium oxide) is then accelerated. The same positive effect has been found for the increase of O_2 concentration: higher O_2 concentration in the gas phase decreases the surface FUR coverage and consequently the condensation of FUR to heavy resins molecules at the surface is inhibited. The addition of water has also been shown to promote the MA formation: supposedly water displaces FUR from surface sites preventing the FUR condensation and additionally inhibiting the homogeneous auto-oxidation of FUR with gaseous O_2. It has also been proposed that the deposition of heavy resin products on the active catalyst sites has a kinetic effect inhibiting the reaction rate [2].

Basically, two routes of transformation have been proposed; they are summarized in Figure 3.11.2. In both cases the oxidation to

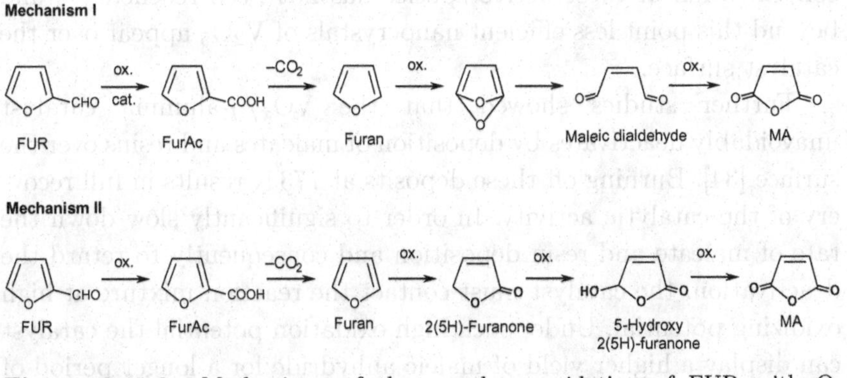

Figure 3.11.2. Mechanisms of the gas-phase oxidation of FUR with O_2 (adapted from Refs. [1, 2]).

furoic acid (FurAc) with subsequent decarboxylation are the first two steps. Milas *et al.* proposed that furan (mechanism I) undergoes the addition of an oxygen atom in the 1,4 positions of the ring [1]. Subsequent ring opening affords maleic dialdehyde, and successive oxidation affords maleic anhydride. A second route (i.e., mechanism II) has been proposed by Slavisnkaya *et al.* [2,32] in which once furan is produced by decarboxylation, it is oxidized to furan-2(5H)-one (also known as 2(5H)-furanone) by insertions of an O atom in position 2 of the ring and the latter oxidized to 5-hydroxy-furan-2(5H)-one (also known as 5-hydroxy-2(5H)-furanone) by O insertion in the position 5. Neither of these proposed mechanisms has been supported by any experimental evidence, either kinetic or spectrometric or by theoretical studies. In this regard, the formation of minor amounts of furan and 2(5H)-furanone have been reported [33], indicating that these compounds must be considered in the reaction mechanism.

More recently a structure-activity study conducted on VO_x/γ-alumina catalysts has showed that surface polyvanadate species present higher intrinsic reaction rates for MA formation than highly dispersed isolated vanadates or crystalline V_2O_5 species [33]. Consequently, the best alumina-supported vanadium oxide catalyst is that with the largest amount of well-dispersed vanadium oxide species. The maximum surface concentration of well-dispersed vanadium oxide species is attained in the range of 8–10 atoms of V per nm^2 of alumina support. At lower V loading the maximum surface concentration of these active species has not been reached yet and beyond this point less efficient nanocrystals of V_2O_5 appear over the catalyst surface.

Further studies showed that the VO_x/γ-alumina catalyst unavoidably deactivates by deposition of maleates and resins over the surface [34]. Burning off these deposits at 773 K results in full recovery of the catalytic activity. In order to significantly slow down the rate of maleate and resin deposition and consequently to retard the deactivation, the catalyst must contact the reaction mixture at high oxidizing potential. Under such high oxidation potential the catalyst can display a higher yield of maleic anhydride for a longer period of time A higher oxidizing potential at a given high temperature can be accomplished by increasing the O_2/furfural mole ratio. An easy rule of thumb is to initially contact the catalyst with the reaction mixture at temperatures at which the furfural conversion is (very close to) full conversion (above 80–90%). The DRIFT studies also demonstrated that if the primary contact occurs at temperatures at which the temperature of the initial contact is low (and so is the furfural conversion), intense deposition of maleates and resins takes place and the catalyst is rapidly deactivated. The increase of the temperature does not result in removal of deposits but, on the contrary, accelerates their deposition. Thus, by initial contacting at 573 K with 1 vol.% furfural and 20 vol.% O_2, the maleic anhydride yield was initially close to 75% and above 60% after 15 h on stream. On the contrary by starting the contact at 523 K and then increasing the temperature, the yield of maleic anhydride never exceeded 30% [34].

3.11.3. Aqueous-phase oxidation with O_2

The oxidation of furfural to MA has also been explored in liquid phase. Lower temperatures than those of the gas phase oxidation are required (363–373 K) but on the other hand higher O_2 pressure (2 MPa) are needed. Both homogeneous and heterogeneous catalysts have been explored. Regarding the homogeneous process, a MAc yield of 49% at ∼100% conversion (368 K, 2 MPa O_2) has been reported by using phosphomolybdic acid, $Cu(NO_3)_2$ is required as co-catalysts [35]. In this case, water was used as solvent and MAc, rather than MA, was formed. The use of organic solvents immiscible with water and with high affinitiy to FUR (tetrachloromethane, nitrobenzene, toluene, cyclohexane, and others) has also been evaluated to overcome the problem of the unselective oxidation of furfural to resins when O_2 is present [9]. The organic solvent extracts FUR, lowering its concentration in the aqueous phase, where oxidation takes place because phosphomolybdic acid is soluble in water. Furfural apparently cannot polymerize in the organic solvent. Product separation and reutilization can in principle be feasible because maleic acid is essentially concentrated in the aqueous layer. The extraction of maleic acid from the aqueous phase would allow reutilization of the catalyst. The organic solvent containing unconverted FUR could also be recycled for further reaction runs. However this protocol has not experimentally been tested. This reaction has also been conducted in pure organic solvents and the highest yield was obtained in acetonitrile/acetic acid mixtures (2/1.3 v/v) [36]. H_2O was not initially present and maleic anhydride (54% yield), rather than MAc, was preferentially formed. It must be remarked that in this medium, 5-acetoxy-2(5H)-furanone, a high added-value product with important pharmaceutical applications, was also detected (7.5% yield). As in the other tests, no reutilization tests were carried out.

The reaction has also been attempted using solid catalysts that present the advantage of an easy recycling of the catalyst by simple filtration or centrifugation. Vanadium based catalysts are those so far attempted: a combined MA and MAc yield of 65% was accomplished with a Mo–V mixed oxide (Mo/V = 4) catalyst (2 MPa O_2, 393 K and 16 h), and a combined yield of 62% with a catalyst based on vanadyl species anchored on graphene oxide (2 MPa O_2, 363 K and 8 h) [37, 38]. Those high yields were only obtained by using acetic acid as solvent, when other solvents like acetonitrile, MIBK, ethanol, DMSO or toluene are used the yield is severely reduced. The same occurs when the reaction is conducted with aqueous solution. The latter feature and the deactivation of both catalysts because of leaching of the metal active components is the major drawback to overcome in the reaction driven by solid catalysts.

It is well accepted that reaction proceeds via a radical mechanism because the reaction is severely suppressed by the incorporation of free radical inhibitors [9, 37]. The following radical mechanism has been proposed for phosphomolybdic acid [9]. Reaction starts with the formation of a furfural radical (Figure 3.11.3, species 1) through H abstraction by either O_2 and/or phosphomolybdic acid. Radical 1 can either be the starting point for polymerization to resin-like compounds or can be transformed to a furfural cation 2 by electron transfer to the catalyst. Cation 2 can undergo a nucleophilic attack by water to form species 3 (5-hydroxy furfural), which can undergo a 1,4 rearrangement to form compound 4. The latter subsequently hydrolyses to compound 5, which is then oxidized to MAc. Compound 3

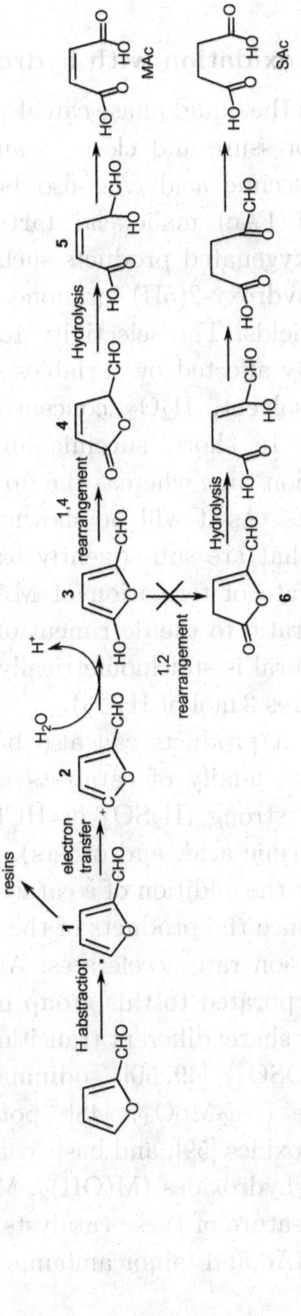

Figure 3.11.3. Proposed mechanism for the oxidation of FUR with O$_2$ in liquid phase (adapted from Ref. [9]).

can also undergo a 1,2 rearrangement to compound 6, which can be successively transformed to SAc. Because SAc is not formed, the 1,2 rearrangement route appears to be blocked.

3.11.4. Liquid-phase oxidation with hydrogen peroxide

The oxidation of FUR in the liquid phase can also be conducted using H_2O_2 at atmospheric pressure and close to ambient temperature. Besides maleic acid, succinic acid can also be obtained in quite high yield. Fumaric acid (FAc), malic acid, tartaric acid, and formic acid, as well as other oxygenated products such as 2(5H)-furanone, 2(3H)-furanone, and 5-hydroxy-2(5H)-furanone, are also formed but in principle, at lower yields. The selectivity to the different acids and furanones is strongly affected by variables such as the reaction temperature, type of solvent, H_2O_2 concentration, and type of catalyst [10, 13, 39–58]. In short, succinic and maleic acids are obtained at longer reaction time whereas the furanones predominate at shorter reaction times. As it will be shown later furanones are intermediate products that are subsequently oxidized to MAc and SAc. In addition, the rate of formation of MAc is accelerated at higher H_2O_2/FUR mol ratio to the detriment of that of SAc (2 mol of H_2O_2 per mol of furfural is stoichiometrically needed to produce SAc whereas MAc requires 3 mol of H_2O_2).

Selectivity to different products can also be modulated by the type of catalysts. A first family of catalysts are those possessing Brønsted acidity, either strong (H_2SO_4 or HCl) or weak (organic acids like acetic acid, formic acid, and others). In fact the reaction can also proceed without the addition of a catalyst because although the initial rate is slow, once the products of the reaction are formed (organic acids) the reaction rate accelerates. Another two types of compounds can be incorporated to this group based on the similar reaction mechanism they share: different transition metal compounds like vanadyl sulfate ($VOSO_4$) [49, 50], sodium vanadate ($NaVO_4$) [43], sodium molybdate (Na_2MoO_4) [48], potassium dichromate ($K_2Cr_2O_7$) [10], and Nb oxides [59], and basic compounds like NaOH and other alkaline earth hydroxides ($M(OH)_2$, M = Mg, Ca, Sr and Ba) [60]. The common feature of these catalysts is that they render a mixture of SAc and MAc and minor amounts of other C_4 diacids and furanones and that, as we will explain it the next paragraph, they possess a common mechanism of reaction featuring a Baeyer–Villiger oxidation of the aldehydic group of furfural as the initial step of the reaction. A higher yield of C_4 diacids can be achieved with those catalysts containing a tolyl group, namely, Amberlyst-15 and p-toluenesulfonic acid [13, 57, 58, 61]. In these sulfonic catalysts SAc yield is higher than MAc for H_2O_2/furfural mol ratio <7.5. Thus, Amberlyst-15 resulted in SAc and MAc yields of 72% and 14%, respectively, using dilute FUR solutions (close to 2.5 wt.%) at 353 K, a H_2O_2/FUR mole ratio of 4, and 24 h of reaction time. But the MAc yield can be improved (46%) by working at H_2O_2/FUR mol ratios higher than 7.5. SAc is unavoidably formed, albeit in lower yields [61]. Solid Amberlyst-15 presented the advantages over the soluble catalysts of facile separation from the reaction medium by decantation and reusability (reusability was demonstrated for three runs) [13, 57, 58]. Very recently betaine hydrochloride (BHC) has been used as acid catalyst and a yield of MA of 60% was obtained (the yield of its isomer, FumAc, was 30%) at 373 K, 0.5 h and at 6.7 wt.% of furfural concentration [62]. A quite large concentration of BHC (ca. 40 wt.%) is required but it was demonstrated that this soluble catalyst can be reutilized for six runs without visible deactivation. The reutilization required the precipitation of BHC with acetone and then subsequent filtration.

For all the above-mentioned catalysts, there is a strong consensus that the first step of the reaction consists of the Baeyer–Villiger oxidation of FUR to the corresponding 2-formyloxyfuran ester (see Figure 3.11.4), which rapidly undergoes hydrolysis in aqueous solution forming the corresponding 2-hydroxyfuran and formic acid. This step results in practice in the removal of the aldehydic group of furfural forming C_4 molecules. The alcohol is in keto-enol-like equilibrium with either furan-2(3H)-one or with the furan-2(5H)-one isomers. SAc is derived from the former furanone through a hydrolysis step that affords 4-oxobutanoic acid (also known as β-formyl propionic acid), which is further oxidized with H_2O_2 to SAc. In contrast, furan-2(5H)-one can be oxidized with H_2O_2 to 5-hydroxy-furan-2(5H)-one, which is in equilibrium with 4-oxo-2-butenoic acid (also known as 4-oxocrotonic acid or cis-β-formyl acrylic acid); the

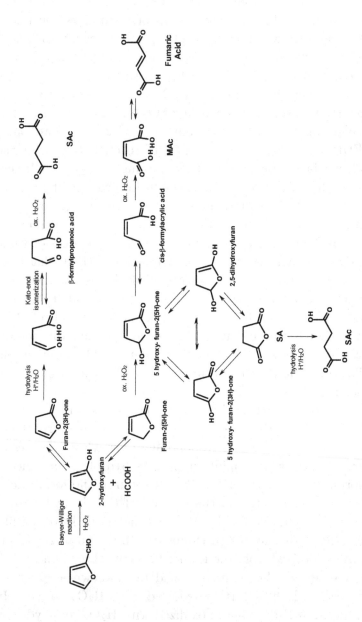

Figure 3.11.4. Proposed mechanism for the oxidation of FUR in aqueous H_2O_2 in the presence of catalysts based either in Brønsted acid groups or in compounds of transition metals (adapted from Refs. [10, 13]).

latter is then further oxidized to MAc (or to its Fumaric acid isomer). Malic acid and Tartaric acid, formed at much smaller yields, are derived from the addition of H$_2$O and H$_2$O$_2$ to the double bonds of MAc and Fumaric acid. More details can be found elsewhere [10, 13, 39–58].

A second group of catalyst is represented by titanium silicalite zeolite (TS-1), in this case MAc predominates and formation SAc is not favored. Thus a MAc yield of *ca.* 80% was reported using a 5 wt.% aqueous solution of FUR at 323 K after 24 h using titanium silicalite (TS-1), whereas only minor yields of 5-hydroxy-furan-2(5H)-one (<10%), HCOOH (<5%), malic acid (<2%), furanones (<2%), and SA (<2%) were observed. It must be remarked that the H$_2$O$_2$/FUR mole ratio used was 7.5, much lower than that needed with the previous groups of catalysts to favor MAc formation [4]. Reutilization studies, conducted under these conditions showed that the catalyst could be reused for six runs without noticeable deactivation [4]. However in longer term reutilization studies, conducted under conditions that favor the detection of causes of deactivation with medium–low impact, leaching of Ti and deposition of by-products were identified as causes of deactivation [3]. Deposits can be removed by burning them off at 773 K and consequently catalytic activity significantly recovered. Leaching is irreversible and threatens the long-term stability of the catalyst.

A different reaction mechanism to that presented in Figure 3.11.4 has been proposed for TS-1 catalyst in order to explain the higher yield of MAc. Reaction does not start with a Baeyer–Villiger oxidation of the aldehydic group but, see Figure 3.11.5, with the epoxidation of the C = C bond at the farthest position of the aldehydic group. Consequently furan-2(3H)-one, the precursor of SAc is not formed. The epoxide so formed is isomerized to (Z)-4-oxopent-2-enedial. The removal of the the original carbonyl group of furfural is accomplished by a Baeyer–Villiger oxidation of the latter compound forming the corresponding ester which is hydrolyzed to *cis-β*-formylacrylic acid, releasing formic acid. The latter C$_4$ compound is in equilibrium with 5-hydroxy-furan-2(5H)-one. Further oxidation of one of these compounds renders MAc.

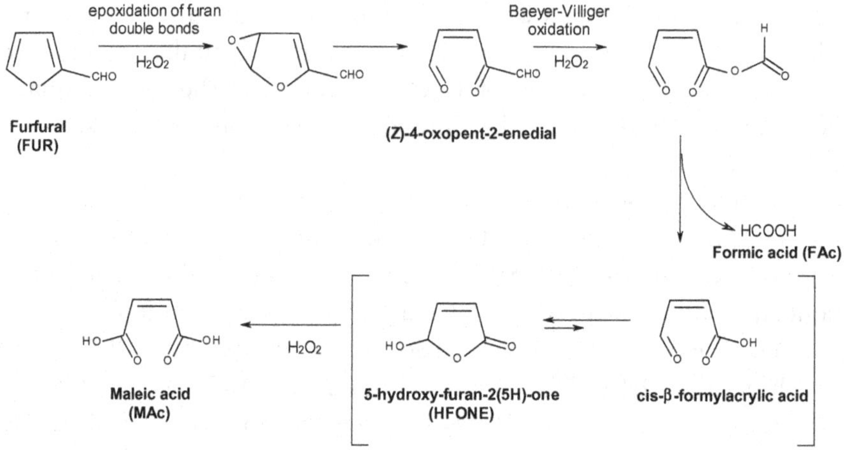

Figure 3.11.5. Proposed mechanism for the oxidation of FUR in aqueous H_2O_2 in TS-1 catalyst (adapted from Refs. [61, 62]).

High yields of MAc has also been accomplished with methyltrioxorhenium either in solution or supported on a variety of polymers [63]. A 70% MAc yield was achieved in the first run after 24 h at 293 K using polystyrene-supported methyltrioxorhenium, 5 wt.% catalyst, 1% HBF_4 as a co-catalyst and 5 equivalents H_2O_2. Succinic and furoic acids were also formed as minor products. The catalyst was reused for four additional runs under the same reaction conditions, and the yield of MAc did not decrease in successive runs (in fact, MAc yields from 77% to 90% were reported).

3.11.5. Oxidation to different furanones

3.11.5.1. *5-hydroxy-furan-2(5H)-one*

5-Hydroxy-furan-2(5H)-one (HFONE) finds applications in the synthesis of pharmaceuticals, insecticides, and fungicides [64, 65]; the synthesis of surfactants has also been proposed [66]. HFONE is the precursor of MAc when oxidizing furfural with H_2O_2 on TS-1 catalyst (see above), therefore relatively high yield of HFONE can be obtained by adjusting the H_2O_2/furfural mol ratio. Thus HFONE yield higher than 60% was achieved at 4.6 wt.% furfural, 323 K, 4.6 wt.%

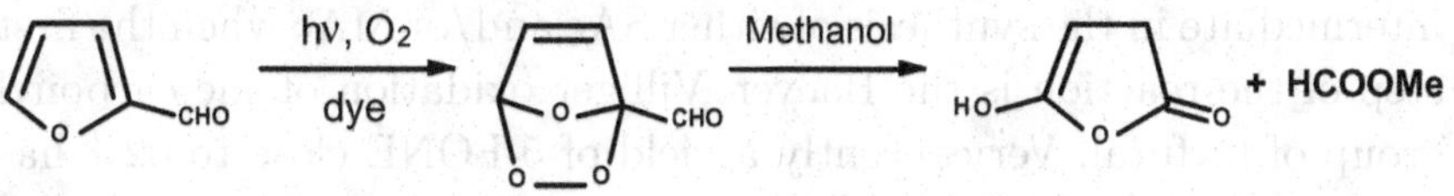

Figure 3.11.6. Scheme of the photo-oxidation of FUR to 5-hydroxy-furan-2(5H)-one.

of TS-1 catalyst after 4 h and using a H$_2$O$_2$/furfural mol ratio of 2.4 [4].

Photochemical oxidation of FUR rendering HFONE has also been technically demonstrated. Photosensitizer dyes, like rose bengal or methylene blue, are required to absorb in the visible light region [64–67]. The reaction of FUR with the singlet oxygen generated during the photochemical sensitization of O$_2$ results in the formation of endo-peroxide (Figure 3.11.6), which is decarbonylated to afford the HFONE. Methanol was selected as the solvent and consequently methyl formate is also formed though its reaction with the formic acid released by the decarbonylation [66]. Although long-term operation assays to test the reutilization of the very expensive sensitizer dyes were not presented, synthesis with sunlight in solar photoreactors has been technically demonstrated [64] laying the first stones for an economical and green synthesis of this fine chemical.

3.11.5.2. *Furan-2(5H)-one*

The interest in the synthesis of furan-2(5H)-one family compounds is justified by their frequent biological activity [68–71] and possible applications as plant growth regulator, antiulcer and fish growth promoter [72]. More recently it has been showed that furan-2(5H)-one (5-FONE) can be readily converted to gamma-butyrolactone (GBL) by hydrogenation [73,74]. The importance of GBL, currently a petrochemical with multiple applications, has been summarized at the beginning of this chapter.

There exist a number of synthetic routes to obtain 5-FONE but all of them suffer of extreme reaction conditions and expensive reactants as well of low yields [68]. A more sustainable process is the oxidation of furfural with H$_2$O$_2$ because as shown above, is an intermediate in the synthesis of either SAc and/or MAc when the first step of the reaction is the Baeyer–Villiger oxidation of the carbonyl group of furfural. Very recently a yield of 5-FONE close to 62% has been reached by using a biphasic water/ethyl acetate system and formic acid as catalyst (formic acid/furfural mol ratio = 0.8) after 2 h at 333 K, H_2O_2/furfural mol ratio = 1.9 and a furfural conc. of around 40% [68]. In the absence of formic acid as catalyst the yield is considerably lower (37%) [72] and the utilization of other solvents like water, isopropanol, THF or GBL results in the deeper oxidation to either SAc or MAc [68]. It must be also remarked that when using basic catalysts like NaOH or alkaline earth hydroxides like M(OH)2 where M = Mg, Ca, Sr and Ba) 5-FONE yield is also larger than SAc yield [60].

3.11.6. Summary and future prospects

This chapter summarizes the main routes to selectively oxidize furfural to important C_4 feedstocks like furanones, maleic anhydride, maleic acid and succinic acid. Catalytic oxidation processes in gas phase (using O_2) and liquid phase (either with O_2 or H_2O_2) have been technically demonstrated with reasonably high yield to the compounds of interest.

However important challenges remain to accomplish the economic viability when confronted with the current commercial processes. Maleic anhydride and maleic acid are important petrochemical commodities and succinic has a petrochemical and biorenewable profile, the latter through fermentation of sugars. In this context the techno-economic evaluations of the chemical processes described in this chapter are essential to assess on their viability and to reveal their bottle-necks and weaknesses and where important technical developments are required. Unfortunately, none has been so far conducted or disclosed.

Even if these analyses are not available yet, the development of more selective and stable catalysts, especially for the liquid-phase oxidation case, seems critical. The research in new technological approaches may also result in substantial improvements in the productivity and in the long-term operation stability, like for example recirculating fluidized catalytic bed for the gas-phase reaction or new solvents or using biphasic reactors in the liquid oxidation case.

Fundamental studies on the routes and mechanisms of reaction are also urgently needed because they are key in the design of more active and stable catalysts. The careful investigation of the causes of deactivation will also have a translation in both the design of more active and stable catalysts and the development of very effective regeneration processes.

Processing low-grade furfural will have a very positive impact in the economic viability because is much cheaper than high-grade furfural, the latter requires a very energy intensive double distillation. However all the investigation accomplished so far has been conducted with high-grade furfural and consequently there is a lack of investigations on the utilization of low-grade furfural.

List of abbreviations

BDO	1,4 Butanediol
GBL	γ-Butyrolactone
HFONE	5-Hydroxy-furan-2(5H)-one
5-FONE	Furan-2(5H)-one
FumAc	Fumaric acid
FUR	Furfural
FurAc	Furoic acid
MA	Maleic anhydride
MAc	Maleic acid
THF	Tetrahydrofuran
SA	Succinic anhydride
SAc	Succinic acid

References

1. N.A. Milas, W.L. Walsh, Catalytic oxidations. I. Oxidations in the furan series, *Journal of the American Chemical Society*, 57 (1935) 1389–1393.
2. V.A. Slavinskaya, D.R. Kreile, E. Dzilyuma, D. Sile, Incomplete catalytic oxidation of furan compounds (review), *Chemistry of Heterocyclic Compounds*, 13 (1978) 710–721.

3. A.C. Alba-Rubio, J.L.G. Fierro, L. León-Reina, R. Mariscal, J.A. Dumesic, M. López Granados, Oxidation of furfural in aqueous H_2O_2 catalysed by titanium silicalite: Deactivation processes and role of extraframework Ti oxides, *Applied Catalysis B: Environmental*, 202 (2017) 269–280.

4. N. Alonso-Fagundez, I. Agirrezabal-Telleria, P.L. Arias, J.L.G. Fierro, R. Mariscal, M.L. Granados, Aqueous-phase catalytic oxidation of furfural with H2O2: high yield of maleic acid by using titanium silicalite-1, *RSC Advances*, 4 (2014) 54960–54972.

5. R.C.A.V. Bridgwater, P.W. Smith, Report "Identification and market analysis of most promising added-value products to be co-produced with the fuels", Deliverable of European project "BIOREF-INTEG", http://www.bioref-integ.eu/fileadmin/bioref-integ/user/documents/D2total_including_D2.1_D2.2_D2._.pdf (2010).

6. H.H.K. Lohbeck, W. Fuhrmann, N. Fedtke, Maleic and fumaric acids, in: *Ullmann's Encyclopedia of Industrial Chemistry*, Weinheim, Germany, 2000, pp. 463–473.

7. J.C.B.T.R. Felthouse, B. Horrell, M.J. Mummey, Yeong-Jen Kuo, Maleic anhydride, maleic acid and fumaric acid, in: *Kirk-Othmer Encyclopedia of Chemical Technology Online*, 2001.

8. R.M. Contractor, Dupont's CFB technology for maleic anhydride, *Chemical Engineering Science*, 54 (1999) 5627–5632.

9. H.J. Guo, G.C. Yin, Catalytic aerobic oxidation of renewable furfural with phosphomolybdic acid catalyst: An alternative route to maleic acid, *Journal of Physical Chemistry C*, 115 (2011) 17516–17522.

10. L.A. Badovskaya, V.M. Latashko, V.V. Poskonin, E.P. Grunskaya, Z.I. Tyukhteneva, S.G. Rudakova, S.A. Pestunova, A.V. Sarkisyan, Catalytic oxidation of furan and hydrofuran compounds. 7. Production of 2(5H)-furanone by oxidation of furfural with hydrogen peroxide and some of its transformations in aqueous solutions, *Chemistry of Heterocyclic Compounds*, 38 (2002) 1040–1048.

11. R. Taylor, L. Nattrass, G. Alberts, P. Robson, C. Chudziak, A. Bauen, I.M. Libelli, G. Lotti, M. Prussi, R. Nistri, D. Chiaramonti, A.M. López-Contreras, H.H. Bos, G. Eggink, J. Springer, R. Bakker, R.V. Ree, From the sugar platform to biofuels and biochemicals: Final report for the European Commission Directorate-General Energy, E4tech/Re-CORD/Wageningen UR, 2015, http://ec.europe.eu/energy/sites/ener/files/documents/ECSugarPlatformfinalreport.pdf.

12. A. Cukalovic, C.V. Stevens, Feasibility of production methods for succinic acid derivatives: A marriage of renewable resources and chemical technology, Biofuels, *Bioproducts and Biorefining*, 2 (2008) 505–529.

13. H. Choudhary, S. Nishimura, K. Ebitani, Highly efficient aqueous oxidation of furfural to succinic acid using reusable heterogeneous acid catalyst with hydrogen peroxide, *Chemistry Letters*, 41 (2012) 409–411.

14. S. Thiyagarajan, H.C. Genuino, M. Śliwa, J.C. Van Der Waal, E. De Jong, J. Van Haveren, B.M. Weckhuysen, P.C.A. Bruijnincx, D.S. Van Es,

Substituted phthalic anhydrides from biobased furanics: A new approach to renewable aromatics, *ChemSusChem*, 8 (2015) 3052–3056.

15. Z. Lin, M. Ierapetritou, V. Nikolakis, Phthalic anhydride production from hemicellulose solutions: Technoeconomic analysis and life cycle assessment, *Aiche Journal*, **61** (2015) 3708–3718.

16. S. Giarola, C. Romain, C.K. Williams, J.P. Hallett, N. Shah, Techno-economic assessment of the production of phthalic anhydride from corn stover, *Chemical Engineering Research and Design*, 107 (2016) 181–194.

17. S. Giarola, C. Romain, C.K. Williams, J.P. Hallett, N. Shah, Production of phthalic anhydride from biorenewables: Process design, in: *Computer Aided Chemical Engineering*, Elsevier, 2015, pp. 2561–2566.

18. E. Mahmoud, D.A. Watson, R.F. Lobo, Renewable production of phthalic anhydride from biomass-derived furan and maleic anhydride, *Green Chemistry*, 16 (2014) 167–175.

19. BioConSept project, Determination of market potential for selected platform chemicals, https://www.cbp.fraunhofer.de/content/dam/cbp/en/docu ments/BioConSepT_Market-potential-for-selected-platform-chemicals_repot t1.pdf, 2013.

20. J.M. Pinazo, M.E. Domine, V. Parvulescu, F. Petru, Sustainability metrics for succinic acid production: A comparison between biomass-based and petrochemical routes, *Catalysis Today*, 239 (2014) 17–24.

21. C.S.K. Lin, R. Luque, J.H. Clark, C. Webb, C. Du, Wheat-based biorefining strategy for fermentative production and chemical transformations of succinic acid, *Biofuels, Bioproducts and Biorefining*, 6 (2012) 88–104.

22. I. Bechthold, K. Bretz, S. Kabasci, R. Kopitzky, A. Springer, Succinic acid: A new platform chemical for biobased polymers from renewable resources, *Chemical Engineering and Technology*, 31 (2008) 647–654.

23. C. Delhomme, D. Weuster-Botz, F.E. Kühn, Succinic acid from renewable resources as a C4 building-block chemical — A review of the catalytic possibilities in aqueous media, *Green Chemistry*, 11 (2009) 13–26.

24. W.V. Sessions, Catalytic oxidation of furfural in the vapor phase, *Journal of the American Chemical Society*, 50 (1928) 1696–1698.

25. F. Zumstein, Process for preparing maleic anhydride and maleic acid, US Patent 1,956,482, 1934.

26. E.R. Nielsen, US Patent 2,464,285, 1949.

27. E.R. Nielsen, US Patent 2,421,428, 1947.

28. E.R. Nielsen, Vapor phase oxidation of furfural, *Industrial and Engineering Chemistry*, 41 (1949) 365–368.

29. K. Rajamani, P. Subramanian, M.S. Murthy, Kinetics and mechanism of vapor phase oxidation of furfural over tin vanadate catalyst, *Industrial & Engineering Chemistry Process Design and Development*, 15 (1976) 232–234.

30. M.S. Murthy, K. Rajamani, P. Subramanian, Mechanism of vapour phase oxidation of furfural on vanadium catalyst, *Chemical Engineering Science*, 30 (1975) 1529.

31. M.S. Murthy, K. Rajamani, Kinetics of vapour phase oxidation of furfural on vanadium catalyst, *Chemical Engineering Science*, 29 (1974) 601–609.

32. D.R. Kreile, V.A. Slavinskaya, M.V. Shimanskaya, E.Y. Lukevits, The reactivity of furan compounds in vapor-phase catalytic oxidation, *Chemistry of Heterocyclic Compounds*, 5 (1972) 429–430.

33. N. Alonso-Fagundez, M.L. Granados, R. Mariscal, M. Ojeda, Selective conversion of furfural to maleic anhydride and furan with VOx/Al_2O_3 catalysts, *ChemSusChem*, 5 (2012) 1984–1990.

34. N. Alonso-Fagúndez, M. Ojeda, R. Mariscal, J.L.G. Fierro, M. López Granados, Gas phase oxidation of furfural to maleic anhydride on V_2O_5/γ-Al_2O_3 catalysts: Reaction conditions to slow down the deactivation, *Journal of Catalysis*, 348 (2017) 265–275.

35. S. Shi, H.J. Guo, G.C. Yin, Synthesis of maleic acid from renewable resources: Catalytic oxidation of furfural in liquid media with dioxygen, *Catalysis Communications*, 12 (2011) 731–733.

36. J. Lan, Z. Chen, J. Lin, G. Yin, Catalytic aerobic oxidation of renewable furfural to maleic anhydride and furanone derivatives with their mechanistic studies, *Green Chemistry*, 16 (2014) 4351–4358.

37. X. Li, B. Ho, Y. Zhang, Selective aerobic oxidation of furfural to maleic anhydride with heterogeneous Mo-V-O catalysts, *Green Chemistry*, 18 (2016) 2976–2980.

38. G. Lv, C. Chen, B. Lu, J. Li, Y. Yang, T. Deng, Y. Zhu, X. Hou, Vanadium-oxo immobilized onto Schiff base modified graphene oxide for efficient catalytic oxidation of 5-hydroxymethylfurfural and furfural into maleic anhydride, *RSC Advances*, 6 (2016) 101277–101282.

39. V.G. Kul'nevich, L.A. Badovskaya, G.F. Muzychenko, 2-Formyloxyfuran as an intermediate in the hydrogen peroxide oxidation of furfuraldehyde, *Chemistry of Heterocyclic Compounds*, 6 (1973) 535–537.

40. G.F. Muzychenko, L.A. Badovskaya, V.G. Kul'nevich, Role of water in the oxidation of furfural with hydrogen peroxide, *Chemistry of Heterocyclic Compounds*, 8 (1972) 1311–1313.

41. L.A. Badovskaya, V.G. Kul'nevich, Influence of the reaction conditions on the nature of the hydrogen peroxide oxidation products of furfural, *Chemistry of Heterocyclic Compounds*, 5 (1972) 146–150.

42. V.V. Poskonin, L.A. Badovskaya, Unusual conversion of 5-hydroxy-2(5H)-furanone in aqueous solution, *Chemistry of Heterocyclic Compounds*, 39 (2003) 594–597.

43. N.K. Strizhov, V.V. Poskonin, O.A. Oganova, L.A. Badovskaya, Polarographic study of reactions in the system 2-furaldehyde-hydrogen peroxide-NaVO3, *Russian Journal of Organic Chemistry*, 37 (2001) 1313–1317.

44. L.A. Badovskaya, T.Y. Kalyugina, V.G. Kul'nevich, Role of the solvent in the oxidation of furfural with hydrogen peroxide, *Chemistry of Heterocyclic Compounds*, 13 (1977) 484–488.

45. L.A. Badovskaya, L.V. Povarova, Oxidation of furans (Review), *Chemistry of Heterocyclic Compounds*, 45 (2009) 1023–1034.

46. R.I. Ponomarenko, L.A. Badovskaya, V.M. Latashko, Catalytic oxidation of furan and hydrofuran compounds. 8. Conversions of 2(5H)-furanone by hydrogen peroxide in media of various pH, *Chemistry of Heterocyclic Compounds*, 38 (2002) 1049–1051.

47. V.V. Poskonin, L.A. Badovskaya, L.V. Povarova, Catalytic oxidation of furan and hydrofuran compounds. 5. Hydroxy- and ethoxydihydrofurans and ethoxyfuran — New products from the reaction of furan with hydrogen peroxide, *Chemistry of Heterocyclic Compounds*, 34 (1998) 900–906.

48. E.P. Grunskaya, L.A. Badovskaya, V.V. Poskonin, Y.F. Yakuba, Catalytic oxidation of furan and hydrofuran compounds. 4. Oxidation of furfural by hydrogen peroxide in the presence of sodium molybdate, *Chemistry of Heterocyclic Compounds*, 34 (1998) 775–780.

49. V.V. Poskonin, L.A. Badovskaya, L.V. Povarova, Catalytic oxidation of furan and hydrofuran compounds. 3. Synthesis of 2,5-diethoxy-2,5-dihydrofuran in the furan-hydrogen peroxide-aqueous ethanol-vanadyl sulfate system, *Chemistry of Heterocyclic Compounds*, 34 (1998) 771–774.

50. V.V. Poskonin, L.A. Badovskaya, Catalytic oxidation of furan and hydrofuran compounds. 2. Oxidation of furfural in the hydrogen peroxide-vanadyl sulfate-sodium acetate system, *Chemistry of Heterocyclic Compounds*, 34 (1998) 646–650.

51. V.V. Poskonin, L.V. Povarova, L.A. Badovskaya, Reactions of catalytic oxidation of furan and hydrofuran compounds. I. General principles of the oxidation of furan in the system hydrogen peroxide-vanadium(IV) compounds depending on the type of solvent and catalyst, *Chemistry of Heterocyclic Compounds*, 32 (1996) 543–547.

52. V.V. Poskonin, L.A. Badovskaya, Reaction of furan compounds with hydrogen peroxide in the presence of vanadium catalysts, *Chemistry of Heterocyclic Compounds*, 27 (1992) 1177–1182.

53. V.V. Poskonin, L.A. Badovskaya, Catalytic oxidation of furan and hydrofuran compounds. 2. Oxidation of furfural in the system hydrogen peroxide vanadyl sulfate sodium acetate, *Khimiya Geterotsiklicheskikh Soedinenii*, (1998) 742–747.

54. E.P. Grunskaya, L.A. Badovskaya, V.V. Poskonin, Y.F. Yakuba, Reactions of catalytic oxidation of furan and hydrofuran compounds. 4. Oxidation of furfural with hydrogen peroxide in the presence of sodium molybdate, *Khimiya Geterotsiklicheskikh Soedinenii*, (1998) 898–903.

55. E.P. Grunskaya, L.A. Badovskaya, T.Y. Kaklyugina, V.V. Poskonin, Kinetics and mechanism of the furan peroxide formation in the reaction of furfural with hydrogen peroxide in the presence or absence of sodium molybdate, *Kinetics and Catalysis*, 41 (2000) 447–450.

56. L.A. Badovskaya, K.A. Latashko, V.V. Poskonin, E.P. Grunskaya, Z.L. Tjukhteneva, S.G. Rudakova, S.A. Pestunova, A.V. Sarkisyan, Reactions of catalytic oxidation of furan and hydrofuran compounds. 7. Preparation of 2(5H)-furanone by oxidation of furfural with hydrogen peroxide and some its conversions in aqueous solutions, *Khimiya Geterotsiklicheskikh Soedinenii*, (2002) 1194–1203.

57. A. Takagaki, S. Nishimura, K. Ebitani, Catalytic transformations of biomass-derived materials into value-added chemicals, *Catalysis Surveys from Asia*, 16 (2012) 164–182.

58. H. Choudhary, S. Nishimura, K. Ebitani, Metal-free oxidative synthesis of succinic acid from biomass-derived furan compounds using a solid acid catalyst with hydrogen peroxide, *Applied Catalysis A: General*, 458 (2013) 55–62.

59. V.V. Poskonin, Catalytic oxidation reactions of furan and hydrofuran compounds 9. Characteristics and synthetic possibilities of the reaction of furan with aqueous hydrogen peroxide in the presence of compounds of niobium(II) and (V), *Chemistry of Heterocyclic Compounds*, 45 (2009) 1177–1183.

60. X. Xiang, B. Zhang, G. Ding, J. Cui, H. Zheng, Y. Zhu, The effect of Mg(OH)2 on furfural oxidation with H2O2, *Catalysis Communications*, 86 (2016) 41–45.

61. N. Alonso-Fagúndez, V. Laserna, A.C. Alba-Rubio, M. Mengibar, A. Heras, R. Mariscal, M.L. Granados, Poly-(styrene sulphonic acid): An acid catalyst from polystyrene waste for reactions of interest in biomass valorization, *Catalysis Today*, 234 (2014) 285–294.

62. N. Araji, D.D. Madjinza, G. Chatel, A. Moores, F. Jérôme, K. De Oliveira Vigier, Synthesis of maleic and fumaric acids from furfural in the presence of betaine hydrochloride and hydrogen peroxide, *Green Chemistry*, 19 (2017) 98–101.

63. A.F.R. Saladino, Process for the oxidation of alcohol and/or aldehyde groups, U.S. Patent (Ed.), 2011.

64. P. Esser, B. Pohlmann, H.D. Scharf, The photochemical synthesis of fine chemicals with sunlight, *Angewandte Chemie International Edition*, 33 (1994) 2009–2023.

65. K.J. Zeitsch, *The Chemistry and Technology of Furfural and its Many By-Products, Sugar Series, Vol. 13*, Elsevier, The Netherlands, 2000.

66. A. Gassama, C. Ernenwein, N. Hoffmann, Photochemical key steps in the synthesis of surfactants from furfural-derived intermediates, *ChemSusChem*, 2 (2009) 1130–1137.

67. A. Gassama, C. Ernenwein, N. Hoffmann, Synthesis of surfactants from furfural derived 2 5H -furanone and fatty amines, *Green Chemistry*, 12 (2010) 859–865.

68. X. Li, X. Lan, T. Wang, Selective oxidation of furfural in a bi-phasic system with homogeneous acid catalyst, *Catalysis Today*, 276 (2016) 97–104.

69. A. Hashem, E. Kleinpeter, The chemistry of 2(5H)-furanones, in: R.K. Alan (Ed.), *Advances in Heterocyclic Chemistry*, Academic Press, 2001, pp. 107–165.

70. Y.S. Rao, Recent advances in the chemistry of unsaturated lactones, *Chemical Reviews*, 76 (1976) 625–694.

71. N.B. Carter, A.E. Nadany, J.B. Sweeney, Recent developments in the synthesis of furan-2(5H)-ones, *Journal of the Chemical Society, Perkin Transactions 1*, (2002) 2324–2342.
72. R. Cao, C. Liu, L. Liu, A convenient synthesis of 2(5H)-furanone, *Organic Preparations and Procedures International*, 28 (1996) 215–216.
73. X. Li, W. Wan, S. Kattel, J.G. Chen, T. Wang, Selective hydrogenation of biomass-derived 2(5H)-furanone over Pt-Ni and Pt-Co bimetallic catalysts: From model surfaces to supported catalysts, *Journal of Catalysis*, 344 (2016) 148–156.
74. X. Li, X. Lan, T. Wang, Highly selective catalytic conversion of furfural to γ-butyrolactone, *Green Chemistry*, 18 (2016) 638–642.

Chapter 3.12

Biofuels and Chemicals from Furfural Condensation Reactions

Irantzu Sádaba* and Manuel López Granados
*Sustainable Chemistry and Energy Group (EQS),
Institute of Catalysis and Petrochemistry (CSIC),
C/Marie Curie 2, Cantoblanco 28049 Madrid, Spain*

3.12.1. Introduction

Reactions of condensation of furfural lead to the formation of new C–C bonds and allow for the transformation of furfural into larger molecules that can be used in different applications, either as chemicals or biofuels. The aldehyde group in furfural is susceptible of aldol condensation reactions with other aldehydes or ketones (Claisen–Schmidt condensation). This is the most studied condensation of furfural, and will be covered in the first section. The second section will focus on the applications of the condensation products as biofuels. Other different types of C–C bond formation reactions from furfural will be addressed in the third section: reductive self-condensation, hydroxyalkylation-alkylation, dehydrogenative cross-coupling, Baylis–Hillman reaction with activated alkenes, Knoevenagel condensations, and Diels–Alder reactions of the double bonds in the furanic ring. Some of these products have also been used to produce biofuels.

---
*Corresponding author: irsz@topsoe.dk

3.12.2. Aldol condensation (Claisen–Schmidt condensation)

A great number of fine chemistry processes use aldol condensation reactions to form C–C bonds [1]. Since furfural does not possess an α-H, the self-aldol condensation with itself is not possible, and the aldol condensation with other carbonyl-containing molecules is denoted Claisen–Schmidt. One of the most studied reactions is the condensation with acetone to form furfurylidene acetone (4-(2-furanyl)-3-buten-2-one, FAc) and difurfurylidene acetone (1,5-bis-(2-furanyl)-1,4-pentadien-3-one, DFAc) as shown in Figure 3.12.1. The reaction occurs through the formation of the aldol adduct (4-(2-furyl)-3-hydroxy-2-butanone) that undergoes further dehydration to yield furfurylidene acetone (FAc). These C8 monomers can further condense with another furfural molecule to form the corresponding C13 dimers like difurfurylidene acetone (DFAc). The final products of the reaction present interesting applications in the preparation of polymers with high thermal and chemical resistance, as well as organo-mineral concretes [2, 3]. Later, other studies tried to obtain the furfurylidene acetone (FAc), as flavoring agent in the food industry [4, 5]. More recently, Dumesic and co-workers pioneeringly proposed this reaction as a way to increase the number of carbon atoms in the molecules in order to produce biofuels [6–12]. Subsequent steps to hydrogenate and hydrodeoxygenate the condensation products and form alkanes will be addressed in following section.

During the 1960s and 1970s, numerous studies covered the condensation of furfural and acetone in water in the presence of NaOH [2, 13–19]. The different parameters affecting the reaction were studied, including the temperature, ratio of reactants, solvent

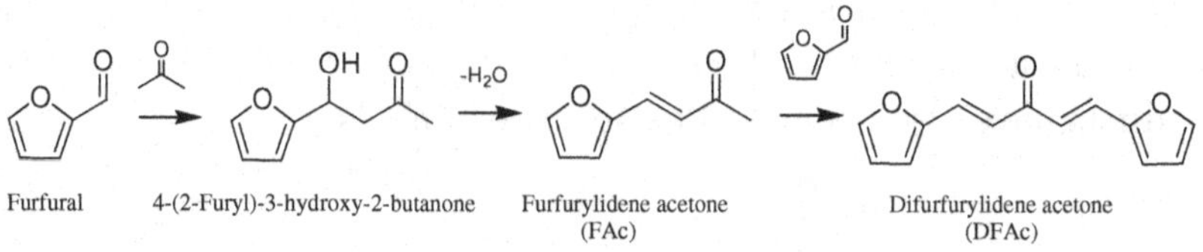

Furfural   4-(2-Furyl)-3-hydroxy-2-butanone   Furfurylidene acetone (FAc)   Difurfurylidene acetone (DFAc)

Figure 3.12.1.   Aldol condensation of furfural with acetone.

and mechanism. As a general summary, FAc (a C8 product) is favored in excess of acetone and water as solvent (60–70% selectivity at 348–358 K) whereas DFAc (a C13 compound) is more readily formed at higher FUR/acetones ratio and in alcohol as solvent (80% selectivity at 348–358 K) [15]. It must be stressed that under these reaction conditions, furfural and the condensation products can further oligomerize, which is a challenge for the selectivity and catalyst stability.

Even though many studies have dealt with homogeneous catalysts, especially NaOH, in more recent publications alternative heterogeneous catalytic processes have been used. The latter option is preferred in industrial applications, allowing for continuous operation either in stirred or in fixed-bed reactors (in the former case, the catalyst is easily recovered by filtration). In pioneering Dumesic's work, a number of solid basic catalysts were tested, including hydrotalcites, MgO, CaO, La–ZrO_2, Y–ZrO_2, MgO–ZrO_2, MgO–TiO_2, ionic exchange resins and functionalized silica with amino groups [9]. The best catalyst in terms of activity and stability was a mixed oxide MgO–ZrO_2 prepared by co-precipitation of the nitrate precursors. Other research groups studied the same reaction employing other solid basic catalysts like nitrogen-substituted zeolite Na–Y [20], Co–Al spinels [21, 22]; mixed oxides like Mg–Al [23–27], Ca–Zr [28], WO_3–ZrO_2 [29], and MgO–ZrO_2 with different compositions and preparation methods [28,30–33]; supported oxides such Mg–Zr oxides on graphite [34], MgO on Na–Y zeolite [35] or Mg oxide on carbon nanotubes [36, 37]; catalysts from natural origin such as dolomites [38] and chitosan [39]; acidic zeolites [40–42] and basic zeolites [43]; nanocrystalline anatase TiO_2 [44]; N-doped carbon nanotubes [45]; amino functionalized silica [46] and Brønsted acidic ionic liquids [7]. Table 3.12.1 shows a comparison of the most representative solid catalysts giving best results. It is clear from the results that catalytic activity of most of the catalysts presented is far from being optimal. With some exceptions, long reaction times (>1000 min) are used, together with relatively high catalyst loadings (FUR/cat ratio <3).

Table 3.12.1. Comparison of the catalytic activity of some of the different solid catalysts tested for the furfural–acetone aldol condensation reaction.

| Catalyst | FUR/cat. (wt.) | Reaction conditions | | | FUR conv. (%) | Selectivity to FAc and DFAc (%) | Ref. |
		Pressure (MPa)	Temp. (K)	Time (min)			
MgO–ZrO$_2$[a]	3.7	Aut. P	393	1560	98	78	[8]
MgO–ZrO$_2$[a,b]	1.25	5 (He)	393	1440	57	95	[20]
Co–Al spinel	6	Aut. P	413	300	96	99	[21]
WO$_3$–ZrO$_2$	1	1 (He)	353	1800	85	n/a	[29]
Carbon nanotubes/MgO	n/a	2.7 (N$_2$)	373	180	16	close to 100	[36]
MgO–ZrO$_2$[c]	1	1 (He)	353	1800	85	n/a	[29]
Chitosan[d]	0.96		423	240	97	100	[39]
MgO–ZrO$_2$[a]	1.5	1 (N$_2$)	323	1440	80	100	[30]
Mg–Zr aerogels	13	0.1	333	180	70	90	[33]
MgO/NaY	0.8	Aut. P	358	480	99	98	[48]
Mg–Al hydrotalcite	3.25	Aut. P	373	120	95	90	[24]
MgO–ZrO$_2$/graphite	1.5	1 (N$_2$)	323	1440	97	85	[34]
Dolomite[e]	11	Aut. P	413	60	90	100	[38]
MCM-22	6.5	Aut. P	373	120	60	89	[42]
H-BEA	3.25	Aut. P	373	480	50	100	[41]

Notes: [a]Preparation method according to Ref. [49].
[b]Reaction in the presence of methanol.
[c]Same result with both catalysts prepared via coprecipitation.
[d]Catalyst dried under supercritical CO$_2$ conditions. Reaction in microwave reactor.
[e]Reaction in water and methanol mixture (1:1.5).

Regarding stability and recyclability of the solid catalyst, the deposition of heavy species is found to be the main cause of deactivation in most of the studies presented in Table 3.12.1 [28,30]. However, in some cases the stability studies were carried out with a bifunctional $Pd/MgO-ZrO_2$ after the consecutive condensation and hydrogenation reactions, as it will be addressed in Section 3.12.3 [8]. The leaching of active species is another main deactivation mechanism in this reaction. In the case of the $MgO-ZrO_2$ catalyst, some authors have proposed that the use of batch type reactions can avoid the leaching of active species, because the decrease in the reaction temperature for stopping the reaction causes reprecipitation of the solubilized species, and consequently, the reutilization of the catalyst is improved [20]. A problem common to all the catalyst possessing strong basic sites is the deactivation caused by interaction with CO_2 from the atmosphere during storage and handling. Besides, during reaction conditions, furfural can be converted via Cannizzaro reaction to produce furoic acid, which can also interact with the strong basic sites causing a decrease in catalytic activity, as described in the case of K-BEA catalyst [43]. Finally, the presence of water can deactivate the catalyst by modifying the phases present, for instance, reacting with MgO to form a much less active phase, $Mg(OH)_2$ [31, 32]. It must be stressed that conducting the reutilization in batch reactors can miss the deactivation because in practice the deactivation may need longer term operation under continuous flow. A more realistic approach, simulating industrial operation, is to study the stability under continuous operation with fixed-bed reactor. Thus, the group of Kubička found that Mg–Al mixed oxides were stable for 40 h but then a sharp decrease in furfural conversion was observed after 40 h on stream. The rapid deactivation beyond 40 h of operation was due to fouling [27]. It was observed by GC-MS analysis that as the reaction progresses, heavier condensation products resulting from condensation of several molecules of furfural and acetone were formed, with molecular weights in the range of 200–350 g/mol. These results are in good agreement with batch studies showing that carbon deposits on the catalyst due to the formation of heavier condensation products are the main cause for the deactivation, but the deactivation may have been unnoticed if only few cycles had been conducted in batchwise studies.

The condensation of furfural with other ketones was also investigated in Dumesic's first studies, such as dihydroxyacetone, hydroxyacetone and glyceraldehyde [50]. Hydroxyacetone provided the best results in terms of furfural disappearance (100% furfural conversion after 4.5 h at room temperature with Mg–Al oxide). In other study, hydroxyacetone was the best suitable candidate for the condensation with furfural when amine-based homogeneous catalysis was used in the condensation (88% yield to condensation product in THF as solvent at room temperature) [51]. This condensation of furfural and hydroxyacetone has been evaluated in the context of upgrading of bio-oils, since both molecules are present and could potentially react to form larger molecules that can be subsequently hydrogenated to hydrocarbons [52]. In this particular example, NiMo/carbon nanotubes (CNT) catalysts were employed for both condensation and hydrogenation, giving a 25% condensation product under tested conditions (473 K). Methyl isobutyl ketone (MIBK) has also been tested as ketone in the aldol condensation with furfural, presenting a 100% yield to condensation product when hydrotalcite and CaO were used as catalysts at 403 K for 8 h [53].

Other biomass-derived molecules containing carbonyl groups can undergo condensation with furfural. The aldol condensation of furfural and levulinic acid or its esters, although known for several decades, is attracting a lot of interest in recent years. Levulinic acid is also a renewable chemical derivable from glucose present in lignocellulose what implies that all the sugars present in lignocellulose are in practice used in this reaction. In an early work by Lange, the aldol condensation of furfural and ethyl levulinate was studied with different solid base catalyst [54]. Different branched products were obtained (Figure 3.12.2), together with trimers and heavier compounds in a reaction at 443–503 K for 2–7 h. When Cs/MgO was used as catalyst, up to 20% yield of dimers was obtained after 1.5 h, but these intermediates further reacted to produce over 80% yield of trimers and heavier after 4 h [54]. When aqueous solution of KOH was used as catalyst and at 323 K and furfural/levulinate

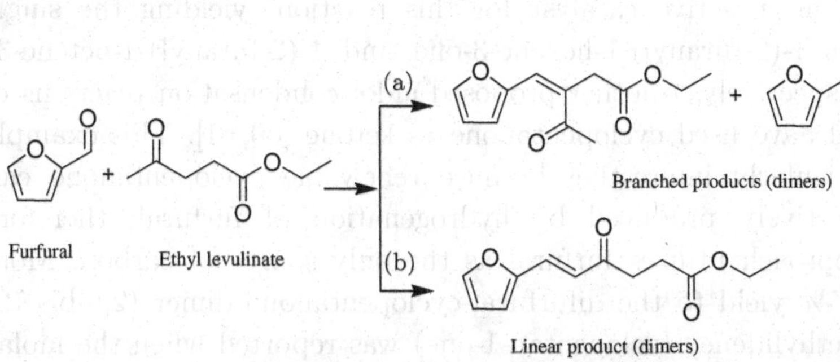

Figure 3.12.2. Condensation products of furfural and ethyl levulinate.

mol ratio = 2, 56% of the aldol adduct formed by condensation of two furfural molecules and one ethyl levulinate was obtained at mild reaction conditions [55]. After a low temperature hydrogenation with Pd/C a more stable C15 product was obtained, which can be further converted to branched alkane. A yield up to 97% of the same C15 product has also been obtained after one hour of reaction in aqueous NaOH at 323 K and furfural/levulinate mol ratio = 2 [56]. Another study of condensation of furfural with ethyl levulinate also used NaOH as catalyst leading to a solid consisting of furanic-keto acid polymers by self-Michael addition of the aldol adducts formed [57]. It must be stressed that a furfural/levulinate mol ratio = 1 was used and the polymer appeared when 1M HCl was used to recover the product of condensation. In an attempt to make the reaction more attractive from an industrial point of view, solid catalysts, similar to those reported in Table 3.12.1 for furfural–acetone aldol condensation have also been employed in this reaction [58]. MgO and ZnO were found to be the most active and selective catalyst for this reaction. A very interesting remark was the fact the MgO, base solid catalyst produced mainly the linear product (Figure 3.12.2(b)), while ZnO was highly selective for the production of the branched product (Figure 3.12.2(a)).

Alternative ketones derived from acetone–butanol-ethanol (ABE) fermentation of lignocellulose, have also been studied [59]. These ketones were 2-pentanone and 2-heptanone, and CaO was found to be the most active catalyst for this reaction, yielding the single adducts 1-(2-furanyl)-1-hexene-3-one and 1-(2-furanyl)-1-octene-3-one, respectively. Another proposed aldol condensation reactions of furfural have used cyclopentanone as ketone [60, 61]. This example is particularly interesting because renewable cyclopentanone can be selectively produced by hydrogenation of furfural; therefore this approach utilizes furfural as the only source of carbon. More than 95% yield to the difurfural-cyclopentanone dimer (2,5-bis (2-furylmethylidene) cyclopentan-1-one) was reported when the molar ratio of furfural to cyclopentanone was 2 to 1. A wide range of temperature, 313–373 K was tested giving very high molar yields of the dimer. A different research group found very consistent results using similar conditions, with successful recovery of the condensation product as solid after reaction [61]. Other authors used 3-pentanone, which is obtained by ketonization of lactic acid-derived propanoic acid [62]. The aldol condensation step was carried out under solvent-free conditions using solid base catalyst, such as CaO. In this same study, 5-nonanone, the ketonization product of valeric acid, was also used as ketone. After hydrodeoxygenation step, the branched aldol adducts lead to jet fuel range branched alkanes.

3.12.3. Applications of the condensation products to biofuels

Two steps are required to obtain liquid alkanes from the aldol products: first hydrogenation to produce alcohols, followed by a dehydration/hydrogenation (also called hydrodeoxygenation) reaction that forms the final alkanes [63].

The hydrotreatment of the furfural–acetone Claisen–Schmidt products was already studied in the late 70s, although at that time the interest was the formation of polymer precursors as it will be explained in the next section. Hydrogenation of the FAc and DFAc produced different alcohols [64,65]. In some cases, the cleavage of the furan ring was achieved, forming octanetriol from FAc using a copper chromite catalyst at 20 MPa and 473 K [64]. Milder hydrogenation conditions (Raney Nickel at 323 K and atmospheric pressure) lead to the hydrogenation of the furan ring and double bonds [65]. The sequence in the hydrogenation of the different functionalities in the FAc molecule was early established: (i) double bonds C=C in the side chain, (ii) C=O bond, (iii) C=C bonds in the furan ring, (iv) hydrogenolysis of the C–O bond in the furan ring, and (v) hydrogenolysis of other C–C bonds. The hydrogenation was also investigated in the case of DFAc under similar reaction conditions and again with Raney Nickel [66]. The effect of the solvent during the hydrogenation was examined at both atmospheric and high pressure (6 MPa) with a Ni–Ti–Al catalyst and it was found that by selecting the solvent, the selectivity to a desired product could be tuned [67].

More recent works have revisited these processes, now aiming at obtaining liquid biofuels, most of them in combination with the aldol condensation of furfural, or even the total conversion of sugars to fuels. In Dumesic's first approach, the previously formed aldol products were hydrogenated using a 3 wt.% Pd/Al_2O_3 catalyst at 393 K and 5.5 MPa [68]. These saturated molecules were finally transformed into alkanes in a four-phase dehydration/hydrogenation (4-PD/H) reactor operating at 523–538 K, 5.2–6.0 MPa with a 4 wt.% Pt/SiO_2–Al_2O_3 catalyst in presence of hexadecane. After this latter reaction, a spontaneous separation of the alkanes from water phase avoids the necessity of any distillation step to separate the products of interest, which favored the energy balance of this process compared to the production of other aqueous soluble biofuels like bioethanol.

Other variations of the hydrogenation and dehydration/ hydrogenation, also considered as hydrodeoxygenation (HDO) reactions have been tested since then. For instance, 99% selectivity to linear alkane was obtained in supercritical carbon dioxide under mild temperature conditions using Pd/Al-MCM-41 catalyst [69]. In this work, the intermediates after condensation reaction were hydrogenated to form the alkanes in supercritical CO_2 in the presence of Pd catalyst at 353 K under 4 MPa H_2 and 14 MPa CO_2. Palladium-based catalyst on SiO_2 promoted by FeO_x mixtures also exhibited high activity for the production of alkanes from furfural–MIBK aldol adducts. The resulting alkanes were mainly composed of 2-methyl-decane and 2-methyl-nonane and account for 90% of the carbon at 623 K and 6 MPa H_2 [70].

An industrial process to produce liquid alkanes from sugars has been envisaged and tested by Dumesic and co-workers [10]. It consists of four reactors in series as depicted in Figure 3.12.3. The first step is the dehydration at 423 K of the sugar (xylose, for instance) to form furfural. This reaction is carried out in a biphasic reactor using aqueous HCl as catalyst and THF as organic solvent. The formed furfural is extracted with a solvent and the homogenous catalyst in the aqueous phase is recycled. The separation of the phases and the extraction are accomplished by the addition of salts, which increases the immiscibility of the phases and improves the partitioning coefficient of the organic molecules formed.

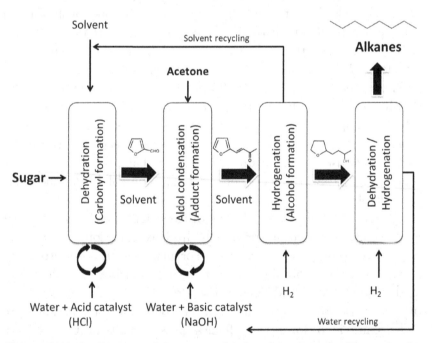

Figure 3.12.3. Diagram of the proposed industrial process for the production of liquid fuel from sugars (adapted from Ref. [10]).

After the dehydration reaction, the organic phase enriched in furfural is transferred to a second reactor, in which an aqueous stream of the basic catalyst (NaOH) and a ketone are mixed. Although in Figure 3.12.3 the formation of FAc is proposed, ketone/furfural ratio can be modified as to target higher yields to the dimer DFAc, resulting in a higher proportion of C13 alkanes (instead of C8 as it is proposed in the figure). The aldol adducts in the organic phase are subsequently transferred to the third reactor and the aqueous phase containing the catalyst is recycled. In the third reactor, the condensation products are hydrogenated over a Pd/γ-Al$_2$O$_3$ catalyst at 2.5–5.2 MPa H$_2$ and 373–413 K. The acyclic C=C bonds are the first to be hydrogenated, followed by the carbonyl group and then the furan ring to form water-soluble alcohols. The organic solvent is then recycled after this step and the water-soluble molecules are transferred to the fourth reactor and subjected to a deep hydrogenation over a bifunctional catalyst (Pt/NbPO$_5$) at 568 K. Finally, the alkanes, spontaneously separated from water, are recovered. The water stream is recycled to the second and third reactor. The total yield from sugar to alkanes was reported to be 58–69% [10].

Other authors have proposed similar processes but using carbohydrate feedstocks [56, 71]. In one case, the feedstock was a by-product stream in wood-processing industries and the authors included a preliminary economic analysis [71]. According to this study, the production cost of the alkanes would lie in the range of 2.0–4.4 U.S. dollars per gallon. The factors that can most significantly affect the OPEX are the price of raw materials, the amount of organic phase used and concentration of xylose in the hemicellulose extract feed. The main challenge is to increase the yield in all of the steps so the obtained yield can be closer to the theoretical yield for this process (0.61 kg of alkanes per kg of xylose) [71]. Other example of study using real biomass was done by the group of Wang and used corncob as feedstock [56]. In this case, an integrated process in which furfural and isopropyl levulinate are prepared from the hemicellulose and cellulose fractions respectively, followed by an aldol-condensation step forming a C15 product containing two molecules of furfural and one of levulinate. The last step is the final hydrodeoxygenation of the C15 oxygenates into branched alkanes.

Other ketones have also been used in the aldol condensation reaction, as explained in the previous section. These new aldol products have the advantage of leading to branched alkanes with supposedly improved properties to be used as jet fuels. The oligomers formed in the first aldol condensation step can be subjected to a hydrodeoxygenation process to lead to diesel fraction.

Finally, some studies have demonstrated the possibility of performing the condensation and the first hydrogenation step in one single reactor to form water soluble molecules. For this purpose, the bifunctional catalysts consist of a support with basic functionalities to conduct the aldol condensation ($MgO-ZrO_2$ mixed oxide) and impregnated with Pd particles, which are able to catalyze hydrogenation and hydrodeoxygenating reactions [8,9]. The hydrogenation was carried out at 393 K and 5.5 MPa with 95% yield to C8 and C13 alcohols [8]. Following this initial approach other authors have studied different basic catalysts functionalized with Pd, such as $Pd/MgO-ZrO_2$ or Pd/Co_2AlO_4 [22, 28]. When using the Pd on cobalt aluminate, 90% selectivity to saturated alcohols was obtained at 393 K and 4 MPa. The authors found, in good agreement with previous studies, that the hydrogenation of furan rings is the rate-determining step, and other by-products were identified, including biciclic spiro compounds formed through the cyclization of the intermediate dihydrofuran alcohol [22]. Consecutive steps to further produce octane have also been proposed, in which a bifunctional Pt/Co_2AlO_4 converts first the FAc to octanediol [72]. Subsequently, a mesoporous $Pt/NbOPO_4$ catalyst produces octane from the diols, at the same time that the catalyst is also able to open the tetrahydrofuran ring present in the feed with 76% octane yield at 2.5 MPa and 448 K.

A series of Pd/WO_3-ZrO_2 bifunctional catalysts have been tested in the total hydrolysis/dehydration/aldol condensation/ hydrogenation of lignocellulosic biomass to fuel precursors [29]. Further HDO of these precursors can lead to alkanes. All the reactions were carried out in a single reactor using Pd/WO_3-ZrO_2

as catalyst with 15–20% total yield to hydrogenated products from various biomass materials. In this case, both furfural and HMF are present in the reaction (derived from xylose and glucose-fructose, respectively) and can condense simultaneously with acetone to form C13 and C15 molecules. The sequence of reactions follows the steps: (i) hydrolysis/dehydration at 523 K for 5 min; (ii) aldol condensation at 353 K for 30 h, and (iii) hydrogenation at 393 K for 6 h [29].

In these examples, bifunctional catalysts are used but the reactions (aldol condensation and hydrogenation) are carried consecutively. The main limitation to the one-step process is the hydrogenation of reactants (furfural and acetone), which prevents condensation reactions to occur [73]. Faba and co-workers observed that only 14% selectivity to aldol products (C8 and C13) was obtained when hydrogen was present during the aldol condensation reaction.

Following the results regarding the stability of the catalyst explained in the previous section, the deposition of carbonaceous species blocking the access to the active sites was identified as one of the main deactivation mechanisms. As mentioned above, after the condensation, heavy insoluble products are formed and they adsorb strongly on the catalyst surface. The removal of these heavy molecules from the surface of the catalyst by partial hydrogenation forming more soluble alcohols can effectively minimize this source of deactivation [74]. Therefore, the stability of the catalyst is improved by using bifunctional catalyst, which allows for hydrogenation of the carbon deposits formed, as in the case of the one-pot reactions. For instance, when using a $MgO–ZrO_2$ catalyst, the loss of selectivity in successive cycles was found to be 90%. This deactivation was reduced to only 25% when the $Pd/MgO–ZrO_2$ catalyst was employed in the one-pot condensation and hydrogenation [74].

One of the main limitations of the aldol adducts in fuels arises as a consequence of the considerable amount of hydrogen required (approximately one molecule of hydrogen per carbon atom) [68]. It is important that the origin of the hydrogen should be renewable as a decisive condition to have a "green" and sustainable process.

3.12.4. Other C–C forming reactions

Even though many studies use aldol condensation to form new C–C bonds from furfural, there are other possible alternatives. For example, Fu *et al.* demonstrated a different two-step route leading to biofuels in the range of linear C8–C10 alkanes [75]. A C–C bond between two furfural molecules is formed in this case through a **reductive self-condensation** (Figure 3.12.4). The reducing step uses metals such as Al, Zn or Mg (in the form of powders) as stoichiometric reductants. C10 dimers are obtained in yields up to 95% using Zn (metal/furfural molar ratio = 1) as reducing agent at room temperature, after 1 h of reaction and using a 10% NaOH aqueous solution of furfural (6 wt.% approx.). Other Al or Mg powders gave lower yields. The authors claimed that the metal hydroxide formed after the reduction step can be recovered and regenerated by smelting to give the reducing agent.

Alkanes can be obtained from the dimers by subsequent dehydration/hydrogenation steps. The best result was obtained by using a mixture of $TaOPO_4$ and Pt/C as catalysts (molar yield of 84% from the dimer), at 4 MPa H_2, 573 K and 3 h of reaction. The use of a mixture of Pt/C and $NbOPO_4$ catalysts resulted in quite similar yields. This reductive condensation route was also demonstrated with other chemical platforms derived from cellulose (methyl furfural from hydroxymethyl furfural) and lignin (anisaldehyde, vanillin and veratraldehyde), yielding C_{11}–C_{12} and branched C_{13}–C_{14} alkanes, respectively.

Huber *et al.* described a different route for forming a C–C bond between furfural and furan (a biomass-derived nucleophile) via **hydroxyalkylation–alkylation** (HAA) [51]. In this study, furfural reacted with furan, affording C9 and C13 molecules (Figure 3.12.5).

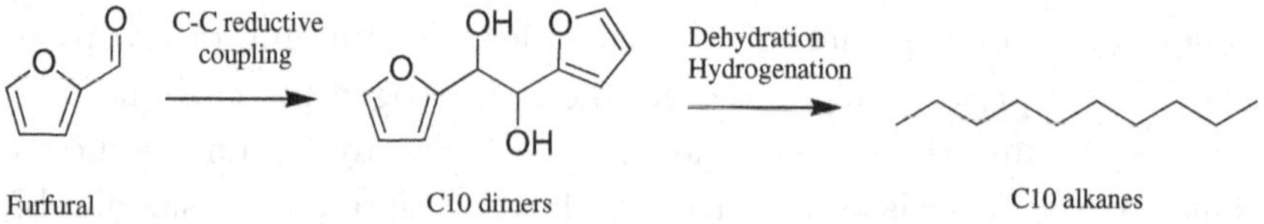

Figure 3.12.4. Formation of C_{10} alkanes via C–C reductive coupling of furfural.

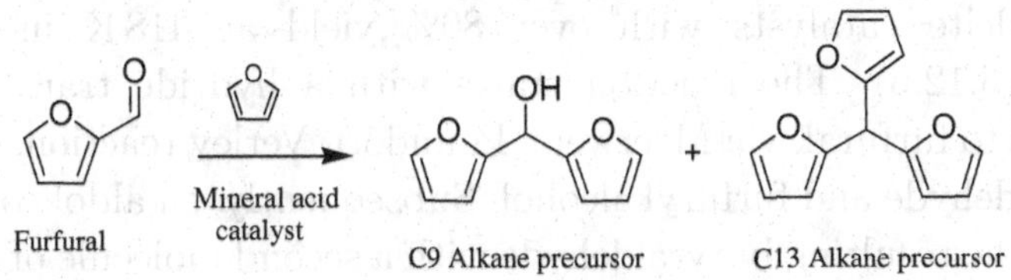

Figure 3.12.5. Formation of C_9 and C_{13} alkane precursors via furfural–furan condensation reactions.

Further hydrodeoxygenation of the C9 and C13 precursors lead to linear and branched diesel and jet fuels. A yield of 79% of C13 dimer was obtained by refluxing a solution of furfural in furan for 1 h in the presence of sulphuric acid. Other carbonyl-containing molecules derived from furfural (such as HMF) were also successfully condensed with furan in high yields. Information is lacking regarding other relevant aspects of the reaction, such as the amount of catalyst used, kinetic studies and the possibility of using other solid acids that can be separated from the reaction mixture and reused for a number of runs.

Following this idea, other authors used 2-methyl furan (2-MF), which in turn is produced from selective hydrogenation of furfural [76] (more insights on this reaction can be found in Chapter 3.5). In this case, solid-acid catalysts were used for the HAA reaction with furfural. Hydrogenation over a Pd/C catalyst and hydrodeoxygenation over a Pt supported on solid-acid were carried out afterwards to lead to branched C15 alkanes. Nafion-212 resin was found to have the highest activity and stability with 75% yield of the trimer alkane precursor (C15, produced from one molecule of furfural and two molecules of 2-MF). Successive reaction over a Pt/zirconium phosphate (Pt/ZrP) catalyst, lead to 94% carbon yield of diesel products, with almost 80% selectivity to C15 hydrocarbons (6-butylundecane as the main component).

The next example of C–C bond formation from furfural, which is a modification of the aldol condensation explained in the previous section, is the **dehydrogenative cross-coupling** reaction with alcohols [77]. A C9 aldol product obtained via transfer hydrogenation and aldol condensation of furfural and 1-butanol is produced using hydrotalcite catalysts with over 80% yield at 418 K in toluene (Figure 3.12.6). The reaction starts with a hydride transfer from butanol to furfural, via Meerwein–Ponndorf–Verley reaction, forming butyraldehyde and furfuryl alcohol. Subsequently, an aldol condensation of the resulting butyraldehyde with a second molecule of furfural renders an C9 aldol product. Furfuryl alcohol so formed is recycled back to furfural by oxidation. Reactions were carried out in toluene at 418 K for 20 h. A final hydrodeoxygenation step allowed for the production of alkanes in the range of jet or diesel fuels with 75–78% overall yield using Pt/NbOPO$_4$ as catalyst.

The **Baylis–Hillman** (BH) reaction between furfural and ethyl acrylate in aqueous medium leads to interesting products with many functionalities [78]. The reaction produced 90% of the adduct (Figure 3.12.7) in mixtures of ethanol/water (vol. 1:1) using 1,4-diazabicyclo(2,2,2)octane (DABCO) as catalyst.

Another type of C–C bond formation is the **Knoevenagel addition**, in which furfural can react with malonitrile to form a product with potential applications in photovoltaic device fabrication [79]. Hydrotalcites were again used as catalyst and different active methylene compounds (for instance, malonitrile, ethyl cyanoacetate, diethylmalonate) were tested in combination with furfural. A yield

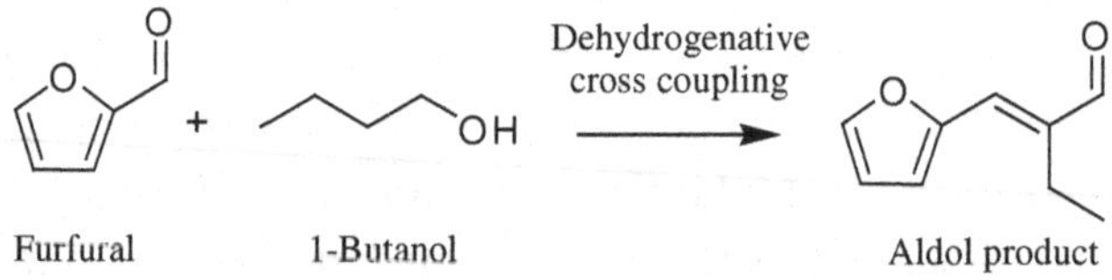

Figure 3.12.6. Dehydrogenative cross-coupling reaction of furfural and 1-butanol.

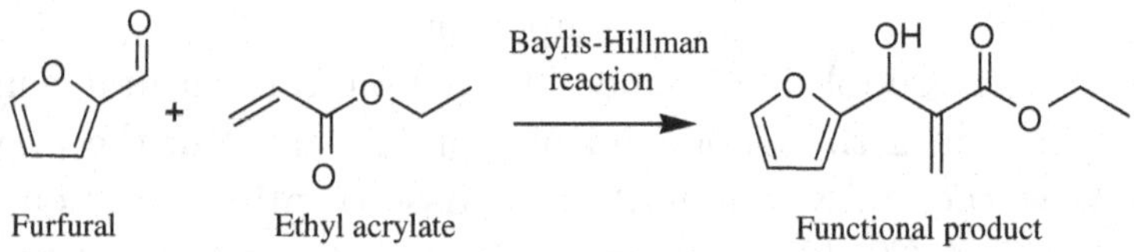

Figure 3.12.7. BH reaction of furfural with ethyl acrylate.

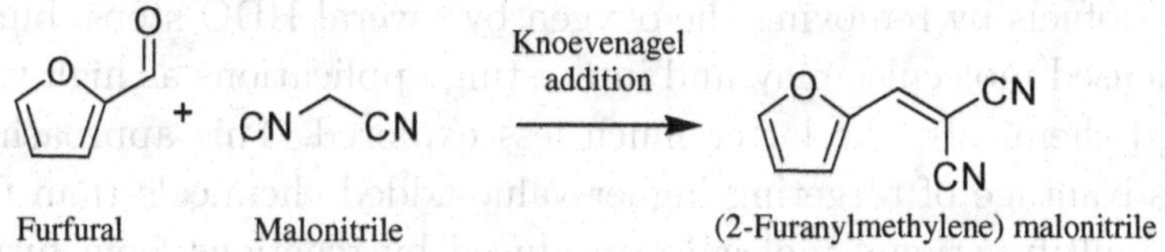

Figure 3.12.8. Knoevenagel reaction between furfural and malonitrile.

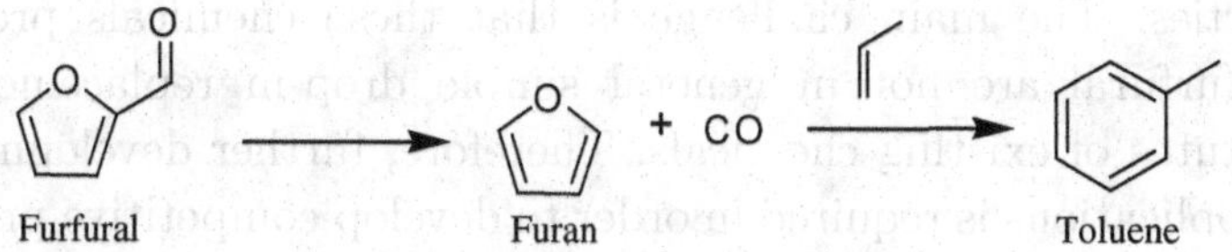

Figure 3.12.9. Diels–Alder reaction of furfural with propylene.

of 99% was obtained with malonitrile in DMF as solvent at 373 K after 15 min (Figure 3.12.8).

Finally, the **Diels–Alder reaction** between furfural and ethylene and propylene was studied by the group of Huber [80]. The authors proposed that the first step of the process was the decarbonylation of furfural to form furan and CO, after which, Diels–Alder reactions with propylene can occur. Overall carbon selectivity of 40% to aromatic was obtained, 60% of which was toluene (Figure 3.12.9). The reaction was carried out at 873 K in a fixed-bed reactor, with a WHSV of $9.0\,h^{-1}$.

3.12.5. Conclusions and future prospectives

Along this chapter, it has been shown that the reactivity of furfural offers multiple possibilities regarding transformations via condensation reactions. Many of these reactions have been known for decades, but nowadays have been revisited due to growing interest in furfural as biobased platform molecule. Several types of condensation reactions have been compiled but so far much of the effort has been concentrated on the aldol condensation with different types of ketones, especially interesting are those using ketones derived from biomass. The investigation has been driven because of the possibility of converting the condensation products into biofuels by removing the oxygen by several HDO steps, but the condensed molecules may find interesting applications as high value-added chemicals, the latter much less explored. This approach has the advantage of targeting higher value-added chemicals than fuels. The multifunctional molecules produced by reactions from furfural could have great potential in a plethora of different applications, from specialty chemicals to polymer and materials with enhanced properties. The main challenge is that these chemicals produced from furfural are not in general simple drop-in replacements or substitutes of existing chemicals. Therefore, further development of new applications is required in order to develop competitive products and processes that can eventually make their way to the market and contribute to the biobased economy.

Despite the fact that furfural condensation reactions seem very promising in the context of biofuels production, it is not mature yet. The main reason behind this is that production of biofuels, in general from biomass-related molecules but also in particular from furfural, is still not economically competitive compared to fossil fuels. The processes required to produce biofuel from biomass involve many steps in which the initial oxygen atoms present in the biomass-derived molecules need to be removed and large amounts of hydrogen are necessary. As a consequence of this, preliminary economic analysis of the final price of the alkanes obtained by aldol condensation with acetone indicated that substantial reduction in the cost of H_2 is required for economic viability. Economic viability was seen to be also very sensitive to the cost of biomass and acetone [71]. In order to ensure positive business cases for the production of biofuels, optimal yields in every step need to be reached. The use of heterogeneous catalysts to make the process as efficient as possible is therefore required. High yields and selectivities to desired products, together with development of robust production processes that can lead to low CAPEX and OPEX for industrial plants will greatly contribute to the economic viability of these processes. Some examples of this approach could be to ensure the recycling of all side-effluents, simple separation and purification processes, while minimizing the number of steps and side-products, and ensuring a long life-time of the catalyst.

References

1. R. Mahrwald, D.A. Evans, *Modern Aldol Reactions*, Wiley-VCH, 2004.
2. D.A. Isacescu, I. Gavat, C. Stoicescu, C. Vass, I. Petrus, Studies on furfural. XXVI compounds resulted from furfural with acetone condensation, *Revuel Roumaine de Chimie*, 10 (1965) 219–231.
3. A.A. Patel, S.R. Patel, Synthesis and characterization of furfural-acetone polymers, *European Polymer Journal*, 19 (1983) 231–234.
4. N. Fakhfakh, P. Cognet, M. Cabassud, Y. Lucchese, M.D. de Los Ríos, Stoichio-kinetic modeling and optimization of chemical synthesis: Application to the aldolic condensation of furfural on acetone, *Chemical Engineering and Processing: Process Intensification*, 47 (2008) 349–362.
5. M. Diaz de los Rios, R.L. Planes, B.H. Cruz, Kinetical study of the condensation reaction of 2-furaldehyde and 2-furfurylidene methylketone, *Acta Polymerica*, 41 (1990) 306–309.
6. H. Olcay, A.V. Subrahmanyam, R. Xing, J. Lajoie, J.A. Dumesic, G.W. Huber, Production of renewable petroleum refinery diesel and jet fuel feedstocks from hemicellulose sugar streams, *Energy & Environmental Science*, 6 (2013) 205–216.
7. R.M. West, Z.Y. Liu, M. Peter, C.A. Gartner, J.A. Dumesic, Carbon-carbon bond formation for biomass-derived furfurals and ketones by aldol condensation in a biphasic system, *Journal of Molecular Catalysis A-Chemical*, 296 (2008) 18–27.
8. C.J. Barrett, J.N. Chheda, G.W. Huber, J.A. Dumesic, Single-reactor process for sequential aldol-condensation and hydrogenation of biomass-derived compounds in water, *Applied Catalysis B: Environmental*, 66 (2006) 111–118.
9. J.N. Chheda, J.A. Dumesic, An overview of dehydration, aldol-condensation and hydrogenation processes for production of liquid alkanes from biomass-derived carbohydrates, *Catalysis Today*, 123 (2007) 59–70.
10. R.M. West, Z.Y. Liu, M. Peter, J.A. Dumesic, Liquid alkanes with targeted molecular weights from biomass-derived carbohydrates, *ChemSusChem*, 1 (2008) 417–424.
11. J.A. Dumesic, G.W. Huber, J.N. Chheda, C.J. Barrett, Method to make alkanes and saturated polyhydroxy compounds from carbonyl compounds, U.S. Patent 7671246 B2, 2010.
12. G.W. Huber, S. Iborra, A. Corma, Synthesis of transportation fuels from biomass: Chemistry, catalysts, and engineering, *Chemical Reviews*, 106 (2006) 4044–4098.
13. D.A. Isacescu, F. Avramescu, Reaction mechanism for the condensation of furfural with acetone catalysed by sodium hydroxide I. Systematization and interpretation of experimental results, *Revue Roumaine de Chimie*, 23 (1978) 661–665.
14. D.A. Isacescu, F. Avramescu, Reaction mechanism for the condensation of furfural with acetone catalysed by sodium hydroxide. II. The activated complex and its catalytic action, *Revue Roumaine de Chimie*, 23 (1978) 865–871.

15. D.A. Isacescu, F. Avramescu, Reaction mechanism for the condensation of furfural with acetone catalysed by sodium hydroxide. III. Solvent influence on the nature of the activated complex and on its catalytic action, *Revue Roumaine de Chimie*, 23 (1978) 873–881.

16. D.A. Isacescu, I. Gavat, I.V. Ionescu, C. Stoicescu, Studies on furfural. XXVII Controlled synthesis of furfural-acetone monomers, *Revue Roumaine de Chimie*, 10 (1965) 223–244.

17. D.A. Isacescu, I. Gavat, C. Stoicescu, C. Vass, I. Petrus, Studies on furfural. XXVI, *Revue Roumaine de Chimie*, 10 (1964) 219–231.

18. D.A. Isacescu, I. Gavat, V. Ursu, Studies on furfural. XXIX, *Revue Roumaine de Chimie*, 10 (1965) 257–267.

19. D.A. Isacescu, I. Rebedea, Studies on furfural. XXXIII The rate of formation of mono- and difurfurylidene acetone during the synthesis of "FA monomer", *Revue Roumaine de Chimie*, 10 (1965) 591–597.

20. W.Q. Shen, G.A. Tompsett, K.D. Hammond, R. Xing, F. Dogan, C.P. Grey, W.C. Conner, S.M. Auerbach, G.W. Huber, Liquid phase aldol condensation reactions with MgO-ZrO2 and shape-selective nitrogen-substituted NaY, *Applied Catalysis A: General*, 392 (2011) 57–68.

21. W.J. Xu, X.H. Liu, J.W. Ren, H.H. Liu, Y.C. Ma, Y.Q. Wang, G.Z. Lu, Synthesis of nanosized mesoporous Co-Al spinel and its application as solid base catalyst, *Microporous and Mesoporous Materials*, 142 (2011) 251–257.

22. W.J. Xu, X.H. Liu, J.W. Ren, P. Zhang, Y.Q. Wang, Y.L. Guo, Y. Guo, G.Z. Lu, A novel mesoporous Pd/cobalt aluminate bifunctional catalyst for aldol condensation and following hydrogenation, *Catalysis Communications*, 11 (2010) 721–726.

23. H.H. Liu, W.J. Xu, X.H. Liu, Y. Guo, Y.L. Guo, G.Z. Lu, Y.Q. Wang, Aldol condensation of furfural and acetone on layered double hydroxides, *Kinetics and Catalysis*, 51 (2010) 75–80.

24. L. Hora, V. Kelbichová, O. Kikhtyanin, O. Bortnovskiy, D. Kubička, Aldol condensation of furfural and acetone over MgAl layered double hydroxides and mixed oxides, *Catalysis Today*, 223 (2013) 138–147.

25. L. Hora, O. Kikhtyanin, L. Čapek, O. Bortnovskiy, D. Kubička, Comparative study of physico-chemical properties of laboratory and industrially prepared layered double hydroxides and their behavior in aldol condensation of furfural and acetone, *Catalysis Today*, 241 (2014) 221–230.

26. S. Ordoñez, E. Diaz, M. Leon, L. Faba, Hydrotalcite-derived mixed oxides as catalysts for different C-C bond formation reactions from bioorganic materials, *Catalysis Today*, 167 (2011) 71–76.

27. O. Kikhtyanin, L. Hora, D. Kubička, Unprecedented selectivities in aldol condensation over Mg-Al hydrotalcite in a fixed bed reactor setup, *Catalysis Communications*, 58 (2015) 89–92.

28. L. Faba, E. Diaz, S. Ordoñez, Performance of bifunctional Pd/MxNyO (M = Mg, Ca; N = Zr, Al) catalysts for aldolization-hydrogenation of furfural-acetone mixtures, *Catalysis Today*, 164 (2011) 451–456.

29. W. Dedsuksophon, K. Faungnawakij, V. Champreda, N. Laosiripojana, Hydrolysis/dehydration/aldol-condensation/hydrogenation of lignocellulosic

biomass and biomass-derived carbohydrates in the presence of Pd/WO3-ZrO2 in a single reactor, *Bioresource Technology*, 102 (2011) 2040–2046.

30. L. Faba, E. Diaz, S. Ordoñez, Aqueous-phase furfural-acetone aldol condensation over basic mixed oxides, *Applied Catalysis B: Environmental*, 113 (2012) 201–211.

31. I. Sadaba, M. Ojeda, R. Mariscal, J.L.G. Fierro, M.L. Granados, Catalytic and structural properties of co-precipitated Mg-Zr mixed oxides for furfural valorization via aqueous aldol condensation with acetone, *Applied Catalysis B: Environmental*, 101 (2011) 638–648.

32. I. Sadaba, M. Ojeda, R. Mariscal, R. Richards, M.L. Granados, Mg-Zr mixed oxides for aqueous aldol condensation of furfural with acetone: Effect of preparation method and activation temperature, *Catalysis Today*, 167 (2011) 77–83.

33. I. Sádaba, M. Ojeda, R. Mariscal, R. Richards, M. López Granados, Preparation and characterization of Mg-Zr mixed oxide aerogels and their application as aldol condensation catalysts, *ChemPhysChem*, 13 (2012) 3282–3292.

34. L. Faba, E. Diaz, S. Ordoñez, Improvement on the catalytic performance of Mg-Zr mixed oxides for furfural-acetone Aldol condensation by supporting on mesoporous carbons, *ChemSusChem*, 6 (2013) 463–473.

35. X.M. Huang, Q. Zhang, T.J. Wang, Q.Y. Liu, L.L. Ma, Production of jet fuel intermediates from furfural and acetone by aldol condensation over MgO/NaY, *Ranliao Huaxue Xuebao/Journal of Fuel Chemistry and Technology*, 40 (2012) 973–978.

36. P.A. Zapata, J. Faria, M. Pilar Ruiz, D.E. Resasco, Condensation/o hydrogenation of biomass-derived oxygenates in water/oil emulsions stabilized by nanohybrid catalysts, *Topics in Catalysis*, 55 (2012) 38–52.

37. M. Li, X. Xu, Y. Gong, Z. Wei, Z. Hou, H. Li, Y. Wang, Ultrafinely dispersed Pd nanoparticles on a CN@MgO hybrid as a bifunctional catalyst for upgrading bioderived compounds, *Green Chemistry*, 16 (2014) 4371–4377.

38. R.E. O'Neill, L. Vanoye, C. De Bellefon, F. Aiouache, Aldol-condensation of furfural by activated dolomite catalyst, *Applied Catalysis B: Environmental*, 144 (2014) 46–56.

39. H. Kayser, C.R. Müller, C.A. García-González, I. Smirnova, W. Leitner, P. Domínguez de María, Dried chitosan-gels as organocatalysts for the production of biomass-derived platform chemicals, *Applied Catalysis A: General*, 445–446 (2012) 180–186.

40. O. Kikhtyanin, V. Kelbichová, D. Vitvarová, M. Kubů, D. Kubička, Aldol condensation of furfural and acetone on zeolites, *Catalysis Today*, 227 (2014) 154–162.

41. O. Kikhtyanin, D. Kubička, J. Čejka, Toward understanding of the role of Lewis acidity in aldol condensation of acetone and furfural using MOF and zeolite catalysts, *Catalysis Today*, 243 (2015) 158–162.

42. O. Kikhtyanin, P. Chlubná, T. Jindrová, D. Kubička, Peculiar behavior of MWW materials in aldol condensation of furfural and acetone, *Dalton Transactions*, 43 (2014) 10628–10641.

43. O. Kikhtyanin, R. Bulánek, K. Frolich, J. Čejka, D. Kubička, Aldol condensation of furfural with acetone over ion-exchanged and impregnated potassium BEA zeolites, *Journal of Molecular Catalysis A: Chemical*, 424 (2016) 358–368.

44. D. Nguyen Thanh, O. Kikhtyanin, R. Ramos, M. Kothari, P. Ulbrich, T. Munshi, D. Kubička, Nanosized TiO_2 — A promising catalyst for the aldol condensation of furfural with acetone in biomass upgrading, *Catalysis Today*, 277 (2016) 97–107.

45. C. Ramirez-Barria, A. Guerrero-Ruiz, E. Castillejos-López, I. Rodríguez-Ramos, J. Durand, J. Volkman, P. Serp, Surface properties of amphiphilic carbon nanotubes and study of their applicability as basic catalysts, *RSC Advances*, 6 (2016) 54293–54298.

46. N.C. Nelson, U. Chaudhary, K. Kandel, I.I. Slowing, Heterogeneous multicatalytic system for single-pot oxidation and C-C coupling reaction sequences, *Topics in Catalysis*, 57 (2014) 1000–1006.

47. A.S. Amarasekara, B. Wiredu, Single reactor conversion of corn stover biomass to C5-C20 furanic biocrude oil using sulfonic acid functionalized Brönsted acidic ionic liquid catalysts, *Biomass Conversion and Biorefinery*, 4 (2014) 149–155.

48. Y.B. Huang, Z. Yang, J.J. Dai, Q.X. Guo, Y. Fu, Production of high quality fuels from lignocellulose-derived chemicals: A convenient C-C bond formation of furfural, 5-methylfurfural and aromatic aldehyde, *RSC Advances*, 2 (2012) 11211–11214.

49. M.A. Aramendia, V. Borau, C. Jimenez, A. Marinas, J.M. Marinas, J.A. Navio, J.R. Ruiz, F.J. Urbano, Synthesis and textural-structural characterization of magnesia, magnesia-titania and magnesia-zirconia catalysts, *Colloids and Surfaces A: Physicochemical and Engineering Aspects*, 234 (2004) 17–25.

50. J.A. Dumesic, G.W. Huber, J.N. Chheda, C.J. Barrett, Stable, aqueous-phase, basic catalysts and reactions catalyzed thereby, Wisconsin Alumni Research Foundation, USA, 2007, 60 pp.

51. A.V. Subrahmanyam, S. Thayumanavan, G.W. Huber, C-C bond formation reactions for biomass-derived molecules, *ChemSusChem*, 3 (2010) 1158–1161.

52. M. Zhou, G. Xiao, K. Wang, J. Jiang, Catalytic conversion of aqueous fraction of bio-oil to alcohols over CNT-supported catalysts, *Fuel*, 180 (2016) 749–758.

53. J. Yang, N. Li, G. Li, W. Wang, A. Wang, X. Wang, Y. Cong, T. Zhang, Solvent-free synthesis of C10 and C11 branched alkanes from furfural and methyl isobutyl ketone, *ChemSusChem*, 6 (2013) 1149–1152.

54. J.P. Lange, E. van der Heide, J. van Buijtenen, R. Price, Furfural: A promising platform for lignocellulosic biofuels, *ChemSusChem*, 5 (2012) 150–166.

55. C. Cai, Q. Liu, J. Tan, T. Wang, Q. Zhang, L. Ma, Synthesis of renewable jet fuel precursors from C-C bond condensation of furfural and ethyl levulinate in water, *Korean Chemical Engineering Research*, 54 (2016) 519–526.

56. C. Li, D. Ding, Q. Xia, X. Liu, Y. Wang, Conversion of raw lignocellulosic biomass into branched long-chain alkanes through three tandem steps, *ChemSusChem*, 9 (2016) 1712–1718.

57. A.S. Amarasekara, T.B. Singh, E. Larkin, M.A. Hasan, H.J. Fan, NaOH catalyzed condensation reactions between levulinic acid and biomass derived furan-aldehydes in water, *Industrial Crops and Products*, 65 (2015) 546–559.

58. G. Liang, A. Wang, X. Zhao, N. Lei, T. Zhang, Selective aldol condensation of biomass-derived levulinic acid and furfural in aqueous-phase over MgO and ZnO, *Green Chemistry*, 18 (2016) 3430–3438.

59. J. Yang, N. Li, S. Li, W. Wang, L. Li, A. Wang, X. Wang, Y. Cong, T. Zhang, Synthesis of diesel and jet fuel range alkanes with furfural and ketones from lignocellulose under solvent free conditions, *Green Chemistry*, 16 (2014) 4879–4884.

60. M. Hronec, K. Fulajtárová, T. Liptaj, M. Štolcová, N. Prónayová, T. Soták, Cyclopentanone: A raw material for production of C15 and C17 fuel precursors, *Biomass and Bioenergy*, 63 (2014) 291–299.

61. Q. Deng, J. Xu, P. Han, L. Pan, L. Wang, X. Zhang, J.J. Zou, Efficient synthesis of high-density aviation biofuel via solvent-free aldol condensation of cyclic ketones and furanic aldehydes, *Fuel Processing Technology*, 148 (2016) 361–366.

62. F. Chen, N. Li, S. Li, J. Yang, F. Liu, W. Wang, A. Wang, Y. Cong, X. Wang, T. Zhang, Solvent-free synthesis of C9 and C10 branched alkanes with furfural and 3-pentanone from lignocellulose, *Catalysis Communications*, 59 (2015) 229–232.

63. G.W. Huber, J.A. Dumesic, An overview of aqueous-phase catalytic processes for production of hydrogen and alkanes in a biorefinery, *Catalysis Today*, 111 (2006) 119–132.

64. C.R. Russell, K. Alexander, W.O. Erickson, L.S. Hafner, L.E. Schniepp, Polyhydroxyalkanes from furfural condensation products, *Journal of the American Chemical Society*, 74 (1952) 4543–4546.

65. Y.M. Mamatov, E.I. Klabunovskii, V.S. Kozhevnikov, Y.I. Petrov, Catalytic hydrogenation of furfurylideneacetone, *Chemistry of Heterocyclic Compounds*, 8 (1972) 263–265.

66. R. Mat'yakubov, Y.M. Mamatov, N.K. Mukhamadaliev, E.G. Abduganiev, Hydrogenation of difurfurylideneacetone on a raney nickel catalyst, *Chemistry of Heterocyclic Compounds*, 15 (1979) 374–377.

67. R. Mat'yakubov, Y.M. Mamatov, Effect of the nature of a solvent on the rate and mechanism of catalytic hydrogenation of difurfurylideneacetone, *Khim. Geterotsikl. Soedin.*, (1981) 889–893.

68. G.W. Huber, J.N. Chheda, C.J. Barrett, J.A. Dumesic, Production of liquid alkanes by aqueous-phase processing of biomass-derived carbohydrates, *Science*, 308 (2005) 1446–1450.

69. M. Chatterjee, K. Matsushima, Y. Ikushima, M. Sato, T. Yokoyama, H. Kawanami, T. Suzuki, Production of linear alkane via hydrogenative ring

opening of a furfural-derived compound in supercritical carbon dioxide, *Green Chemistry*, 12 (2010) 779–782.

70. J. Yang, S. Li, L. Zhang, X. Liu, J. Wang, X. Pan, N. Li, A. Wang, Y. Cong, X. Wang, T. Zhang, Hydrodeoxygenation of furans over Pd-FeO$_x$/SiO$_2$ catalyst under atmospheric pressure, *Applied Catalysis B: Environmental*, 201 (2017) 266–277.

71. R. Xing, A.V. Subrahmanyam, H. Olcay, W. Qi, G.P. van Walsum, H. Pendse, G.W. Huber, Production of jet and diesel fuel range alkanes from waste hemicellulose-derived aqueous solutions, *Green Chemistry*, 12 (2010) 1933–1946.

72. W. Xu, Q. Xia, Y. Zhang, Y. Guo, Y. Wang, G. Lu, Effective production of octane from biomass derivatives under mild conditions, *ChemSusChem*, 4 (2011) 1758–1761.

73. L. Faba, E. Díaz, S. Ordóñez, One-pot aldol condensation and hydrodeoxygenation of biomass-derived carbonyl compounds for biodiesel synthesis, *ChemSusChem*, 7 (2015) 2816–2820.

74. L. Faba, E. Diaz, S. Ordoñez, Improvement of the stability of basic mixed oxides used as catalysts for aldol condensation of bio-derived compounds by palladium addition, *Biomass Bioenergy*, 56 (2013) 592–599.

75. Y.-B. Huang, Z. Yang, J.-J. Dai, Q.-X. Guo, Y. Fu, Production of high quality fuels from lignocellulose-derived chemicals: A convenient C-C bond formation of furfural, 5-methylfurfural and aromatic aldehyde, *RSC Advances*, 2 (2012) 11211–11214.

76. G.Y. Li, N. Li, Z.Q. Wang, C.Z. Li, A.Q. Wang, X.D. Wang, Y. Cong, T. Zhang, Synthesis of high-quality diesel with furfural and 2-methylfuran from hemicellulose, *ChemSusChem*, 5 (2012) 1958–1966.

77. S. Sreekumar, M. Balakrishnan, K. Goulas, G. Gunbas, A.A. Gokhale, L. Louie, A. Grippo, C.D. Scown, A.T. Bell, F.D. Toste, Upgrading lignocellulosic products to drop-in biofuels via dehydrogenative cross-coupling and hydrodeoxygenation sequence, *Chemsuschem*, 8 (2015) 2609–2614.

78. J.N. Tan, M. Ahmar, Y. Queneau, Bio-based solvents for the Baylis–Hillman reaction of HMF, *RSC Advances*, 5 (2015) 69238–69242.

79. M. Shirotori, S. Nishimura, K. Ebitani, One-pot synthesis of furfural derivatives from pentoses using solid acid and base catalysts, *Catalysis Science and Technology*, 4 (2014) 971–978.

80. Y.T. Cheng, G.W. Huber, Production of targeted aromatics by using Diels-Alder classes of reactions with furans and olefins over ZSM-5, *Green Chemistry*, 14 (2012) 3114–3125.

Chapter 3.13

Fuel Additives by Furfural Acetalization with Glycerol

Manuel López Granados*

Sustainable Chemistry and Energy Group (EQS),
Institute of Catalysis and Petrochemistry (CSIC),
C/Marie Curie 2, Cantoblanco 28049 Madrid, Spain

3.13.1. Introduction

The intense growth in the global production of biodiesel has resulted in an excess of glycerol, the main by-product of the biodiesel industry. This surplus of glycerol cannot be consumed by the conventional routes, among other reasons, because of the presence of impurities derived from the biodiesel process. Managing this excess of glycerol poses a serious problem but, on the other hand, it also comes with the reduction of the price of glycerol. The ease of access to worldwide available low-cost glycerol has triggered the exploration of new applications for the crude glycerol produced in the biodiesel process. Consequently the development of new chemical routes to transform the glycerol in chemicals, biofuels and/or fuel additives is currently a vivid topic of research. Thus, carboxylation, oxidation hydrogenolysis, etherification, reforming, dehydration, chlorination, etc. have been explored as possible means of glycerol valorization [1].

In this context, the transformation of glycerol into diesel and biodiesel additives via acetalization of glycerol with aldehydes or

---

*Corresponding author: mlgranados@icp.csic.es

ketones has been identified as a promising valorization route. When compared with conventional diesel, reductions in the emissions of hydrocarbons, carbon monoxide, unregulated aldehyde and particulate has been detected in the fuels that have been additivated with glycerol acetals [2, 3]. Improvements in the flash point, oxidative stability, cloud point, pour point and viscosity, have also been confirmed when incorporated into biodiesel [4–7]. The transformation of glycerol into biodiesel additives means in practice the complete transformation of the vegetable oil into biofuels because glycerol, approximately 10 wt.% of the triglycerides, ends up fully incorporated into the biofuel.

3.13.2. Acetalization of furfural with glycerol

Besides the acetalization of glycerol with acetone [3, 5, 8] and formaldehyde, butyraldehyde or benzaldehyde [1,6] the acetalization with furfural has also been demonstrated. Figure 3.13.1 presents the scheme for the glycerol acetalization with furfural, resulting in the formation of the dioxane and dioxolane regioisomers (in practice a mixture of the *cis* and *trans* isomers of each regioisomers can be formed) [9]. The results reported so far are compiled in Table 3.13.1 Although soluble $ZnCl_2$ salt (with potential Lewis acid sites) resulted in high yield (91%) at very short contact time [10], its practical

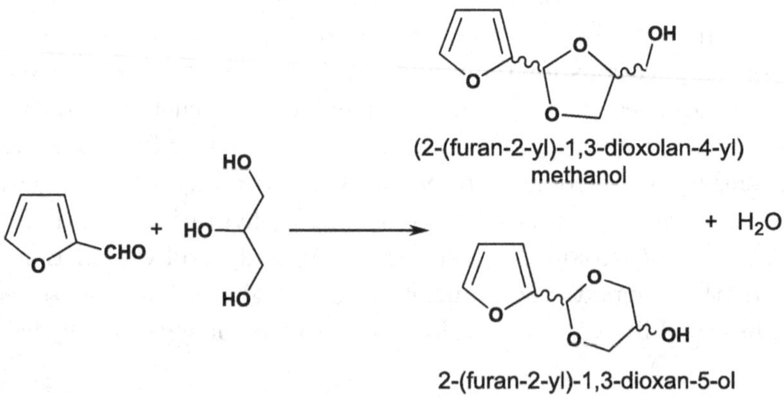

(2-(furan-2-yl)-1,3-dioxolan-4-yl)
methanol

+ H_2O

2-(furan-2-yl)-1,3-dioxan-5-ol

Figure 3.13.1. Simplified scheme of the glycerol acetalization reaction with furfural.

Table 3.13.1. Summary of results reported on the acetalization of glycerol with furfural.

Catalyst	Temp. (K)	Gly/FUR (molar)	Time (min.)	Cat. (wt.%)[a]	Conv. (mol%)	Yield (mol%)[b]	Ref.
PTSA[c]	b.p.[d]	1:1	180	0.06	n.r.[e]	70-75	[9]
Oxorhenium(V) oxazoline	373	1:5	240	0.2	100	77	[17]
ZnCl$_2$	373	1:5	40	1 (mol.%)	n.r.	91[f]	[10]
Al-MCM41 (3% Al)	373	1:5	40	10 (mol.%)	n.r	90[f]	[10]
MoO$_3$–SnO$_2$	298	1:1	120	5	100	100	[12,13]
SO$_4^{2-}$–SnO$_2$		1:1	120		100	100	
Amberlyst-15	343	1.1:1	180	6	n.r.	80	[7]
Al-SBA-15	373	1:1	720	54	69	n.r.	[14]
Fe oxide/Al-SBA-15	373	1:1.5	720	54	95	n.r.	[15]
Bentonite	333	2:1	60	5	69	n.r.	[18]
MCM-41 supported Phosphotungstic acid	Boiling toluene	1:1.05	80	5	n.r.	89	[11]
Graphene	373	1:10	120	27	n.r.	85	[16]
Microwave heating (no catalyst)	413	2:1	15	—	75	n.r.	[19]

Notes: [a]Weight or mole percentage with respect to glycerol.
[b]Yield of all isomers.
[c]PTSA: p-toluenesulphonic acid.
[d]b.p.: boiling point of the mixture (benzene, furfural and glycerol).
[e]n.r.: not reported.
[f]stripping with N$_2$.

use presents the serious drawback that its reutilization in successive runs is very complicated. On the contrary the acid solid Al-MCM-41 is also very active and its reutilization was demonstrated for four cycles. A yield as high as 90% was achieved by preventing the backward reaction through stripping the water produced during the course of the reaction with dry N_2. Remarkably the utilization of crude glycerol obtained from a conventional biodiesel reaction was explored and it was concluded that NaCl present in crude glycerol plays contradictory roles in the activity. The protons released through the exchange between Cl^- with acid sites in Al-MCM-41 partially contributes to the reaction but also results in the inefficient resinification of furfural observed when crude glycerol is used. Once the exchangeable protons are exhausted this side reactions are depleted.

For other catalysts, reutilization has not been explored, as in the case of MCM-41 supported Phosphotungstic acid [11] and the quite active Amberlyst-15 [7] or, as in the case of S- and Mo-promoted SnO_2, Al-SBA-15, Fe oxide/Al-SBA-15 and graphene catalysts, the catalysts deactivated [12–16].

Interestingly the acetalization has also been accomplished without catalysts (75% furfural conversion and 45 and 55% selectivity to dioxolane and dioxane isomers, respectively) but under microwave heating, in 15 min under 600 W of microwave power, equivalent to 313 K.

Unfortunately, solubility of the dioxolane and dioxane acetals in diesel and biodiesel is very limited. The hydrogenation of the acetals to produce the more soluble tetrahydrofuryl-1,3-dioxacyclanes has been proposed as a solution [10, 11]. Selective hydrogenation of these acetals to the corresponding tetrahydrofuryl-1,3-dioxane and dioxolane isomers has been selectively accomplished under mild conditions using 5% Pd/C catalyst (2.76 MPa hydrogen at 22°C) [10]. Up to a 5 wt.% of hydrogenated acetals were incorporated to biodiesel with no negative effects on the cloud point, density and flash point. Higher concentrations were not possible due to insolubility problems. Acetylation with acetic anhydride of the latter hydrogenated derivatives further improves the solubility. The dioxane and dioxolane acetates formed can be incorporated to biodiesel up to 10 wt.% and no negative effect on the tested properties was observed.

3.13.3. Summary and future challenges

Acetals derived from the reaction between glycerol and furfural or ketones have been proposed as additives of biodiesel and diesel to improve their properties as fuel. This strategy is relevant because implies the valorization of the surplus of glycerol produced by the biodiesel industry and it could increase significantly the furfural market. This chapter briefly summarizes the chemical technologies (catalytic and microwave irradiation) so far described to produce the mixture of dioxane and dioxolane regioisomers obtained from the acetalization of glycerol with furfural.

Catalytic technologies have been basically described but the reaction can also be rapidly and selectively accomplished by microwave irradiation. The difficulties to scale the microwave technologies up to production levels favors so far the catalytic technologies, but this can change in the near future if new advances are accomplished. On the other hand, in the case of catalytic technologies, the deactivation is a serious issue, especially when considering the utilization of crude glycerol, the cheaper feedstock for this process. There is a lack of information regarding this aspect that must be investigated in detail.

References

1. C.H. Zhou, H. Zhao, D.S. Tong, L.M. Wu, W.H. Yu, Recent advances in catalytic conversion of glycerol, *Catalysis Reviews — Science and Engineering*, 55 (2013) 369–453.
2. B. Delfort, I. Durand, A. Jaecker, T. Lacome, X. Montagne, F. Paille, Diesel fuel compounds containing glycerol acetals, InstitutFrancais du Petrole, 2005.
3. G. Vicente, J.A. Melero, G. Morales, M. Paniagua, E. Martín, Acetalisation of bio-glycerol with acetone to produce solketal over sulfonic mesostructured silicas, *Green Chemistry*, 12 (2010) 899–907.
4. J.D. Puche, Procedure to obtain biodiesel fuel with improved properties at low temperature, Industrial Management SA, 2003.
5. E. García, M. Laca, E. Pérez, A. Garrido, J. Peinado, New class of acetal derived from glycerin as a biodiesel fuel component, *Energy and Fuels*, 22 (2008) 4274–4280.

6. M.B. Güemez, J. Requies, I. Agirre, P.L. Arias, V.L. Barrio, J.F. Cambra, Acetalization reaction between glycerol and n-butyraldehyde using an acidic ion exchange resin. Kinetic modelling, *Chemical Engineering Journal*, 228 (2013) 300–307.

7. J.R. Dodson, T. Avellar, J. Athayde, C.J.A. Mota, Glycerol acetals with antioxidant properties, *Pure and Applied Chemistry*, 86 (2014) 905–911.

8. G.S. Nair, E. Adrijanto, A. Alsalme, I.V. Kozhevnikov, D.J. Cooke, D.R. Brown, N.R. Shiju, Glycerol utilization: Solvent-free acetalisation over niobia catalysts, *Catalysis Science and Technology*, 2 (2012) 1173–1179.

9. E.V. Gromachevskaya, F.V. Kvitkovsky, E.B. Usova, V.G. Kulnevich, Investigation in the area of furan acetal compounds. 13. Synthesis and structure of 1,3-dioxacyclanes based on furfural and glycerol, *Chemistry of Heterocyclic Compounds*, 40 (2004) 979–985.

10. B.L. Wegenhart, S. Liu, M. Thom, D. Stanley, M.M. Abu-Omar, Solvent-free methods for making acetals derived from glycerol and furfural and their use as a biodiesel fuel component, *ACS Catalysis*, 2 (2012) 2524–2530.

11. D. Bombos, S. Velea, M. Bombos, G. Vasilievici, E.E. Oprescu, Ecological component for motor fuels based on furfural derivates, *Revista de Chimie*, 67 (2016) 745–750.

12. B. Mallesham, P. Sudarsanam, B.M. Reddy, Eco-friendly synthesis of bio-additive fuels from renewable glycerol using nanocrystalline SnO2-based solid acids, *Catalysis Science and Technology*, 4 (2014) 803–813.

13. B. Mallesham, P. Sudarsanam, G. Raju, B.M. Reddy, Design of highly efficient Mo and W-promoted SnO2 solid acids for heterogeneous catalysis: Acetalization of bio-glycerol, *Green Chemistry*, 15 (2013) 478–489.

14. C. Gonzalez-Arellano, R.A.D. Arancon, R. Luque, Al-SBA-15 catalysed cross-esterification and acetalisation of biomass-derived platform chemicals, *Green Chemistry*, 16 (2014) 4985–4993.

15. C. Gonzalez-Arellano, S. De, R. Luque, Selective glycerol transformations to high value-added products catalysed by aluminosilicate-supported iron oxide nanoparticles, *Catalysis Science and Technology*, 4 (2014) 4242–4249.

16. N. Oger, Y.F. Lin, E. Le Grognec, F. Rataboul, F.X. Felpin, Graphene-promoted acetalisation of glycerol under acid-free conditions, *Green Chemistry*, 18 (2016) 1531–1537.

17. B.L. Wegenhart, M.M. Abu-Omar, A Solvent-Free Method for Making Dioxolane and Dioxane from the Biorenewables Glycerol and Furfural Catalyzed by Oxorhenium(V) Oxazoline, *Inorganic Chemistry*, 49 (2010) 4741–4743.

18. R.R. Pawar, K.A. Gosai, A.S. Bhatt, S. Kumaresan, S.M. Lee, H.C. Bajaj, Clay catalysed rapid valorization of glycerol towards cyclic acetals and ketals, *RSC Advances*, 5 (2015) 83985–83996.

19. R.R. Pawar, S.V. Jadhav, H.C. Bajaj, Microwave-assisted rapid valorization of glycerol towards acetals and ketals, *Chemical Engineering Journal*, 235 (2014) 61–66.

Chapter 3.14

Furanic Resins and Polymers

Thomas J. Schwartz* and Sikander H. Hakim[†,‡]

*Department of Chemical and Biological Engineering,
University of Maine, Orono, ME 04469-5737, USA

†Glucan Biorenewables LLC, 505 South Rosa Road,
Suite 112, Madison, WI 53719, USA

3.14.1. Introduction

Furfural can serve as a building block for yet another class of products — *furan resins*. Furan resins are thermosetting condensation resins with the characteristic feature of a furan ring in the molecular structure. Furan resins can be obtained from both furfural as well as its hydrogenated derivative, 2-furfuryl alcohol. Currently, furan resins are the single most important terminal downstream product of furfural. Of all the furfural produced worldwide, nearly 90% is converted to 2-furfuryl alcohol, and about 85–90% of worldwide demand for 2-furfuryl alcohol in 2015 was used to produce furan resins [1].

Furan resins have been known for over a century. As early as 1840 they were recognized as resinous masses obtained from furfural [2]. Initially, little attention was paid to understanding their physical properties or applications but soon after, resins soluble in organic solvents such as benzene and acetone were obtained from furfural and used as varnishes and stains varying in shades from light brown

‡Corresponding author: shakim@glucanbio.com

to black. In 1921, a process was patented to produce phenol–furfural resins for hot molding [3, 4]. From 1935 to 1945, the production of industrial furan resins increased more than 16-fold [5]. While the chemistry of furanic structures is versatile, industrial progress remained slow in subsequent decades due in part to widespread availability of cheaper fossil-based phenolic resins, which remain the main competitor to furan resins in the industry today [6]. However, in recent decades, these sustainable biomass-sourced resins have enjoyed a resurgence because of more stringent environmental and health regulations and the global drive to replace petroleum-based materials [7].

The tremendous versatility of furans as starting material for resin formation is illustrated by the wide variety of reactions that are used for their conversion in the plastics industry: (1) direct aldehyde condensation, as used in phenol–furfural resins; (2) polymer formation via etherification, as in the reactions of furfuryl alcohol with dimethyl urea; (3) methylene bridging, as in the formation of furfuryl alcohol resins; (4) addition polymerization through the conjugated ring structure of the furan molecule, as in acid-catalyzed furfuryl alcohol resinification; and (5) chemical modifications, as in the production of polyamide resins [5].

The addition of furan groups in polymeric structures can provide unique properties for the synthesis of novel materials. A variety of versatile materials with excellent chemical and thermal resistance can be obtained [7, 8]. Furan resins have greater chemical and heat resistance than polyesters, epoxides, and phenolic or amino-plastic resins. For example, corrosion-resistant furan resins are available and used as lining of chemical equipment. The corrosion-resistance behavior has been studied in detail [9, 10]. Similarly, furan-glass-fiber-reinforced composites are more heat-resistant than phenol-aldehyde-glass-fiber-reinforced composites such that the former can be used at elevated temperatures for extended periods of time.

Furfural and 2-furfuryl alcohol possess qualities that make them attractive starting materials for the resin and plastics industry: (1) they are globally available in large quantities and are consistent in quality; (2) they are highly reactive and suitable for direct conversion to synthetic resins and polymers; (3) they have a high degree of compatibility with other resins; and (4) they are marketable at a reasonable price. Despite these advantages, there has been limited success in the application of such renewable polymers. Although comparable to many fossil-based plastics, the price point for furan resins remains higher than their petrochemical cousins. Moreover, extended industrial scale production is hindered because detailed chemical mechanisms of resinification of either furfural or 2-furfuryl alcohol have not yet been elucidated. A higher-value material is poly(tetrahydrofuran) which can also be obtained from furfural via tetrahydrofuran [11] (see Figure 3.14.1).

There are many advantages to obtaining polymers from renewable resources: the potential to obtain unique structures and chemistries, reduced demand for petroleum-derived monomers in the

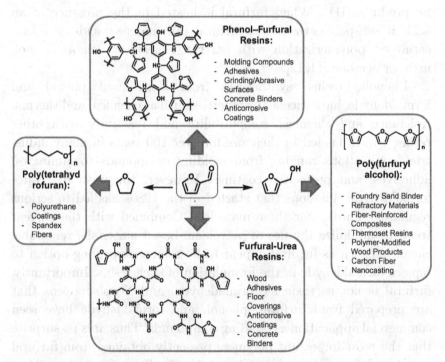

Phenol–Furfural Resins:
- Molding Compounds
- Adhesives
- Grinding/Abrasive Surfaces
- Concrete Binders
- Anticorrosive Coatings

Poly(tetrahydrofuran):
- Polyurethane Coatings
- Spandex Fibers

Poly(furfuryl alcohol):
- Foundry Sand Binder
- Refractory Materials
- Fiber-Reinforced Composites
- Thermoset Resins
- Polymer-Modified Wood Products
- Carbon Fiber
- Nanocasting

Furfural–Urea Resins:
- Wood Adhesives
- Floor Coverings
- Anticorrosive Coatings
- Concrete Binders

Figure 3.14.1. Overview of the application of furfural and its derivatives: 2-furfuryl alcohol and tetrahydrofuran in various furan resins.

manufacture of polymers, reduced environmental and health concerns, and the potential to obtain technologically superior properties. Thus, there is tremendous future potential in these materials as seen in the recent increased interest in both industry and academia. The application for furan resins has expanded from its initial use in the foundry sector and as linings of tanks and piping. The most recent emerging and exciting developments for furan resins are in applications in nanotechnology as well as smart and high performance materials, among many others [8].

3.14.2. Furan resins from furfural

Furfural is a highly reactive molecule, with its aldehyde group being particularly attractive given the well-developed organic chemistry of aldehydes. There are several ways in which furfural resins can be produced [11]. When furfural is heated in the presence of an acid, it self-polymerizes, but furfural can undergo acid- or alkali-catalyzed polymerization with partner molecules such as phenol, urea, or acetone [11, 12].

Phenolic resins, synthesized from fossil-based phenol and formaldehyde, have excellent properties such as chemical and thermal resistance and chemical compatibility and reactivity with other resins, which has led to their use for over 100 years in many industrial applications ranging from molding compounds to laminates, adhesives, and protective coatings. However, the push to reduce formaldehyde emissions that started in the 1980s has led to serious regulatory hurdles for these materials. Combined with the current trend to minimize the use of petroleum-based materials, renewable chemicals such as furfural appear to be a more appealing option to replace formaldehyde in the formulation of such resins. Importantly, furfural is not as toxic as formaldehyde, and Novolac resins that are prepared from furfural, phenol, and formaldehyde have seen commercial application as molding compounds. Thus, it is no surprise that the most important polymers presently obtained from furfural are phenol–furfural thermosetting resins [7]. These furan resins have desirable properties such as resistance to water and acids, which make them suitable for application in grinding and abrasive surfaces, and as adhesives to concrete binders and anticorrosive coatings [13, 14]. However, furfural is a slower-reacting aldehyde than formaldehyde, which results in a lower curing rate that can be undesirable in some industrial applications. This issue has been remedied by the use of additives such as boric acid and hexamethylenetetramine [15–17]. It has been reported that the boric acid-hexamethylene tetramine catalyzed phenol–furfural resin cures as rapidly as a commercial phenol–formaldehyde formulation.

The reaction of furfural with phenol is analogous to the formation of phenol/formaldehyde resins [5, 16, 18]. Here the aldehyde group of the furfural reacts with the reactive hydrogen of the phenol in the *ortho* and *para* positions (Scheme 3.14.1). As a result of this reaction, a furfurylol group is formed. An α-substituted (*o*- or *p*-hydroxyphenyl)-furfuryl alcohol (furfurylol) is generated. The –OH group in the phenol does not react but instead activates the three hydrogen atoms on the benzene ring as a result of which one, two, or all three can react with furfural to form phenol monofurfurylol, phenol difurfurylol, or phenol trifurfurylol, respectively [18]. The reaction is exothermic and requires proper control to prevent the resinification from reacting beyond a workable stage. The condensation products of this general type are usually compounded or mixed with modifying ingredients and used for molding compounds, impregnating solutions, bonding agents, coating materials, or adhesives.

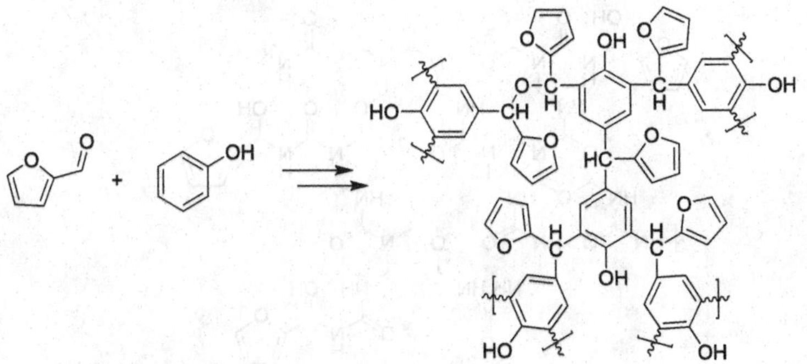

Scheme 3.14.1. Representative scheme of a furan–phenol resin (adapted from Ref. [7]).

The main advantages of phenol–furfural resins are: (1) added functionality that allows for new material properties and, potentially, new commercial applications, (2) unique reactivity with acids and unusual solubility, (3) improved flow/cure characteristics, (4) high tolerance to hydrocarbon solvents (often an advantage in the preparation of resin solutions for the impregnation of paper or cloth for laminating), (5) reduced brittleness and enhanced flexibility, (6) low melt viscosity and a low viscosity index, (7) reduced emissions relative to phenol–formaldehyde polymerization processes, and (8) the renewable nature of furfural. Instead of phenol, other substituted phenols, including renewable phenolics such as o-cresol, m-cresol, or p-cresol or cardanol [14, 16], can be used to produce Novolac resins by condensation with furfural. The main disadvantages associated with the use of furfural in this process are: (1) slow cure rates, (2) increased expense of furfural relative to formaldehyde, (3) weak chemical stability, (4) low flexibility, (5) weak impact resistance, and (6) red-colored resins that are undesirable in certain applications (Scheme 3.14.2).

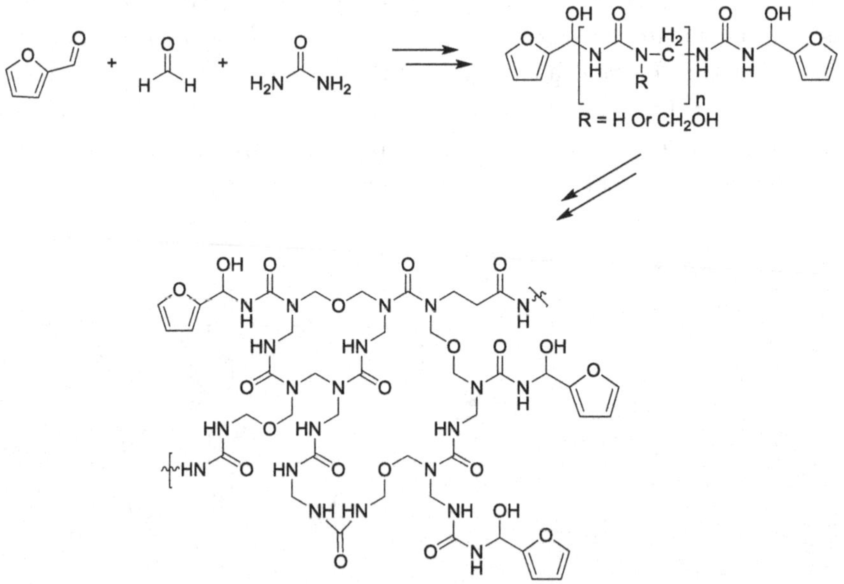

Scheme 3.14.2. Formation of urea–formaldehyde–furfural resins (adapted from Ref. [19]).

The reaction of furfural and urea produces another important class of furan resin. Similar to phenolic resins, urea–furfural resins have chemistry analogous to urea–formaldehyde resins [18]. It was observed for urea–formaldehyde–furfural co-condensed resins (UFFR) that the substitution of furfural in place of formaldehyde (see Scheme 3.14.2) reduced the free formaldehyde content effectively but at the expense of prolongation of the curing time. The latter challenge can be addressed by adding suitable curing agents, such as $(NH_4)_2S_2O_8$ [19]. Similarly resins from furfural can also be synthesized with acetone as a partner and were reported to be obtained in aqueous alkaline medium and then cured by acid catalysis [8]. These types of resins find several applications such as corrosion-resistant coatings, concrete binders, wood adhesives, floor coverings for the chemical industry, and, with the addition of dialkylphosphites, as flame-resistant materials. In spite of the excellent properties reported, they have yet to be used commercially due to their high cost.

Furfural has also been explored for its application as a building block for ion exchange resins. Furfural can be copolymerized with styrene because aldehydes and styrenes can undergo cationic polymerization, similar to the reaction between styrene and divinylbenzene used to form cross-linked styrene-divinylbenzene copolymers (S-DVB) that are used as ion-exchange resins for various applications such as lubrication, space fillers, and adsorption [20, 21]. There were some efforts to explore this application, reported in a body of the literature in the 1970s and 1980s [20]. While new work on this application has been scarce, this might change given the increasing need to provide feedstocks for the chemical industry by using renewable resources. Besides, the structure of conjugated furan rings provides unique properties. For example, Budinova *et al.* demonstrated that furfural-based resins exhibit superior performance for the adsorption of heavy metals, such as mercury, from aqueous solution [22]. It was found that the adsorption capacities of the carbons produced from furfural are significantly higher than those of commercial activated carbons due to the prevalence of oxygen-containing functional groups that are ideal for chelation (i.e., chemisorption) of Hg(II) ions, while only a small fraction of the Hg(II) ions are bound by physical adsorption as would have occurred using traditional activated carbons. This is an important finding in addressing the issue of metal ion contamination in potable water and wastewaters. Such materials can increase the efficiency of heavy metal chelation relative to the most common process that uses activated carbon for the elimination of heavy metals from wastewater. Additionally, there is a clear motivation for obtaining carbon adsorbents from non-conventional materials (e.g., agricultural byproducts, biomass materials, etc.) that are not only renewable but also result in materials with a high adsorptive capacity, considerable mechanical strength, and low ash and sulfur content.

Finally, furfural can serve as a feedstock to selectively produce well-ordered polymers via furan and tetrahydrofuran (THF) as intermediates. Currently these materials are obtained primarily from petroleum [11], however, furfural provides a renewable non-fossil based alternative and the chemistry to obtain these from furfural is well known. First, furfural must be decarbonylated, which has long been reported to occur via treatment with NaOH at elevated temperatures [23]. The decarbonylation can also be catalyzed by metals, with commercial production now occurring over a supported Pd catalyst [18, 24, 25]. Further hydrogenation of this furan leads to THF, which is practiced commercially [26, 27]. Cationic ring-opening polymerization of THF using strong acid catalysts (e.g., triflurormethanesulfonic acid) leads to high molecular weight poly(tetrahydrofuran) [28] (Scheme 3.14.3). The material is characterized by a very low glass transition temperature of $-84°C$ and a low melting point around $43°C$ [29]. The polyether functionality of the material makes it a suitable resin for use in polyurethane coatings [30]. Importantly, the low glass transition and melting temperatures make low-molecular-weight poly(THF)

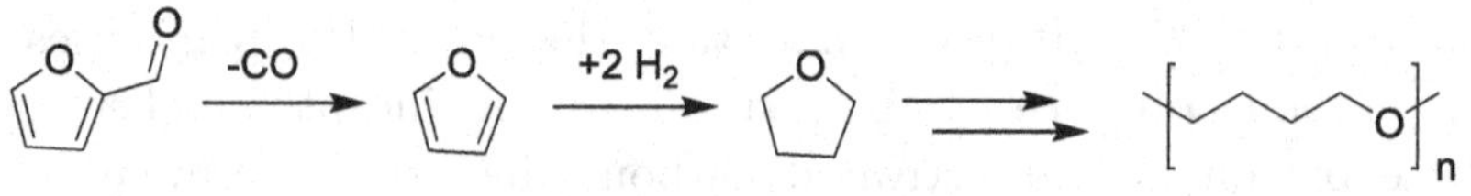

Scheme 3.14.3.   Reaction steps to convert furfural to poly(tetrahydrofuran).

highly suitable for use in spandex-type materials used for clothing manufacture [29].

3.14.3. Furan resins from 2-furfuryl alcohol

The most important furan resins are those that are prepared from 2-furfuryl alcohol, as evidenced by the fact that 80–90% of commercially produced furfural goes into making 2-furfuryl alcohol, the majority of which is converted to resins. There are many ways in which 2-furfuryl alcohol resins can be obtained. Similar to furfural resins, 2-furfuryl alcohol can produce resins by either self-condensation polymerization or via polymerization with a partner molecule [8, 18] (Scheme 3.14.4).

Poly(furfuryl alcohol) is a thermally cross-linked polymer used as a thermosetting resin, and it is synthesized by self-polymerization of 2-furfuryl alcohol via cationic condensation [31]. Poly(furfuryl alcohol)-based (PFA) resins have found a range of useful applications in the foundry industry, as wood adhesives and binders, as polymers, in concretes, and as precursors to carbonaceous products. Poly(furfuryl alcohol) consists of furan rings joined together by methylene bridges, formed via the reaction of the alcohol group of one furfuryl alcohol reacting with the active hydrogen from the ring of an adjacent furfuryl alcohol molecule and subsequent elimination of a water molecule [18]. Alternatively, the polymerization of furfuryl alcohol can also be carried out with a partner molecule that contains reactive hydrogen, such as formaldehyde. This reaction gives a series of thermosetting resins, which contains more methylene bridges as compared to a polymer obtained by self-polymerization of furfuryl alcohol. Such resins are used in bonding of wood [18].

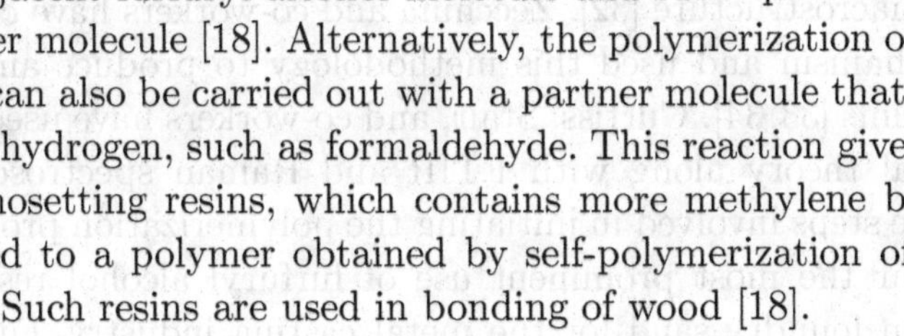

Scheme 3.14.4. Self-polymerization of furfuryl alcohol to poly(furfuryl alcohol).

This discussion mostly focuses on the self-polymerization of 2-furfuryl alcohol, which is a highly exothermic reaction that can be explosive in nature if poorly controlled. However, when adequate heat transfer is provided, along with careful pH control and neutralization of the reaction solution at the desired point, this reaction can be well-controlled, and polymers of almost any desired viscosity can be obtained. A continuous process can reduce the hazards of batch operation and circumvents the storage of large batches of the resin. Cross-linking or setting reactions are carried out at elevated temperatures and transform these liquid resins into solid products. During these final setting reactions, the double bonds in the furan rings in one chain react with the double bonds in the other chains to yield a cross-linked solid. Thus, when mixed with various fillers, this process forms the basis to obtain various products ranging from cements to composites to produce chemically resistant equipment [18].

The self-polymerization reaction is often carried out in the presence of an acid catalyst. Several groups have investigated the mechanism for the polymerization reaction. According to Gandini and co-workers, the resins are formed first by sequential condensation reactions leading to poly(furfuryl) chains that are conjugatively unsaturated. After sufficient unsaturation is achieved, the chains begin to branch and cross-link, leading to the formation of color bodies and macrostructure [32]. Zecchina and co-workers have confirmed this mechanism and used this methodology to produce amorphous carbon films [33,34]. Curtiss, Stair, and co-workers have used density functional theory along with FTIR and Raman spectroscopies to probe the steps involved in initiating the polymerization process [35].

By far the most prominent use of furfuryl alcohol resins is as binders in foundry sand for the metal casting industry. Furan-type resin binders were originally introduced in 1958 as an acid-catalyzed, no-bake alternative to phenolic resins. The primary competitor to furfuryl alcohol resins remains phenolic resins. The main advantage of furfuryl alcohol resins over phenol/formaldehyde resins is that the former does not produce hazardous formaldehyde, released from phenol/formaldehyde resins at high temperatures by degradation of the methylene bridges. By the early 1980s, furan resins had become the largest resin binder consumed, and they are still used extensively today, especially with progressively stricter limits on formaldehyde release. They also offer unparalleled advantages such as higher tensile strength, good dimensional accuracy during casting, predictable polymerization shrinkage, excellent shakeout, and lower smoke and odor during casting than organic solvent-based systems. Additionally, wide ranging binder selection offers alternatives to balance cost versus performance [36].

While having an application history of more than 50 years in the traditional foundry and refractory industry that continues to evolve and improve, current advancements in formulation technology render these resins appropriate for composite production. User-friendly furan resin systems have been prepared by TransFurans Chemicals [37], which is the largest producer of furfuryl alcohol in the world with a capacity of 30,000 MT/year. For example, their new range of furanic resins, BioRez^{TM} and Furolite^{TM} are suitable in a wide range of formulations as the matrix for binding fiberglass, rockwool and carbon fiber, as well as natural fibers such as wood, flax, sisal, and jute or as impregnating agent for porous substrates. The biobased natural fiber-reinforced composites (NFRC) from furan resins produced using hot compression molding are suitable for producing automotive interior trim parts.

An alternative to polymerization followed by cross-linking, is to produce a cross-linked material in a single polymerization step. Goldstein and co-workers in the 1960s showed that impregnating a porous material with furfuryl alcohol or with a solution of 2-furfural alcohol resins yields an inert thermoset upon acid treatment [38]. This *in situ* polymerization method enables the stabilization of fine-pored substances (e.g., brick, wood, and carbon), while the use of pre-polymerized furfuryl alcohol resins could be used only with material having large pores. After impregnation of these fine-pored materials with furfuryl alcohol solutions and subsequent heat curing, enhanced properties are obtained, such as modified bending properties, flexibility, impact strength, and chemical resistance. The treated carbon, unlike the untreated, does not chip or flake when cut.

Impact strength, hardness, and resistance to oxidation were increased by the resin treatment. This treatment has been often used to obtain materials or surfaces where chemical resistance is desirable.

More recently developing this technology, a Norwegian company, Kebony, has commercialized an award winning and environmentally friendly, patented process to produce attractive, long lasting and environmental friendly furan polymer modified wood [39]. The first step is impregnating wood with furfuryl alcohol, which is then followed by heating to cure the polymer such that it is permanently locked into the wood cells. Once locked, it becomes stable and does not leak out resulting in a permanently modified wood, which has outstanding stability, the maximum amount of hardness and a guaranteed long life. Kebony wood also provides a high level of safety as the wood does not splinter and contains no toxins or chemicals, nor does the wood get too hot in the summer. This is important because Kebony® technology can transform more sustainable woods to perform similar to or better than more precious tropical hardwoods. Additionally, this unique process is also a superior alternative to traditional wood treatment with biocides that are both expensive and environmentally damaging.

Recently, nanostructured carbons have been produced by a novel nanocasting process. The key in the process is the use of 2-furfuryl alcohol that has a molecular dimension of $8.43 \times 6.44 \times 4.28$ Å, which is smaller than the channel sizes of most of the hard templates. To prepare porous carbon nanocasts, a 2-furfuryl alcohol solution is impregnated into the pores of a hard template and then polymerized *in situ* either by heating or exposure to acid. After carbonizing the PFA-template in an inert gas atmosphere at high temperatures (e.g., $>400°C$), the template is removed by dissolution, leaving a porous carbon structure that is a negative of the template. In this way, unique materials with pores ranging from the micro- to meso-scale, as well as hierarchical porous materials with pores at multiple scales can be produced. These materials have found use in a wide range of applications as adsorbents, separation membranes, catalysts, and electrodes of fuel cells, lithium batteries, and electric double-layer capacitors [31]. Additionally, doping with other components via co-condensation reactions would be a significant step toward large-scale, cost-effective production of mesoporous and hierarchical porous carbons that are functionalized to improve their performance for hydrogen storage and greenhouse CO_2 capture from power plants and motorized vehicles. The progress of research into poly(furfuryl alcohol)-based nanocomposites has been quite encouraging, and has demonstrated improved performance. For example, PFA-Nafion membranes showed low methanol crossover and high proton conductivity thereby significantly improving the performance of direct methanol fuel cells.

Poly-furfural alcohol is expected to continue having an important role in the fabrication of nanostructured materials. Poly-furfuryl alcohol is compatible with many organic polymers and inorganic materials, and it gives high carbon yield upon pyrolysis and allows for desired semiconductor properties for use in electronic devices such as lithium ion batteries.

3.14.4. Final outlook and future perspective

Despite the advanced properties, benefits, applications, and potential of furanic resins, projections for global market growth have not been strong [1, 40]. This is due to slumping crude oil prices as a result of which phenolic resins retain a price advantage. In the years since the 1960s, due to low profit margins and a difficult competitive situation, most of the 2-furfuryl alcohol and furan resin business has migrated from the US and Europe to China. A small number of western producers are still selling significant volumes into the resin market, such as Illovo Sugar (South Africa) and IFC (Rotterdam/Geel Belgium). However, it has also been suggested that if furfural were more widely available with a consistent quality and lower price point, the market could expand significantly. Many recent efforts both in industry [41, 42], as well as in academia [43], have focused on the co-production of furfural alongside other revenue generating streams such that the produced furfural can meet all the criteria mentioned above for market growth [44]. Meanwhile, new applications of furan resins continue to emerge. 2-Furfuryl alcohol resins have also been reported as appropriate polymers for consolidating and strengthening petroleum reservoirs while retaining or enhancing their permeability [7]. Their use has been illustrated in thermoset matrices reinforced by fibers to produce composites for automotive and even the aerospace industry [45]. The ultimate holy-grail has been thought to be a "100% green resin" by combination of biophenol (obtained, for example, from lignin) and furfural [46–48] to replace traditional petroleum based phenol formaldehyde resins targeting a current global market that is approximately $10 billion USD.

References

1. https://www.ihsmarkit.com/products/furfural-chemical-economics-handbo ok.html, https://www.ihsmarkit.com/products/furfuryl-alcohol-and-furan-chemical-economics-handbook.html.
2. J.P. Trickey, C.S. Miner, H.J. Brownlee, Furfural resins, *Industrial & Engineering*, 15(1) (1923) 65–66.
3. E.E. Novotny, Phenol-furfural resin and method of making same. US Patent 1,737,121, 1929.
4. L.H. Brown, Oil soluble furfural-phenol resins, US Patent 2,601,498, 1952.
5. A.J. Norton, Furan resins, *Industrial & Engineering Chemistry*, 40(2) (1948) 236–238.
6. http://www.furan.com/furfural_historical_overview.html.
7. G. Rivero, L.A. Fasce, S.M. Ceré, L.B. Manfredi, Furan resins as replacement of phenolic protective coatings: Structural, mechanical and functional characterization, *Progress in Organic Coatings*, 77(1) (2014) 247–256.
8. L.B. Manfredi, G. Rivero, Furan resins: Synthesis, characterization and applications, *Chemical Physics Research Journal*, 5(1/2) (2012) 45–91.
9. E. Machnikova, K.H. Whitmire, N. Hackerman, Corrosion inhibition of carbon steel in hydrochloric acid by furan derivatives, *Electrochimica Acta*, 53(20) (2008) 6024–6032.
10. S. Vishwanatham, N. Haldar, Furfuryl alcohol as corrosion inhibitor for N80 steel in hydrochloric acid, *Corrosion Science*, 50(11) (2008) 2999–3004.
11. R. Mariscal, P. Maireles-Torres, M. Ojeda, I. Sádaba, M. López Granados, Furfural: A renewable and versatile platform molecule for the synthesis of chemicals and fuels, *Energy & Environmental Science*, 9(4) (2016) 1144–1189.
12. R. Roque-Malherbe, *The Physical Chemistry of Materials*, CRC Press, Taylor & Francis Group, 2009.
13. L.H. Brown, Resin forming reactions of furfural and phenol, *Industrial & Engineering Chemistry*, 44(11) (1952) 2673–2675.
14. A.U. Patel, S.S. Soni, H.S. Patel, Synthesis, characterization and curing of o-cresol — Furfural resins, *International Journal of Polymeric Materials and Polymeric Biomaterials*, 58(10) (2009) 509–516.
15. L. Brown, D. Watson, Curing phenol-furfural resins, *Industrial & Engineering Chemistry*, 51(5) (1959) 683–684.

16. R. Srivastava, D. Srivastava, Mechanical, chemical, and curing characteristics of cardanol–furfural-based novolac resin for application in green coatings, *Journal of Coatings Technology and Research*, 12(2) (2015) 303–311.

17. G. Rivero, V. Pettarin, A. Vázquez, L.B. Manfredi, Curing kinetics of a furan resin and its nanocomposites, *Thermochimica Acta*, 516(1–2) (2011) 79–87.

18. K.J. Zeitsch, *The Chemistry and Technology of Furfural and its Many By-Products*, Elsevier, 2000.

19. J. Zhang, H. Chen, A. Pizzi, Y. Li, Q. Gao, J. Li, Characterization and application of urea-formaldehyde- furfural co-condensed resins as wood adhesives, *Bioresources*, 9(4) (2014) 6267–6276.

20. A.K. Dalal, R.N. Kapadia, Synthesis and physicochemical properties of phenol-furfural type ion-exchange resins, *Indian Journal of Technology*, 22(2) (1984) 75–76.

21. B.D. Dasare, N. Krishnaswamy, Furfural-based ion-exchange resins. Part I. Preparation and properties of a cation exchanger from furfural–styrene reaction product, *Journal of Applied Polymer Science*, 9(8) (1965) 2655–2659.

22. T. Budinova, D. Savova, N. Petrov, M. Razvigorova, V. Minkova, N. Ciliz, *et al.* Mercury adsorption by different modifications of furfural adsorbent, *Industrial & Engineering Chemistry Research*, 42(10) (2003) 2223–2229.

23. C.D. Hurd, A.R. Goldsby, E.N. Osborne, Furan reaction II. Furan from furfural, *Journal of the American Chemical Society*, 54(6) (1932) 2532–2536.

24. C.L. Wilson, Reactions of furan compounds. Part V. Formation of furan from furfuraldehyde by the action of nickel or cobalt catalysts: Importance of added hydrogen, *Journal of the Chemical Society*, (1945) 61–63.

25. P. Lejemble, A. Gaset, P. Kalck, From biomass to furan through decarbonylation of furfural under mild conditions, *Biomass* 4(4) (1984) 263–274.

26. R.H. Kottke, Furan derivatives, in: *Kirk-Othmer Encyclopedia of Chemical Technology*, 2000.

27. K. Yan, G. Wu, T. Lafluer, C. Jarvis, Production, properties and catalytic hydrogenation of furfural to fuel additives and value-added chemicals, *Renewable and Sustainable Energy Reviews*, 38 (2014) 663–676.

28. G. Pruckmayr, T.K. Wu, Polymerization of tetrahydrofuran by proton acids, *Macromolecules*, 11(4) (1978) 662–668.

29. G. Pruckmayr, P. Dreyfuss, M.P. Dreyfuss, Polyethers, tetrahydrofuran and oxetane polymers, in: *Kirk-Othmer Encyclopedia of Chemical Technology*, 2000.

30. J.R. Harrison, The use of PTMEG in polyurethane coatings, *Journal of Elastomers & Plastics*, 17(1) 6–23.

31. H. Wang, J. Yao, Use of Poly(furfuryl alcohol) in the fabrication of nanostructured carbons and nanocomposites, *Industrial & Engineering Chemistry Research*, 45(19) (2006) 6393–6404.

32. C. Mekki, N.M. Belgacem, A. Gandini, Acid-catalyzed polycondensation of furfuryl alcohol: Mechanisms of chromophore formation and cross-linking, *Macromolecules*, 29(11) (1996) 3839–3850.

33. S. Bertarione, F. Bonino, F. Cesano, A. Damin, D. Scarano, A. Zecchina, Furfuryl alcohol polymerization in H–Y confined spaces: Reaction mechanism

and structure of carbocationic intermediates, *Journal of Physical Chemistry B*, 112(9) (2008) 2580–2589.

34. S. Bertarione, F. Bonino, F. Cesano, S. Jain, M. Zanetti, D. Scarano, A. Zecchina, Micro-FTIR and Micro-Raman studies of a carbon film prepared from furfuryl alcohol polymerization, *Journal of Physical Chemistry B*, 113(31) (2009) 10571–10574.

35. T. Kim, R.S. Assary, C.L. Marshall, D.J. Gosztola, L.A. Curtiss, P.C. Stair, Acid-catalyzed furfuryl alcohol polymerization: Characterizations of molecular structure and thermodynamic properties, *ChemCatChem*, 3 (2011) 1451–1458.

36. P. Carey, M. Lott, Furan no-bake, *Foundry Management & Technology*, 123(7) (1995) 26–31.

37. https://www.polyfurfurylalcohol.com/products.

38. I. Goldstein, W. Dreher, Stable furfuryl alcohol impregnating solutions, *Industrial & Engineering Chemistry*, 52(1) (1960) 57–58.

39. http://kebony.com/en/content/technology.

40. http://www.cnchemicals.com/Press/87143-CCM:%20Export%20volume%20 of%20furfural%20goes%20down%20in%20China%20in%202016.html.

41. http://www.naylornetwork.com/ppi-otw/articles/index-v2.asp?aid=354439 &issueID=42783.

42. http://glucanbio.com/why-glucanbio/.

43. D.M. Alonso, S.H. Hakim, S. Zhou, W. Won, O. Hosseinaei, J. Tao, *et al.*, Increasing the revenue from lignocellulosic biomass: Maximizing feedstock utilization, *Science Advances*, 3(5) (2017).

44. J.-P. Lange, E. van der Heide, J. van Buijtenen, R. Price, Furfural — A promising platform for lignocellulosic biofuels, *ChemSusChem*, 5(1) (2011) 150–166.

45. F.B. Oliveira, C. Gardrat, C. Enjalbal, E. Frollini, A. Castellan, Phenol-furfural resins to elaborate composites reinforced with sisal fibers-molecular analysis of resin and properties of composites, *Journal of Applied Polymer Science*, 109(4) (2008) 2291–2303.

46. J. Liu, J. Wang, Y. Fu, J. Chang, Synthesis and characterization of phenol-furfural resins using lignin modified by a low transition temperature mixture, *RSC Advances*, 6(97) (2016) 94588–94594.

47. J.-M. Pin, N. Guigo, A. Mija, L. Vincent, N. Sbirrazzuoli, J.C. van der Waal, E. de Jong, Valorization of biorefinery side-stream products: Combination of humins with polyfurfuryl alcohol for composite elaboration, *ACS Sustainable Chemistry & Engineering*, 2(9) (2014) 2182–2190.

48. N. Guigo, A. Mija, L. Vincent, N. Sbirrazzuoli, Eco-friendly composite resins based on renewable biomass resources: Polyfurfuryl alcohol/lignin thermosets, *European Polymer Journal*, 46(5) (2010) 1016–1023.

Chapter 4

Future Prospects and Main Challenges

Manuel López Granados* and David Martín Alonso[†,‡,§]

*Sustainable Chemistry and Energy Group (EQS),
Institute of Catalysis and Petrochemistry (CSIC),
C/Marie Curie 2, Cantoblanco 28049 Madrid, Spain

[†]Department of Chemical and Biological Engineering,
University of Wisconsin–Madison,
Madison, WI 53706, USA

[‡]Glucan Biorenewables LLC,
505 South Rosa Road, Suite 112,
Madison, WI 53719, USA

Furfural is considered a key renewable chemical platform in the valorization of lignocellulosic biomass [1]. As it has been described in the previous chapters of the book, furfural is currently being transformed in industry into a number of chemicals. Additionally, other transformations of furfural have been technically explored, like to biofuels, renewable fuel additives (2-methylfuran, methyltetrahydrofuran, furfuryl acetate, etc.), industrially relevant oil-derived chemicals (maleic acid, succinic acid, cyclopentanone, etc.) and finally to other chemicals with interesting properties and potential for application (levulinic acid, γ-valerolactone, 1,5-pentanediol, etc.). Even though its great potential as platform molecule is currently

§Corresponding author: david@glucanbio.com

limited in terms of economic, technical, and/or environmental feasibility, their interesting properties attract the attention of industrial and academic researchers, and it is expected that new technologies will allow to expand the furfural market soon. Two particular molecules that can also be key in a future biorefinery are levulinic acid and γ-valerolactone. They can potentially be produced from both hemicellulose (via furfural and/or furfuryl alcohol) and cellulose (acid-catalyzed dehydration), what implies that all the sugars present in lignocellulose can be valorized to these other chemical platforms [2–4].

Furfural platform is underdeveloped at the moment and its utilization to supply biofuels and chemicals is clearly very limited and restricted to a few applications. In order to expand its industrial application many difficulties must still be tackled. There are still room for improvements and for the discovery of new technologies. Some of these have been revealed in previous chapters but in this final chapter we will summarize what we consider the most important lines of actions to speed up the deployment of furfural biorefineries. This description is not intended to be exhaustive but to provide the most relevant directions.

One of the reasons why furfural has not reached all its potential is its price, which is mostly dominated by the feedstock cost (50–75%) and by the energy required to purify the furfural (25–35%). Besides, furfural price has shown periods of volatility and instability like the shortage of furfural in 2011 that resulted in a price rise up to approximately 2000 USD/ton [5]. The price of furfural in the near future is expected to remain around 1500 USD/tonne [5] but periods of instability in the price and in furfural market cannot be fully ruled out. This instability is derived from the cost of the feedstock, practically limited to corn cobs or bagasse. Any shortage with them affect directly the production of furfural. The development of flexible and effective technologies to process different lignocellulose feedstocks possessing disparate physical shapes (pieces, branches, straw, aquatic biomass, etc.) and composition (regarding structural carbohydrates, lignin and mainly contaminants, etc.) is essential to stabilize the furfural market. In the case of using agro-residues industries, this is essential to guarantee a constant supply of feedstock independent of seasonal accessibility. In addition, significant technological breakthroughs and the deployment of new and larger furfural plants are required to further decrease (or at least maintain) the price and give stability to the furfural trade. Thus, a representative example of a large industry that could benefit from a lower furfural price could be the process to produce jet and diesel fuel range alkanes by aldol condensation of furfural with acetone (and further deoxygenation of the oxygenated adducts so formed to the alkanes) [6]. A preliminary economic assessment has revealed that this process still requires substantial improvements to be economically viable [6]. The same can be applied to many other processes described in this book. Lowering the price of furfural will accelerate the deployment of a bioindustry based on furfural.

Another option to reduce the cost of the chemicals derived from furfural is to use low-grade furfural as feedstock. High-grade furfural (~99%; methylfuran and furfural methyl ketone are the most concentrated impurities) needs two energy intensive distillation steps [7]. The development of robust technologies that allow the processing of cheaper low-grade furfural and tolerate the presence of impurities would have a direct impact in the economic viability of not-commercial yet processes. A first approach could be the direct use of the dilute aqueous FUR solution (ca. 5 wt.%) obtained primarily from lignocellulosic biomass. The challenge is that these raw solutions typically contain some carboxylic acids, aldehydes and other organics besides light oligomers (resins) that can interfere in the downstream processing of furfural. A second approach is to use the concentrated FUR stream (~95% FUR) obtained after the first distillation step. This is a less problematic feedstock because of the reduced presence of contaminants (such as carboxylic acids and resins) [7] and consequently technically more favorable. However, it must be kept in mind that the utilization of either of these two solutions is in practice only possible if the furfural derivative can be simply and economically purified downstream the process. More research should focus on the utilization of this low-grade furfural.

Economic analysis and Life Cycle Analysis (LCA) are indispensable tools, first to assess on the economic and environmental viability of the chemical process, and second to detect where the most important weaknesses of the process are, and consequently to focus the improvements in these critical features. These analyses are complex and require meticulous and complicated compilation of data. Many issues are difficult to be accounted for, such as the logistics and costs involved in supplying the raw materials and shipping the products, the quality and quantity of competitors, the predictions of the price evolution and fluctuations, and in the potential synergy effects with other industries, among others. Actually, there is a remarkable lack of this type of analysis in the open literature for the processes described in this book. Remarkably some studies have already been conducted in the frame of biomass transformation to biofuels and chemicals and represent outstanding examples to encourage this type of evaluations. In this context, the inspiring analyses conducted for the production of 5-nonanone from levulinic acid [8], conversion of lignocellulosic biomass to liquid hydrocarbon via decarboxylation of GVL to butenes [9], the production of alkanes in the range of jet and diesel fuel via aldol condensation of FUR with acetone [6] or the acrylonitrile production from glycerol [10] must mark the direction to follow.

The accomplishment of economic feasibility of not viable yet processes requires of significant technical advances. In this context, an important area is the search of co-solvents and (or in combination with) cost competitive advanced separation technologies (pervaporation, ultrafiltration, ceramic membranes, etc.) to facilitate the purification of the reaction products from the solvent and from the unselective by-products. Another important field for progress is the catalyst design, especially developing more selective and durable catalysts and replacing the very costly catalysts based on noble metals (Ru, Rh, Ir, Pd, Pt, etc.) by others based on more affordable metals. The co-feeding of gases (such as H_2 or water vapor) to improve the selectivity and/or prevent catalyst deactivation are also strategies that may have a very positive outcome.

The causes of catalyst deactivation in several of the transformations depicted in this book are well-known, especially for those already commercial, and effective practices for the rejuvenation of the spent catalyst have also been implemented. However, for most of the processes, the detailed inventory of the likely causes of deactivation and of their real impact on the deterioration of the catalyst is far from being understood. This knowledge is of primary importance for inventing more robust catalysts and for implementing methodologies either to prevent and/or reduce the deactivation phenomena or to regenerate the catalyst.

Regarding the investigation of catalyst deactivation, it must be emphasized that using high-grade furfural may result in overlooking some deactivation phenomena experienced when real lower-grade feedstocks are used in industry. The presence of omitted impurities in the real streams may seriously affect the long-term stability of the catalyst. Actually, long-term studies under real reaction conditions is lacking for many of the processes reported. On the other hand, it will be also very informative and relevant to assess on the deactivation using low-grade furfural like that obtained immediately downstream the biomass processing reactor (raw furfural before any distillation) or the furfural solution obtained after the first distillation step.

The development of more active, selective, and durable processes needs of detailed information of the kinetics of the reaction pathway and the comprehension at molecular level of the surface species involved in the catalytic cycle. This information is scarce, even for those processes that have demonstrated industrial viability. Within this context, tools like theoretical calculations (quantum chemistry or density functional theory) have already demonstrated its utility in the hydrogenation of furfural to furfuryl alcohol, decarbonylation of furfural to furan, and hydrogenation of furan to tetrahydrofuran [11–14] and paves the way for future studies.

Concerning the production of biofuel, an important drawback is the use of large amounts of H_2 involved in the reduction, hydrogenation, and hydrodeoxygenation steps required to remove the O present either in the furfural molecule or in the biofuel precursors formed in condensation reactions. In the case of using non-renewable H_2 the economic and environmental viability of these processes are seriously threatened. The same may apply to chemicals synthesized using H_2 as reactant. In this context catalytic hydrogenation transfer reactions (for instance, Meerwein–Ponndorf–Verley) or the use of other molecules as H_2 sources (for instance, formic acid) are optional routes worth of investigation.

The electrochemical or photochemical conversion of furfural [15, 16] along with the utilization of new reaction environments (supercritical fluids, new green solvents, ionic liquids, etc.) can cause significant advances. The former are environmentally friendly and green methodologies, especially if sunlight or renewable electricity (e.g., produced by wind turbines) are used. Besides in the case of electrochemical conversion the energy (electricity) is stored in the form of biofuels or chemicals [16]. Some examples of photo-oxidation and electrochemical hydrogenation and oxidation of fur-fural to different products have been presented in this book. Further innovations are still demanded in order to increase the time yield productivity. Economic analyses are also urgently needed: the cost of the photosensitizers or of the electricity needed for the photo-oxidation and electrochemical processes, respectively, threaten their industrial implementation.

With regard to hydrogenation reactions, electrochemical cells are particularly attractive because H_2 can be *in situ* produced in the cathode, generally at atmospheric pressure and near ambient temperatures. The scale-up of a continuous electrochemical process should not present more complications than that of the conven-tional catalytic process. Besides the electrochemical hydrogenation reactions can be coupled with an oxidation reaction in the anode of the galvanic cell. Consequently the simultaneous production of two valuable products can be accomplished enhancing the electric efficiency of the whole process.

Finally, it must be stated that the deployment of furfural-based bioindustries is complicated because they compete against the conventional oil-refinery processes. Oil market is very unstable and volatile but it uses very well-demonstrated technologies with relatively lower risk. It is plausible that biorefineries can be competitive in the very long term because of the depleting of oil reserves but in the shorter term, the development of furfural industry must be driven by environmental and energy/chemical supply security issues rather than economic advantages. Furfural biorefineries still demand very large investments and exhibit high technological risks. Accordingly, long-term mandatory policies regarding the use of renewable chemicals and biofuels are clearly demanded to promote the economic viability and to foster biorefineries implementation.

References

1. J.J. Bozell, G.R. Petersen, Technology development for the production of biobased products from biorefinery carbohydrates — The US Department of Energy's "top 10" revisited, *Green Chemistry*, 12 (2010) 539–554.

2. S.G. Wettstein, D. Martin Alonso, E.I. Gürbüz, J.A. Dumesic, A roadmap for conversion of lignocellulosic biomass to chemicals and fuels, *Current Opinion in Chemical Engineering*, 1 (2012) 218–224.

3. D.M. Alonso, S.G. Wettstein, J.A. Dumesic, Gamma-valerolactone, a sustainable platform molecule derived from lignocellulosic biomass, *Green Chemistry*, 15 (2013) 584–595.

4. D.M. Alonso, J.Q. Bond, J.A. Dumesic, Catalytic conversion of biomass to biofuels, *Green Chemistry*, 12 (2010) 1493–1513.

5. G. Marcotullio, The chemistry and technology of furfural production in modern lignocellulose-feedstock biorefineries, Process and Energy Department, Technische Universiteit Delft, 2011.

6. R. Xing, A.V. Subrahmanyam, H. Olcay, W. Qi, G.P. van Walsum, H. Pendse, G.W. Huber, Production of jet and diesel fuel range alkanes from waste hemicellulose-derived aqueous solutions, *Green Chemistry*, 12 (2010) 1933–1946.

7. K.J. Zeitsch, *The Chemistry and Technology of Furfural and its Many By-Products*, Sugar Series, Vol. 13, Elsevier, The Netherlands (2000).

8. A.D. Patel, J.C. Serrano-Ruiz, J.A. Dumesic, R.P. Anex, Techno-economic analysis of 5-nonanone production from levulinic acid, *Chemical Engineering Journal*, 160 (2010) 311–321.

9. D.J. Braden, C.A. Henao, J. Heltzel, C.C. Maravelias, J.A. Dumesic, Production of liquid hydrocarbon fuels by catalytic conversion of biomass-derived levulinic acid, *Green Chemistry*, 13 (2011) 1755–1765.

10. D. Cespi, F. Passarini, G. Mastragostino, I. Vassura, S. Larocca, A. Iaconi, A. Chieregato, J.L. Dubois, F. Cavani, Glycerol as feedstock in the synthesis of chemicals: A life cycle analysis for acrolein production, *Green Chemistry*, 17 (2015) 343–355.

11. S. Wang, V. Vorotnikov, D.G. Vlachos, A DFT study of furan hydrogenation and ring opening on Pd(111), *Green Chemistry*, 16 (2014) 736–747.

12. V. Vorotnikov, G. Mpourmpakis, D.G. Vlachos, DFT study of furfural conversion to furan, furfuryl alcohol, and 2-methylfuran on Pd(111), *ACS Catalysis*, 2 (2012) 2496–2504.

13. J.W. Medlin, Understanding and controlling reactivity of unsaturated oxygenates and polyols on metal catalysts, *ACS Catalysis*, 1 (2011) 1284–1297.

14. S.H. Pang, J.W. Medlin, Adsorption and reaction of furfural and furfuryl alcohol on Pd(111): Unique reaction pathways for multifunctional reagents, *ACS Catalysis*, 1 (2011) 1272–1283.

15. P. Esser, B. Pohlmann, H.D. Scharf, The photochemical synthesis of fine chemicals with sunlight, *Angewandte Chemie International Edition*, 33 (1994) 2009–2023.

16. S.K. Green, J. Lee, H.J. Kim, G.A. Tompsett, W.B. Kim, G.W. Huber, The electrocatalytic hydrogenation of furanic compounds in a continuous electrocatalytic membrane reactor, *Green Chemistry*, 15 (2013) 1869–1879.

Annex 1

Introduction to Chemical Reactors

José Miguel Campos-Martín*

*Sustainable Chemistry and Energy Group (EQS),
Institute of Catalysis and Petrochemistry (CSIC),
C/Marie Curie 2, Cantoblanco 28049 Madrid, Spain*

A.1.1. Introduction

The purpose of this annex is to introduce the reader to the fundamentals aspects of the engineering of chemical reactors. This book is targeted at a wide audience, including undergraduate or postgraduate students not familiarized with the basic concepts of chemical reactors. This annex is directed at briefly introducing the most important notions of the chemical reactors. For deeper information of the points treated in this chapter, the recommended readings included at the end of the annex can be consulted.

A.1.2. General concepts

A *chemical reaction* is a process that results in the conversion of chemical substances. The substance or substances initially involved in a chemical reaction are called *reactants*, which yield one or more products.

A brief representation of the chemical change in terms of the symbols and formulae of the reactants and products is called a *chemical equation*. A chemical equation therefore must fulfill

*Corresponding author: jm.campos@csic.es

the following conditions: (a) represents a true chemical change, (b) balances and (c) has a molecular form. A chemical equation has both qualitative and quantitative significance. Qualitatively, a chemical equation gives the names of the various reactants and products. Quantitatively, it expresses the relative numbers of molecules of the reactants and products taking part in the reaction. Chemical equations give the quantitative relationship between the reactants and products. This quantitative information can be utilized to perform various calculations, which are often required to assess the economic viability of the chemical process.

When more than one reaction occurs in the reactor, they are referred to as multiple reactions. There are four main types of multiple reactions (Figure A.1.1): parallel reactions, reactions in series, consecutive reactions and independent reactions.

Reaction rate is defined as an expression of the speed with which a reaction occurs. The rate of a reaction is always positive. A negative sign is present to indicate that a reactant concentration is decreasing. Reaction rates are represented as algebraic expressions that indicate the effect of each reaction parameter on the behavior of the reaction. The global reaction rate expression is a mathematical function that expresses the actual rate of a chemical reaction, accounting for all the phenomena and mechanisms that occur. The global reaction rate is essential for designing and operating chemical reactors. For most homogeneous chemical reactions, the global rate is the same as the intrinsic kinetic rate. However, for many heterogeneous chemical reactions, a priori determination of the global reaction rate is extremely difficult.

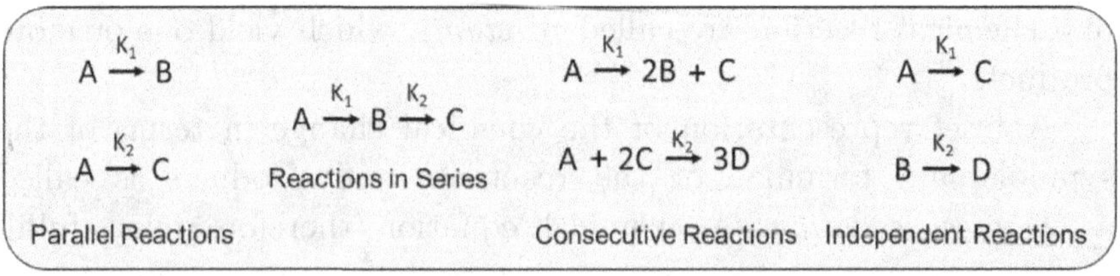

$$A \xrightarrow{K_1} B$$

$$A \xrightarrow{K_2} C$$

Parallel Reactions

$$A \xrightarrow{K_1} B \xrightarrow{K_2} C$$

Reactions in Series

$$A \xrightarrow{K_1} 2B + C$$

$$A + 2C \xrightarrow{K_2} 3D$$

Consecutive Reactions

$$A \xrightarrow{K_1} C$$

$$B \xrightarrow{K_2} D$$

Independent Reactions

Figure A.1.1. Examples of multiple reactions.

Conversion is defined as the fraction of reactant that has been consumed. Three points concerning the conversion should be noted:

1. Conversion is defined only for reactants, and by definition, its value is between zero and one.
2. Conversion is related to the composition (or flow rate) of a reactant and is not defined on the basis of any specific chemical reaction. When multiple chemical reactions occur, a reactant may be consumed in several chemical reactions. However, if reactant A is produced by an independent chemical reaction, its conversion is not defined.
3. Conversion depends on the initial state selected, the initial amount of material (for batch reactors) and the boundaries of the system, both "in" and "out" (for flow reactors).

When several simultaneous chemical reactions produce both desired and undesired products, it is convenient to define parameters that indicate what portion of the reactant is converted to valuable products. Below, we define and discuss two quantities that are commonly used: yield and selectivity of the desired product.

Selectivity is a measure of the portion of a reactant converted to the desired product by the desired chemical reaction. It indicates the amount the desirable product, for instance X, produced relative to the amount of X that could have been produced if only the desirable reaction occurred. The **Selectivity** is defined such that its value is between zero and one.

Product **Yield** indicates the amount of product P produced relative to the theoretical amount of P that could be produced if all reactant A consumed were to react through the desired chemical reaction.

Chemical reactors are vessels designed to contain chemical reactions. They are the site of conversion of raw materials into products and are also known as the heart of a chemical process. Chemical reactors are a vital step in the overall design of a process. Designers ensure that the reaction proceeds with the highest efficiency towards the desired output, producing the highest yield of product in the most cost-effective way.

Chemical reactions can be divided into homogenous reactions and heterogeneous reactions. **Homogeneous reactions** are those in which the reactants, products and any catalyst used form a continuous phase; for example, *gaseous* or *liquid*. **Heterogeneous reactions** are those that contain two or more phases. If there is more than one phase, mass and energy transfer phenomena must be taken into account. The *overriding* problems in reactor design is the promotion of mass transfer between phases.

The reactors used for established processes are usually complex designs that have been developed and have evolved over a period of years to suit the requirements of the process and are unique designs. However, it is convenient to classify reactor designs into the following broad categories.

Stirred tank reactors: Stirred tank reactors consist of a tank fitted with a mechanical agitator and a cooling or heating jacket or coils. They operate as batch reactors or continuous reactors. Several reactors can be used in series. They are used for homogeneous and heterogeneous liquid–liquid and liquid–gas reactions and for reactions that involve freely suspended solids, which are held in suspension by agitation. As the degree of agitation is under the designer's control, stirred tank reactors are particularly suitable for reactions where good mass transfer or heat transfer is required. When operated as a continuous process, the composition in the reactor is constant and the same as the product stream. Except for very rapid reactions, this will limit the conversion that can be obtained in a single stage.

Tubular reactors: These reactors consist of a tube or series of tubes where a flow of reagents is subjected to the reaction conditions. If high heat transfer rates are required, small diameter tubes are used to increase the surface area to volume ratio. Several tubes may be arranged in parallel, connected to a manifold or fitted into a tube sheet in a similar arrangement to shell and tube heat exchangers. For high temperature reactions, the tubes may be arranged in a furnace.

In some cases, the tubular bed is filled with a solid, which are known as packed bed reactors, and the solid can be a reactant or a catalyst.

Fluidized-bed reactors: A fluidized-bed reactor is a combination of the two most common continuous flow reactors: packed-bed and stirred tank. These types of reactors are very important to chemical engineering because of their excellent heat and mass transfer characteristics. The essential feature of a fluidized-bed reactor is that the solids are held in suspension by the upward flow of the reacting fluid. This promotes high mass and heat transfer rates and good mixing. The solids may be a catalyst, a reactant in a fluidized combustion process or an inert powder that is added to promote heat transfer. Though the principal advantage of a fluidized bed over a fixed bed is the higher heat transfer rate, fluidized beds are also useful when it is necessary to transport large quantities of solids as part of the reaction processes, e.g., when catalysts are transferred to another vessel for regeneration. The limitation of the process is the size of particles than can fluidize, i.e., smaller than $300\,\mu$m.

The selection of the kind of reactor is not an easy issue, but in general:

Stirred batch reactors are used on a laboratory scale in investigations to determine reaction kinetics and on a commercial scale in multiproduct production situations and/or when small production volumes (e.g., in the pharmaceutical industry) are required. However, this kind of reactors cannot be used for large volumes because the production is limited by the cyclical nature and the cost of operation is high. For large volumes production continuous reactors are employed, the first option is tubular reactors which have the small size and operation complexity, but stirred tank and fluidized-bed reactors are usually employed when the reaction are very exothermic or endothermic or the catalyst suffers a quick deactivation for instance. Continuous reactors have low cost of operation but the capital expenditure is high.

For liquid-phase reaction, usually a stirred reactor was employed, but lately is changing and the presence of tubular reactors is growing, this is also frequently labeled as flow chemistry. For gas phase reactions are almost exclusively performed in tubular or fluidized-bed reactors.

A.1.3. Design of chemical reactors

The design of a chemical reactor incorporates multiple aspects of chemical engineering. Chemical reactions, chemical energetics and equations/laws of thermodynamics play an important role in the selection and design of chemical reactors.

The two main equations to design a reactor are mass balance (Figure A.1.2) and energy balance (Figure A.1.3). As the chemical reaction rate depends on temperature, it is often necessary to calculate both energy balance (often heat balance rather than full-fledged energy balance) as well as mass balance to fully describe the system. A different reactor model might be needed for energy balance: a system that is closed with respect to mass may be open with respect to energy because heat may enter the system through conduction.

Here, F_{A0} is the molar flow feed to the reactor, F_A is the molar flow at the exit of the reactor, G_A is the molar flow generated in the reactor, N_a is the molar accumulation inside the reactor, V is the reactor volume and r_A is the reaction rate of compound A. Applying the mass balance, Eq. (A.1.1) represents

$$\text{Inlet} - \text{Outlet} + \text{Generation} = \text{Accumulation}$$

$$F_{A0} + F_A + \int r_A dV = \frac{dN_A}{Dt} \qquad \text{(A.1.1)}$$

Equation (A.1.1.) Mass balance of a reactor.

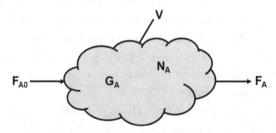

Figure A.1.2.   Schematic representation of mass balance.

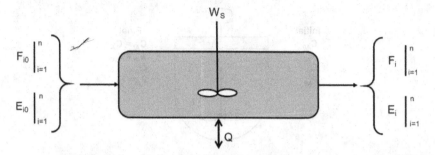

Figure A.1.3. Schematic representation of energy balance.

Here, F_{i0} is the molar flow feed to the reactor for component i, E_{i0} is the molar energy of component i at the inlet conditions, F_i is the molar flow at the exit of the reactor, E_i is the molar energy of component i at the outlet conditions, Q is the heat exchange with the surroundings, W_s is the work exchanged with the surroundings and E_{reactor} is the energy accumulated in the reactor. Applying energy balance, Eq. (A.1.1) represents

$$\text{Inlet} - \text{Outlet} + \text{Generation} = \text{Accumulation}$$

$$Q - W_s + \sum_{i=1}^{n} F_{i0} E_{i0} - \sum_{i=1}^{n} F_i E_i = \frac{dE_{\text{reactor}}}{dt} \qquad (A.1.2)$$

Equation (A.1.2.) Energy balance of a reactor.

To facilitate calculations in the design of chemical reactors, three main basic models are used to estimate the most important process variables for various chemical reactors. These models are ideal and facilitate calculations for reactor design. These three models are as follows:

- Discontinuous Stirred-Tank Reactor (DSTR) or Batch Reactor Model (Batch)
- Continuous Stirred-Tank Reactor (CSTR)
- Plug Flow Reactor (PFR)

In reality, reactors are non-ideal. Some are quasi-ideal and can be described with models. However, in general it is necessary to include in the design equations not only chemical reaction rates but also

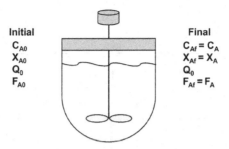

Figure A.1.4.   Scheme of a DSTR.

mass transfer rates, which may be important in the mathematical description of a system, especially heterogeneous systems.

A.1.3.1. *Discontinuous stirred-tank reactor* (*DSTR*) *or batch reactor*

This reactor is simple and consists of a closed vessel with a perfect mixture reactor, in which all properties are identical in the whole volume (temperature, pressure, concentration, etc.). This type of reactor is similar to laboratory vessels or reactors used in batch processes. Batch processes consist of filling the reactor with the reactants, recovering the reaction products, cleaning and restarting of the cycle. These reactors are suited to small production rates. They are conducted in tanks with stirring of the contents by internal impellers, gas bubbles or pumps. Temperature control is achieved with the help of jackets, reflux condensers or pumps through an exchange (Figure A.1.4).

A DSTR is a closed vessel; if no mass addition or extraction occurs, the two first terms of Eq. (A.1.1) can be discarded.

$$ \cancel{F_{A0}^0} - \cancel{F_A^0} + \int r_A dV = \frac{dN_A}{dt} \qquad (A.1.3) $$

where mass balance is applied to a DSTR.

As in an ideal model with perfect mixing, the r_a is constant for the entire reactor volume, and Eq. (A.1.3) can be simplified to the equation described in Figure A.1.5. The amount profile of component A has a shape similar to Figure A.1.5 if is a reagent, or an opposite shape if it is a product.

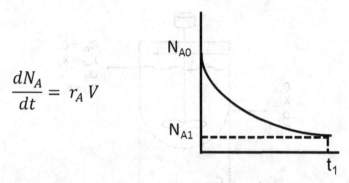

$$\frac{dN_A}{dt} = r_A V$$

Figure A.1.5. Simplified equation and concentration profile of a DSTR.

If the energy involved in the reaction is very small or the reactor operates isothermally (constant temperature), this equation is enough to describe and design the reactor.

A.1.3.2. *Continuous stirred-tank reactor (CSTR)*

This type of reactor is very similar to DSTR, but there are one or more fluid reagents introduced into the reactor tank equipped with an impeller while reactor effluent is recovered. In this case, we assume perfect mixing in the model reactor. Although the composition is uniform in individual vessels, a stepped concentration gradient exists in the system as a whole (Figure A.1.6).

The average amount of time spent by a discrete quantity of reagent inside the tank, or *residence time*, can be obtained by simply dividing the volume of the tank by the average volumetric flow rate through the tank. The reaction proceeds at the reaction rate associated with the final (output) concentration (Figure A.1.6).

A CSTR operates in steady state; there is no accumulation and the last term of Eq. (A.1.1) can be discarded (Eq. (A.1.4)). As it is an ideal model with perfect mixing, the r_a is constant in the entire reactor volume, and Eq. (A.1.4) can be simplified.

$$F_{A0} - F_A + \int r_A dV = \cancel{\frac{dN_A}{dt}}$$

$$F_{A0} - F_{A1} + r_A V = 0 \tag{A.1.4}$$

where mass balance is applied to a CSTR.

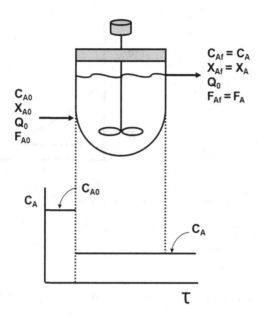

Figure A.1.6.   Scheme of a CSTR.

These reactors proceed at the reaction rate associated with the final (output) concentration and require a large volume (residence time) to reach high productivity. This model reactor is applied to a fluid-bed reactor from a mass and energy balance point of view.

A.1.3.3. *Plug flow reactor (PFR) or tubular flow reactors (TFR)*

In a PFR model reactor, reagents flow through a pipe or tube. It is assumed that no upstream or downstream mixing occurs, as implied by the term "plug flow". Additionally, no axial changes of properties are taken into account; the flow advances uniformly as a plug, without any back mixing or resistance of the walls or filler. Large-diameter vessels with packing or trays may approach plug flow behavior and are widely employed.

The chemical reaction proceeds as the reagents travel through the PFR (Figure A.1.7). In this type of reactor, the reaction rate is a gradient. That is, at the inlet to the PFR, the rate is very high, but as the concentrations of the reagents decrease and the concentration

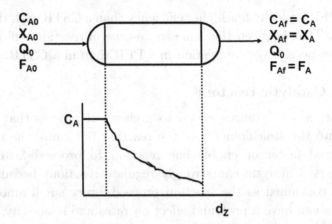

Figure A.1.7. Scheme of a PFR.

of product(s) increases, the reaction rate slows. PFR may be filled with solid particles, either catalytic (if required) or inert, to improve interphase contact in heterogeneous reactions. PFR can operate in horizontal or vertical mode depending of reaction needs. When heat transfer is required, individual tubes are jacketed or a shell-and-tube construction is used. In the latter case, the reactants may be in either the shell or the tube. Reagents may be introduced into the PFR at locations other than the inlet. In this way, higher efficiency may be obtained, or the size and cost of the PFR may be reduced.

A PFR operates in the steady state; there is no accumulation and the last term of Eq. (A.1.1) can be discarded (Eq. (A.1.5)). As is an ideal model for plug flow, each dV travels through the reactor without mixing, and Eq. (A.1.5) can be simplified.

$$F_{A0} - F_A + \int r_A dV = \frac{dN_A}{dt}$$

$$0 - \frac{dF_A}{dV} = -r_A \longrightarrow \boxed{\frac{dF_A}{dV} = r_A} \qquad \text{(A.1.5)}$$

where mass balance is applied to a PFR.

A PFR typically has higher efficiency than a CSTR with the same volume. That is, given the same space-time, a reaction will proceed to higher percentage completion in a PFR than in a CSTR.

A.1.4. Catalytic reactors

Catalysts are substances added to a chemical process that do not enter into the stoichiometry of the reaction but cause the reaction to proceed faster or enable one reaction to proceed faster than the others. Catalysts can strongly regulate reactions because they are not consumed as the reaction proceeds; very small amounts of catalysts can have a profound effect on rates and selectivity. At the same time, catalysts can undergo changes in activity and selectivity as the process proceeds. We distinguish between *homogeneous* and *heterogeneous* catalysts. Homogeneous catalysts are molecules in the same phase as the reactants (usually a liquid solution), and heterogeneous catalysts exist in another phase (usually solids whose surfaces catalyze the desired reaction). Acids, bases, and organometallic complexes are examples of homogeneous catalysts, while solid powders, pellets and reactor walls are examples of heterogeneous catalysts.

Catalysts provide an alternate path for the reaction to occur. If the catalyst provides a lower energy barrier path, this rate enhancement can be described using a potential energy diagram (Figure A.1.8). The energy barrier for reaction (the reaction

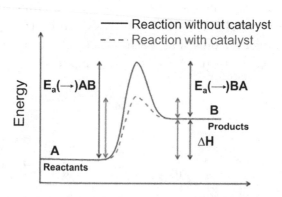

Figure A.1.8. Potential energy diagram.

activation energy) is lowered from E_1 to E_{cat} by the use of an appropriate catalyst, and the reaction rate is enhanced. If the selectivity to form a desired product is more important than reactivity (which is usually the case), a catalyst can provide a lower barrier for the desired reaction, leaving the undesired reaction rate unchanged. The search for active and selective catalyst in biology or chemical synthesis is perhaps the central goal in most biological and physical science research.

The use of heterogeneous catalysts implies the need to account for the essential transport steps for reactants and products to and from the catalyst. In practice, catalytic reaction rates can be thoroughly disguised by mass transfer rates. In fact, in many industrial reactors the kinetics of individual reactions are quite unknown, and some engineers regard knowledge of their rates as unimportant compared to the need to prepare active, selective, and stable catalysts. The role of mass transfer in reactions is therefore essential in describing most reaction and reactor systems (Figure A.1.9).

A catalytic reaction with a heterogeneous catalyst may contain some or all of the following steps (Figure A.1.9):

1. Mass transfer (diffusion) of the reactant(s) (e.g., species A) from the bulk fluid to the external surface of the catalyst pellet
2. Diffusion of the reactant from the pore mouth through the catalyst pores to the immediate vicinity of the internal catalytic surface
3. Adsorption of reactant A onto the catalyst surface
4. Reaction on the surface of the catalyst
5. Desorption of products (e.g., B) from the surface
6. Diffusion of products from the pellet interior to the pore mouth at the external surface
7. Mass transfer of products from the external pellet surface to the bulk fluid

The overall rate of reaction is limited by the rate of the slowest step in the mechanism. When the diffusion steps (1, 2, 6, and 7 in Figure A.1.9) are very fast compared to the reaction steps (3, 4, and 5), the concentrations in the immediate vicinity of the active sites are indistinguishable from those in the bulk fluid. In this situation,

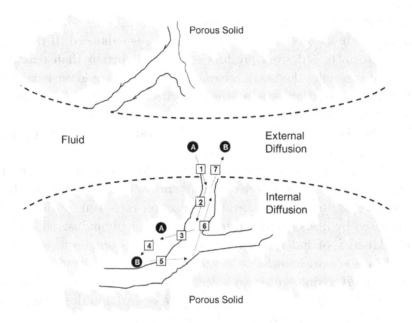

Figure A.1.9. Steps in the reaction with heterogeneous catalysts.

the transport or diffusion steps do not affect the overall rate of the reaction. In other situations, if the reaction steps are very fast compared with the diffusion steps, mass transport affects the reaction rate. In systems where diffusion from the bulk gas or liquid to the catalyst surface or to the mouths of catalyst pores affects the rate, changing the flow rate should change the overall reaction rate. In porous catalysts, however, diffusion within the catalyst pores may limit the rate of reaction, and as a result, the overall rate will be unaffected by external flow conditions even though diffusion affects the overall reaction rate. The reaction occurs on the surface, but the species involved in the reaction must be transported to and from the surface. The external diffusion can be reduced increasing the lineal velocity of the flow in the reactor, increasing the mass transfer between the bulk and surface of the catalyst. The internal diffusion can be reduced decreasing the particle size, but small particle size can produce an increase in the pressured drop in the reactor, for this reason the absence of internal diffusion can only reached in laboratory scale reactors or slurry of fluidized bed reactors at industrial scale.

Based on the characteristics for all reactions that are catalyzed by heterogeneous catalysts, we should expect

1. Non-integral orders and even negative orders, with respect to reactant species in catalytic reactions;
2. Rate expressions obtained with one set of pressures and temperatures may not describe a reactor in another situation because the rate expressions of catalytic reactions do not obey the power law;
3. Traces impurities drastically alter catalytic reaction rates even when they do not enter into the reaction.

As a consequence, many times the reaction rate is described by an experimentally deduced expression that fits these conditions and do not correspond to a complete description of all steps involved in the reaction process.

A.1.5. Photoreactors

A photoreactor is any vessel in which a photochemical reaction takes place. A photochemical reaction is a process that must be preceded by absorption of radiation of the appropriate energy by one molecule. Upon radiation absorption the excited molecule can be transformed, in one or more steps, into a product or it may be converted into an intermediate species that can participate in subsequent reactions of thermal nature as happens, for example, in chain reactions. Sometimes, radiation absorption occurs in a particular molecule but definitive changes occur in others, as in the case of photosensitized and photocatalyzed reactions.

The design of is more difficult than traditional thermal reactors because plant size photoreactors operate under non-uniform concentrations, temperatures, and — particularly important — radiation (light) distributions. In other words, point values of the reaction rates are intrinsically not uniform and in many cases are utterly different from the global ones. In some cases, good mixing can be achieved and nonuniformities in concentrations and temperature may be less severe. However, under no circumstances can stirring render the photon distribution uniform if geometric effects (distance from the radiation source to each different point inside the reactor, particularly for non-collimated beams of radiation) or attenuation effects (due to radiation absorption by the intervening species) have produced a spatial distribution of the absorbed photons.

The modeling of photoreactor based requires the solution of the energy and mass balance equations, like other reactors described in this annex, but is necessary the inclusion of the photon balance (radiation energy) in the case of a photoreactor. The radiation balance can be treated separately from the thermal energy balance because thermal effects of the photochemically useful energy are generally negligible and, consequently, one is mainly concerned with the kinetics effects of the employed radiation, because the absorption of photons produces or initiate the reaction.

Geometry of the photoreactors depends mainly on the application as well as on the available irradiation source. Additionally, the following factors also need to be considered during the design of photoreactors: (1) type and particle size of the photocatalyst; (2) distribution of the photocatalyst (fixed or suspended); (3) type, content, and distribution of pollutants; (4) mass transfer; (5) fluid dynamics (laminar or turbulent flow); (6) temperature control; (7) reaction mechanism; and (8) reaction kinetics.

Recommended reading

1. Uzi Mann, M.D. Morris, advisory editor — 2nd ed., *Principles of Chemical Reactor Analysis and Design: New Tools for Industrial Chemical Reactor Operations*, John Wiley & Sons, Inc., Hoboken, NJ, USA, 2009.
2. Lanny D. Schmidt, *The Engineering of Chemical Reactions*, Oxford University Press, Inc., New York, NY, USA, 1998.
3. Mark E. Davis, Robert J. Davis, *Fundamentals of Chemical Reaction Engineering*, The McGraw-Hill Companies, Inc., New York, NY, USA, 2003.
4. Octave Levenspiel, *Chemical Reaction Engineering*, 3rd ed., John Wiley & Sons, Inc., New York, NY, USA, 1998.
5. H. Scott Fogler, *Essentials of Chemical Reaction Engineering*, Pearson Education, Inc., Boston, MA, USA, 2011.

Annex 2

Introduction to Electrochemistry

María Retuerto and Sergio Rojas*

Sustainable Chemistry and Energy Group (EQS),
Institute of Catalysis and Petrochemistry (CSIC),
C/Marie Curie 2, Cantoblanco 28049 Madrid, Spain

Electrosynthesis is becoming a real possibility for the transformation of biomass-derived chemical products in valuable chemicals. In a similar manner to the previous annex, this chapter is aimed at introducing those readers not familiar with the electrochemical techniques so they can understand those sections of this book describing electrochemical transformations. This annex serves as a brief introduction to some basic electrochemistry concepts useful for the study of electrocatalytic reactions. We encourage the interested reader to consult the references included at the end of the annex for comprehensive details.

Electrochemistry deals with the chemical reactions caused by the passage of an electric current or the generation of an electrical current by chemical reactions. Electrochemical processes are characterized by the *crossing of charge across the interface* between different phases, say an electronic conductor (electrode) and an ionic conductor (electrolyte) thus resulting in the *oxidation* and *reduction* of substances. Charge transport in the electrodes occurs due to the motion of electrons (or holes) whereas ions (anions or cations) are

---

*Corresponding author: srojas@icp.csic.es

responsible for charge motion in the electrolyte. When an electrode is immersed in an electrolyte two type of processes can take place. One kind of such process involves electrons (charge) flowing across the interface causing reduction or oxidation reactions. These processes are governed by Faraday's law, which relates the amount of chemical transformation with the amount of electricity passed per time unit (the charge). Because electrochemistry deals with the study of the flow of current (reaction rate) across the interface these so-called faradaic processes are the most interesting ones for electrocatalysis. The electrode at which the reduction takes place is the cathode whereas the electrode at which the oxidation reaction takes place is the anode. The direction of the flow of electrons defines the oxidation and reduction processes so that cathodic currents are associated with the flow of electrons from the electrode to species in the electrolyte. Conversely, electrons flowing from a species in the electrolyte to the electrode are anodic currents. This convention applies to both types of electrochemical cells; electrolytic and galvanic.

When an electrode is immersed in an electrolyte, a two-dimensional surface (interphase) separating both phases is created as the result of the reorganization of the surface composition of electrode and electrolyte under the effect of each other [1,2]. The point of zero charge (PZC) is the potential at which the charge of the electrode surface drops to zero. The electrode surface is positively charged at potentials more positive than the PZC and is negatively charged at potentials more negative than the PZC.

Several models have been developed to describe the distribution of charges around the interface [2]. A first layer of solvent molecules and ions free from their solvation layer, usually anions (*specifically adsorbed ions*) placed at the surface of the electrode. This layer is known as inner Helmholtz layer (IHP). A second layer formed by the solvated ions is known as outer Helmholtz plane (OHP). In the inner layer mechanism, the electron transfer proceeds through a ligand shared by the electronic conductor and the electroactive species. This mechanism corresponds to specifically adsorbed reactants and kinetics are affected by the nature of the electrode surface. Conversely, in the outer sphere mechanism the reactive species are located in the

OHP so that reactants are not coordinated to (adsorbed onto) the electrode surface [3, 4].

Since electrocatalytic processes are heterogeneous in nature, adsorption and desorption of chemical species (reactants, ions. . .) at the electrode surface can take place along with the passing of electron flow. When the charge does not cross the interface, it is called non-faradaic processes, the charge of the electrode is compensated by the reorganization (adsorption) of the ions in the vicinity of the electrode resulting in the so-called "double layer" (phenomena similar to the charge of a capacitor). Non-faradaic processes can be neglected when evaluating the reaction kinetics if the concentration of electroactive species is sufficiently high. On the other hand, faradaic processes are characterized by the crossing of electrons through the interface, being the case of study of electrocatalytic reactions.

A.2.1. Electrolytic and galvanic cells

Two types of faradaic processes can be conceived. Electrolytic cells are those in which a substance transforms into another through the passage of a current in a process known as electrolysis. This type of cell corresponds to the conversion of electrical energy into chemical energy. In order for the chemical transformation to proceed, an external potential difference should be applied between the electrodes greater than the open circuit potential. In this type of cells, the cathode is the negative electrode. Examples of these cells are electroplating or electrolytic synthesis.

On the other hand, galvanic cells are those in which the chemical reactions take place spontaneously at two different electrodes connected by a conductor, generating electrical current by a potential difference. The open circuit potential decreases with the increasing demand of electric current. In galvanic cells, the anode is the negative electrode. Examples are batteries (discharge) or fuel cells.

A.2.2. Electrochemical cell. Two electrodes are needed

As stated above, electrochemistry studies the current that flows across an electrode|electrolyte interface when a potential between two electrodes is applied or vice versa, i.e., the variation of the potential when a current is passed across the interface. It is not possible to study an individual interface; therefore, in a typical electrochemical cell two (or more) interfaces exist. This feature implies that the absolute potential for a reaction cannot be determined, being the potentials for redox reactions reported (tabulated) relatively to a reference electrode. Nernst proposed to use the hydrogen electrode as standard and defined (arbitrarily) its value as $E^0 = 0.0$ V. This electrode, known as Standard Hydrogen Electrode (SHE), consists of a platinized platinum sheet immersed in an aqueous solution of unit activity of H_{aq}^+ in contact with hydrogen gas at a pressure of one atmosphere. The standard reversible electrode potentials for electrode reactions, reported as reduction reactions, are referred to this value and tabulated in many works [5–7].

In order to form an electrochemical two-electrode cell, reduction and oxidation are needed. The electrochemical reaction would take place, i.e., current will flow spontaneously, when the two electrodes (half-reactions) of the cell are connected through an external electrical circuit and the cell has a positive cell potential value. The cell potential is calculated by subtracting the equilibrium standard potential of the cathode reaction from that of the anode reaction, i.e., $E_{cell} = E_{cathode} - E_{anode}$ If E_{cell} is positive, the reaction would take place in the direction both half-reactions are written. On the other hand, if E_{cell} is negative, the reaction as written is thermodynamically unfavorable and half-reactions should be reversed in order to be spontaneous.

Equilibrium potential is related to the free energy of the reaction by equation $\Delta G = -nFE_{cell}$ where n is the number of electrons exchanged and F is the Faraday constant. Positive values of E_{cell} result in negative values of ΔG, thus spontaneous reactions. The fact that the reaction is thermodynamically possible or spontaneous ($\Delta G < 0$) does not imply that it would take place at appreciable rates. In order to accelerate the rate of a reaction, the potential is driven away from the equilibrium potential so that the reaction takes place at measurable rates. This shift of the potential from the equilibrium potential is known as overpotential (η) so that

$\eta = E - E_{eq}$. This overpotential contains contributions from different effects, including mass-transfer and charge-transfer. In addition, when current is flowing in a resistive solution, e.g., the electrolyte in the electrochemical cell, a potential drop is generated in any point in the solution (electrolyte). This ohmic resistance is referred to as the ohmic drop (iR_{drop}) or uncompensated resistance and it obeys Ohm's law [8]. Therefore, iR_{drop} is characteristic of the cell (not of the reaction) and can be minimized by the correct designing of the cell, for instance by placing the tip of the reference electrode close to the surface of the working electrode by using a Luggin capillary [9]. In any case, the effect of iR_{cell} should be considered in the overall E_{cell} required to drive a reaction so that

$$E_{\text{cell}} = E_{\text{cat}} - E_{\text{an}} + \eta - iR_{\text{drop}} \qquad \text{(A.2.1)}$$

A.2.3. Reaction rates in electrochemistry

Reaction rates are expressed in units of moles of substance produced or consumed per time unit (moles/s). Electrochemical cells do not allow measuring the amount of reactants consumed during the electrochemical experiment. However, it is very simple to measure the current flow passing during the experiment. The unit of current is ampere (A).

$$i(A) = \frac{c}{s} \qquad \text{(A.2.2)}$$

Faraday's law of electrolysis correlates the amount of charge (q), in coulombs, passed during an electrochemical experiment with the amount of moles transformed (m), the number of electrons exchanged (n) and the Faraday's constant (F)

$$q = \int_0^t i\,dt = mnF \qquad \text{(A.2.3)}$$

so that the reaction rate (r) in an electrochemical process is proportional to the current measured in an electrochemical experiment.

$$\text{rate}(r) = \frac{\text{mol}(m)}{\text{time unit }(s)} = \frac{q}{nFs} = \frac{is}{nFs} = \frac{i}{nF} \qquad \text{(A.2.4)}$$

The quantity nF relates the amount of electrical current (flow of electrons) with the amount of chemical transformation.

Since electrochemical reactions are heterogeneous, i.e., occur at the electrode/electrolyte interface, surface (S) normalization of the current is a more suitable descriptor for reaction rate,

$$\frac{r}{S} = \frac{i}{nFS} = \frac{j}{nF} \tag{A.2.5}$$

where j is the current density in A/cm^2.

A.2.4. Factors affecting electrode reaction rate and current

Electron transfer in electrochemical processes takes place at the interface electrode/electrolyte so adsorption and transport of reactants and products to (or from) the electrocatalyst should be considered. In addition, the presence of electrolyte (and the transport of charge by ions) and the presence of electric potential at the interface should be taken into account. In this chapter, we will not address the (crucial) role of charge transport by ions in electrochemical reactions [2,7,10]. It is also important to clarify that the reaction rate indicates the time needed by the reactants to arrange themselves (and their environments) so that electron transfer across the electrode is possible. According to the Franck–Condon principle, the rate of electron transfer itself takes place very rapidly, in the order of 10^{-16} s (Figure A.2.1).

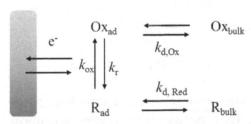

Figure A.2.1.   Pathway of an electrode reaction. k_d are diffusion constants and k are kinetic constants. Note that this scheme does not consider chemical reactions that can occur during the transport of species from the bulk to the electrode.

A.2.5. Reversible and irreversible reactions

Notice that the concept of reversible and irreversible reactions in electrochemistry is related to the relative rates of mass transfer and reaction kinetics. The rate of electrode reactions is affected by mass transport and by electron transfer effects, which can be related to the mass transfer coefficient (k_d) and to the standard rate constant (k^0), respectively. Two extreme situations can be envisaged: if $k_d \gg k^0$ the system is irreversible whereas if $k^0 \gg k_d$ the system is reversible. As a rule of thumb, very fast reactions with $k^0 > 10^{-1}\,\mathrm{cm/s}$ will be reversible regardless of the stirring intensity. On the other hand, very slow reactions with $k^0 < 10^{-5}\,\mathrm{cm/s}$ will always be irreversible [2].

The Nernst equation (Eq. (A.2.6)) applies for the reversible systems, i.e., where the rate of the electrode reaction (current) is limited by the rate at which the reactants are transported to the electrode surface. On the other hand, under conditions where the electron transfer rate is the rate-determining step, and when working at low current and efficient stirring, it is possible to measure kinetic parameters. In these irreversible systems, it has been observed that the current is related exponentially to the overpotential.

A.2.6. Reaction kinetics

Reaction rates are affected by the rate at which the electroactive species are transported to the electrode. The transport of material in solution is driven by gradients of concentration. If the overpotential applied is sufficiently high (positive or negative), then the kinetics of the electrode reaction is sufficiently fast so that the rate of the electrode process is controlled by mass transport, i.e, by the rate at which the electroactive species reach the electrode surface, thus reaching limiting currents.

Consider the reaction $\mathrm{O} + ne^- \leftrightarrow \mathrm{R}$ taking place at the surface of an electrode. If the kinetics of electron transfer is fast, the concentrations of O and R species at the electrodes can be assumed to be at the equilibrium with the electrode potential. The relationship between the potential and the bulk concentration of the redox species at the equilibrium is described by the Nernst equation

$$E = E^0 + \frac{RT}{nF} \ln \frac{a_O}{a_R} \tag{A.2.6}$$

where E^0 is the standard potential for the reaction, R is the gas constant, and a_i is the activity of the species i in solution. At the equilibrium potential both oxidation and reduction reactions take place at identical rates, consequently no net current is observed. In terms of current densities this situation can be expressed as $j_r = -j_{ox} = j_0$, where j_r and j_{ox} are the reduction and oxidation current densities, respectively. They have different signs to indicate that the flow of electrons in oxidation and reduction reactions are opposite. The term j_0 is the exchange current density, which is a key parameter in the description of electrode kinetics. j_0 measures the electron-transfer rate at the equilibrium. High j_0 values are indicative of fast kinetics of the reaction under study whereas low j_0 values indicate slow electrode reactions. Generally, the presence of an electrocatalysts increases reaction rates by increasing the value of j_0 (actually of k_0).

The overall reaction rate will be defined by the contributions of the reduction and oxidation reactions:

$$j = j_r - j_{ox} = nF[C^0_{ox}k_r - C^0_r k_{ox}] \tag{A.2.7}$$

At this point one of the key elements of the heterogeneous transfer process emerges; the rate constant k depends on the potential region. It is possible to increase the reactions rates even for reactions with low k^0 by using high overpotentials.

The relationship between current density and overpotential is described by the Butler–Volmer equation in the form (see [2, 7, 15] for further details):

$$j = j_0 \left[\exp\left(-\frac{\alpha nF}{RT}\eta\right) - \exp\left(\frac{(1-\alpha)nF}{RT}\eta\right) \right] \tag{A.2.8}$$

This equation shows that the measured current density is a function of the overpotential (η), the exchange current density (j_0) and the charge transfer coefficient α. The latter coefficient is an indicator of the symmetry of the activation barrier, i.e., preference of the reaction to proceed in one direction; α values of 0.5 indicate that both directions proceed with equal probability [7].

The equations above show that the total current density of an electrochemical reaction is the sum of the anodic and cathodic currents. By conducting the reaction at sufficiently large overpotentials, and considering the exponential relationship between current density and overpotential, it is possible to ignore the contribution of one of the components of the Butler–Volmer equation (Eq. (A.2.8)). For instance, if the overpotential is driven sufficiently negative, greater than ca. 120 mV, the cathodic current increases while the anodic one becomes negligible and the current density could be expressed as:

$$j = j_0 \left[\exp \left(\frac{-\alpha nF}{RT} \eta \right) \right] \qquad (A.2.9)$$

Equation (A.2.9) can be rewritten as

$$\eta = \frac{RT}{\alpha nF} \ln j_0 - \frac{RT}{\alpha nF} \ln j \qquad (A.2.10)$$

In this form, this equation resembles to the Tafel equation, which predicted that the current is related exponentially to the overpotential

$$\eta = a + b \log j \qquad (A.2.11)$$

Plotting η *vs.* $\ln j$ will render straight lines with slopes $-\alpha F/RT$ and $(1-\alpha)F/RT$ for the cathodic and anodic branches, respectively. Obviously, α can be calculated from the slope of these lines. In addition, the extrapolation of the straight segment of the lines intercepts y-axis at $\ln j_0$. Note that to extract activation energies, the value of i_0 should be used instead of the current density values at any overpotential (Figure A.2.2).

As stated above, at the equilibrium, the rates of the oxidation and reduction process are equal, that is, the net current is zero and j_0 cannot be measured directly experimentally. Instead, j_0 is usually calculated by extrapolating the current obtained at high overpotentials

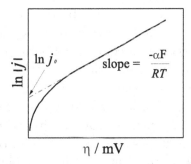

Figure A.2.2. Tafel plot of the cathodic current–overpotential curve for $O + ne^- \leftrightarrow R$.

as explained above. It is equivalent to the standard rate constant, k_0. Note that the activation energy should be calculated by using the i_0 value.

A.2.7. Description of the electrochemical cell

A.2.7.1. *Electrodes*

As discussed above, two electrode reactions are needed to form an electrochemical cell. However, in many cases the study of individual oxidation or reduction processes taking place at a single electrode is pursued (convenient). For that purpose, the most used configuration at lab scale is the so-called three-electrode cell. This cell consists of a working electrode (W.E.) which is the electrode where the reaction under study takes place; a reference electrode (R.E) which is the electrode against the potential of the W.E. is measured; and an auxiliary or counter electrode (C.E.) which is the electrode at which the reaction complementary to that of the W.E. takes place. In order to decrease the ohmic drop (iR_{ohmic}), the R.E. should be placed very close to the W.E. via a Luggin capillary.

The electrodes are immersed in the same electrolyte and potential and current are controlled/measured with a potentiostat/galvanostat. In a typical experiment, a potential (or a potential range) is imposed into the W.E. and the current that passes through the W.E. (and the C.E.) is recorded and represented *vs.* potential or *vs.* time. According to the IUPAC convention, positive currents identify oxidation currents whereas negative ones are for reduction processes.[a]

Several materials can be used as W.E., C.E. and R.E. In order to study electrocatalytic reactions, or more likely the performance of novel electrocatalysts for a given reaction (say oxidation or reduction of molecules) the most used extended approach is to deposit a thin film of the electrocatalyst under study on a carbon disk (or any other inert, electron conducting material) both conforming the W.E. Caution: the actual area of the electrocatalyst under study could be very different to the geometric area of the electrode; the real electroactive area of the catalysts should be determined to establish proper comparisons.

A.2.8. Methods in electrocatalysis

In a typical electrochemical experiment, the system is altered by imposing either potential (potentiostatic control) or current (galvanostatic control) programs (constant, steps, sweeps...) which result in a perturbation of the system. A brief description of the methods more commonly used at laboratory scale is given below. For comprehensive details the reader is referred to Refs. [4, 7].

A.2.8.1. *Potential sweep methods, linear sweep voltammetry and cyclic voltammetry*

In a typical potentiostat method, the electrode potential imposed is maintained either constant or variable and the current that flows through the W.E. is recorded and represented as function of time or potential, respectively. In a potential sweep method the potential applied to the W.E. varies linearly from E_1 to E_2 (linear sweep potential, LSV) at a constant sweep rate (v). Generally, but not always, the value E_1 is chosen so that the current flow is negligible, i.e., the electroactive species in neither oxidized nor reduced. In

---

[a]Note that the convention of taking positive currents as anodic (or oxidation) currents is not always followed, and the convention of referring positive currents to cathodic process can be used.

a cyclic voltammetry method, the potential varies from E_1 to E_2 and back to E_1 (other potentials can be considered), resulting in a cyclic program which can be recorded consecutive times. Cyclic voltammetry is *almost routinely* employed for conducting initial studies of the electrochemical performance of new systems (electroactive species and/or electrocatalysts). This technique allows for a rapid evaluation of the potentials at which the redox reactions take place. Faradaic and capacitive currents are registered within the potential range where the electrode reactions occur. The capacitive current is proportional to sweep rate (v) whereas Faradaic currents are proportional to $v^{1/2}$. As a consequence, capacitive currents are more significant at high sweep rate and should be subtracted in order to attain accurate information of the Faradaic processes under study. The shape of the voltammogram will depend on the following features; the rate constant (k^0; irreversible, quasireversible or reversible behavior), the formal potential ($E^{0'}$), the diffusion coefficients of the electroactive species, and the sweep rate (v) [4].

Figure A.2.3 shows typical shape of a cyclic voltammogram for a reversible system. The most informative features that can be obtained from voltammograms are the onset potential (E_{onset}), the $E_{p/2}$ (half-wave potential) which is the potential at which the current is half of the maximum current), i_p (which is the maximum current) and the peak potential (E_p) which is the potential at which the current is maximum (highest rate).

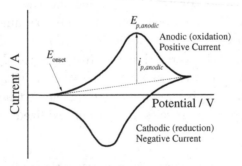

Figure A.2.3. Characteristic cyclic voltammogram for a reversible redox couple.

The two following criteria can be used to discriminate irreversible and reversible processes [15]:

1. Peak to peak separation $\Delta E_p = |E_{\text{panodic}} - E_{\text{pcathodic}}| \approx 57\,\text{mV}$ (at 298 K) for reversible (nernstian) reactions, and it is independent of scan rate. ΔE_p depends on scan rate for irreversible and quasireversible processes.
2. $|i_{\text{panodic}}/i_{\text{pcathodic}}| = 1$ is reversible.

A.2.8.2. *Identification of products*

The identification of the products formed during electrolysis is usually accomplished via *ex situ* techniques such as chromatography (gas or liquid), infrared spectroscopy, NMR or mass spectrometry. It is also possible to identify the products formed during an electrochemical experiment using *in situ* techniques. In addition, it is also possible to analyze the electrode surface and its evolution during the electrochemical reaction. There are several spectroelectrochemical techniques than ca be used to determine the electrode surface and the products formed during electrolysis [16]. One of such techniques uses infrared spectroscopy coupled to an electrochemical cell. The infrared beam passes through a window and a thin layer of liquid electrolyte, reflects on the electrode and it is detected. In this configuration, the layer of the electrolyte must be thin (within the micrometer range) since molecules contained in the electrolyte, typically H_2O, adsorb in the infrared region. Since the concentration of the species of interest is usually orders of magnitude smaller than that of the electrolyte, several approaches are used to avoid the interference from the electrolyte, including modulation techniques or using polarized light, further details can be found in Ref. [20]. A suitable technique to detect the product formed during an electrochemical reaction is the so-called differential electrochemical mass spectrometry (DEMS) [17, 18]. In this setup, the outlet of the electrochemical cell is connected directly to a mass spectrometer through a porous hydrophobic membrane (typically a Teflon membrane) of controlled thickness and pore size using a differential pumping system to ensure a fast transfer of the reaction products to the mass spectrometer. Response can be in the order of few milliseconds so that real-time analysis of products is possible in step potential methods.

A.2.9. Electrochemistry at the industry

Despite the benefits of using electrochemical routes in terms of energy efficiency and product selectivity are acknowledged, the implementation of electrochemistry at the industrial scale for the production of commodities or specialty products is low [19, 20]. One of the main drawbacks of electrochemistry is the high price of electricity. However, this would only apply to large-scale processes such as the chlor-alkali and the aluminum production processes. In the production of specialties and fine chemicals (via electrosynthesis) the consumption of electricity is a minor fraction of the total cost [21].

Electrochemistry is present in several other industrial (or pilot scale) processes including, water purification, metal finishing, corrosion, electrochemical sensors, batteries and fuel cells and organic electrosynthesis [22]. The production of adiponitrile (an intermediate for the synthesis of Nylon 66) from acrylonitrile is the most successful organic electrosynthesis at the industry [19].

The most successful processes based upon electrochemical routes are the aluminum and the chlor-alkali industries, which account to more than 90% of the electricity used in electrolytic processes [20].

Aluminum production from Al_2O_3 is conducted in molten cryolite (Na_3AlF_6) cells operating at $\sim 1000°C$ using current densities as high as $1\,A\ cm^{-2}$. In the anodic reaction, carbon is used to consume oxygen ions to produce CO_2.

$$Al_2O_3 + 3C \rightarrow 4Al + 3CO_2$$

This process results in lower overpotentials than that of the formation and evolution of oxygen from Al_2O_3, and the reversible cell voltage is $-1.18\,V$ compared to $-2.21\,V$.

The chlor-alkali process is the largest of the electrochemical industrial processes. It consists on the electrolysis of aqueous sodium chloride (usually as brine) to yield chlorine, sodium hydroxide and hydrogen. Due to the very large amounts of chlorine and NaOH produced globally, the chlor-alkali industry consumes ca. 125 × 10^9 kWh/year of electric power [21]. The electrode reactions are the production of chlorine gas at the anode electrode and hydrogen gas and hydroxide anions at the cathode. Three main reaction configurations are available, mercury, diaphragm, and membrane cells. In the diaphragm cell reactor, an asbestos separator is deposited onto a steel gauze that acts as cathode whereas anodes are based on metal (Ni) coated titanium. However, the use of asbestos as separator has several drawbacks such as poor selectivity to the passage of ions and generation of large overpotentials (due to high iR drop) of ca. 1.0–1.5 V, resulting in actual voltages ∼3.5 V. In membrane reactors, a cation exchange membrane is used as separator. Perfluorinated membranes allow the selective passage of Na^+ ions (preventing the passage of protons) from the anode to the cathode thus forming sodium hydroxide. Voltages of ca. −2.95 V with current densities of ca. 0.4 A cm^{-2} are characteristic of this technology [22].

Water electrolysis is currently considered among the most promising technologies for the storage of renewable energy (wind, solar, hydraulic, etc.) as hydrogen. During the electrolysis of water, pure H_2 and pure O_2 are produced at the cathode and anode of electrolytic cells, respectively. It is necessary to keep both gases apart, so separators such as asbestos are used. Several types of electrolytic cells can be found ranging from small scale in the order of few kilowatt to large reactors in the range of several MW. Tank cells are the most used ones due to their intrinsic simplicity. In a typical electrolytic tank cell operated in the monopolar manner, metallic electrodes are contained within a rectangular stainless-steel cell. Electrodes with the same polarity are connected in parallel. Highly concentrated KOH or NaOH is used as electrolyte. Recently, a great interest has been devoted to the development of solid polymer electrolyte cells. In these electrolyzers, a thin proton-conducting membrane is used to separate anode and cathode compartments. Several potential advantages of this technology include the higher stability of the electrolyte, the minimization of corrosion issues, and the optimum separation of the gases therefore high gas purity.

However, the electrodes usually contain platinum group metals, which increase their price considerably. Since polymeric cells can be operated at temperatures as high as 150°C lower voltages ($\sim$–2 V) and higher current densities ($>$1 A cm^{-2}) are possible. Hydrogen purity is very high, $>$99.995% with O_2 content below 0.005% volume.

References

1. S. Trasatti, R. Parsons, Interphases in systems of conducting phases, *Pure and Applied Chemistry*, 58 (1986) 437–454.
2. V.S. Bagotsky, *Fundamentals of Electrochemistry*, 2nd edn., John Wiley & Sons, Inc., 2006.
3. L.M. Torres, A.F. Gil, L. Galicia, I. González, Understanding the difference between inner- and outer-sphere mechanisms: An electrochemical experiment, *Journal of Chemical Education*, 73 (1996) 808.
4. R.G. Compton, C.E. Banks, *Understanding Voltammetry*, Imperial College Press, 2011.
5. A.J. Bard, R. Parsons, J. Jordan, *Standard Potentials in Aqueous Solution*, Marcel Dekker, Inc., 1985.
6. C.H. Hamann, A. Hamnett, W. Vielstich, *Electrochemistry*, 2nd ed., Wiley 2007.
7. A. Bard, L. Faulkner, *Electrochemical Methods: Fundamentals and Applications*, 2nd ed., John Wiley & Sons, Inc., 2001.
8. J.C. Myland, K.B. Oldham, Uncompensated resistance. 1. The effect of cell geometry, *Analytical Chemistry*, 72 (2000) 3972–3980.
9. K.B. Oldham, N.P.C. Stevens, Uncompensated resistance. 2. The effect of reference electrode nonideality, *Analytical Chemistry*, 72 (2000) 3981–3988.
10. J. Koryta, J. Dvorak, K.L., *Principles of Electrochemistry*, 2nd ed., John Wiley & Sons Ltd., 1993.
11. A. Zalineeva, S. Baranton, C. Coutanceau, G. Jerkiewicz, Electrochemical behavior of unsupported shaped palladium nanoparticles, *Langmuir*, 31 (2015) 1605–1609.
12. S. Trasatti, O.A. Petrii, Real surface area measurements in electrochemistry, *Journal of Electroanalytical Chemistry*, 327 (1992) 353–376.
13. C.C.L. McCrory, S. Jung, J.C. Peters, T.F. Jaramillo, Benchmarking heterogeneous electrocatalysts for the oxygen evolution reaction, *Journal of the American Chemical Society*, 135 (2013) 16977–16987.
14. J. Suntivich, H.A. Gasteiger, N. Yabuuchi, Y. Shao-Horn, Electrocatalytic measurement methodology of oxide catalysts using a thin-film rotating disk electrode, *Journal of the Electrochemical Society*, 157 (2010) B1263.
15. C.M.A. Brett, A.M.O. Brett, *Electrochemistry: Principles, Methods, and Applications*, Oxford University Press, 1993.
16. R.C. Alkire, D.M. Kolb, J. Lipkowski, P.N. Ross, *Diffraction and Spectroscopic Methods in Electrochemistry*, Wiley, 2006.

17. H. Baltruschat, Differential electrochemical mass spectrometry, *Journal of the American Society for Mass Spectrometry*, 15 (2004) 1693–1706.
18. S.J. Ashton, *Design, Construction and Research Application of a Differential Electrochemical Mass Spectrometer (DEMS)*, Springer, Berlin, Heidelberg, 2012.
19. G.G. Botte, Electrochemical manufacturing in the chemical industry, *The Electrochemical Society Interface*, 23 (2014) 49–55.
20. C.A.C. Sequeira, D.M.F. Santos, Electrochemical routes for industrial synthesis, *Journal of the Brazilian Chemical Society*, 20 (2009) 387–406.
21. D. Pletcher, Guide to electrochemical technology for synthesis, separation and pollution control, Electrosynthesis Company Inc., 1999.
22. D. Pletcher, F.C. Walsh, *Industrial Electrochemistry*, 2nd ed., Blackie Academic & Professional, 1993.

Index

Printed in the United States
By Bookmasters